Fluorides in the Environment

Fluorides in the Environment

Effects on Plants and Animals

Professor L.H. Weinstein

Boyce Thompson Institute for Plant Research
Tower Road
Ithaca
NY 14853
USA

and

Professor A. Davison

School of Biology
Ridley Building
University of Newcastle
Newcastle upon Tyne NE1 7RU
UK

CABI Publishing

CABI Publishing is a division of CAB International

CABI Publishing
CAB International
Wallingford
Oxon OX10 8DE
UK

CABI Publishing
875 Massachusetts Avenue
7th Floor
Cambridge, MA 02139
USA

Tel: +44 (0)1491 832111
Fax: +44 (0)1491 833508
E-mail: cabi@cabi.org
Web site: www.cabi-publishing.org

Tel: +1 617 395 4056
Fax: +1 617 354 6875
E-mail: cabi-nao@cabi.org

A catalogue record for this book is available from the British Library, London, UK.

Library of Congress Cataloging-in-Publication Data

Weinstein, Leonard H., 1926-
  Fluorides in the environment / by L.H. Weinstein and A. Davison.
    p. cm.
Includes bibliographical references and index.
  ISBN 0-85199-683-3 (alk. paper)
  1. Fluorides--Environmental aspects. I. Davison, Alan. II. Title.
TD196.F54W44   2004
363.738--dc21

                                                        2003013590

ISBN 0 85199 683 3

Typeset by AMA DataSet Ltd, UK.
Printed and bound in the UK by Biddles Ltd, King's Lynn.

Contents

The colour plate section can be found following p. 118

Preface

Fluoride emissions from volcanoes and the natural occurrence of excessive amounts of fluoride in drinking water have affected the health of humans and livestock for centuries, if not millennia. In addition to these natural sources, the use of fluoride as a flux, as a feed material in the chemical industry, its occurrence as a contaminant in phosphates and coal, and its use in a myriad of organic compounds have led to widespread air pollution and environmental damage. For over a century inorganic fluoride emissions damaged crops, forests and natural vegetation, and they caused fluorosis in factory workers, livestock and wild mammals. There is no doubt that inorganic fluoride was one of the major air pollutants of the 20th century but in the last 40 years there have been enormous improvements in the containment and scrubbing of emissions, so at the present time modern fluoride-emitting industries generally have little or no environmental impact outside the factory fence. Very few organofluorides were manufactured until the 1930s then CFCs (chlorofluorocarbons) transformed refrigeration and air conditioning, leading to enormous social changes in regions with hot climates. Later they were found to be the cause of ozone depletion in the stratosphere and contributors to climate forcing. They are being phased out but over a million organofluorides are known, including agrochemicals, surfactants, insulating materials, fabrics and life-saving pharmaceuticals such as anaesthetics and antibiotics. They are a vital part of modern society but their effects on the environment are still largely unexplored.

There are thousands of publications that relate to fluorides in the environment but most are specialized, and there is no recent text that attempts to provide an introduction to this large, complex subject; thus the main aim of this book is to provide a source of information for engineers, regulators, academics, students and the public. However, the literature is dominated by publications on industrial air pollution and fluoridation of water supplies so it has a bias towards the problems encountered in the more developed countries. There is much less information available about the immense problems caused by excessive amounts of natural fluoride in water supplies in the less developed countries and about the effects of organofluorides, so a second aim is to redress the balance and provide a more global picture.

Finally, anyone looking at the literature for the first time, and especially if they use the Internet, will immediately become aware of the controversy surrounding fluoridation of public water supplies. Since fluoridation was introduced in 1945, there has been strong support from its proponents but strenuous opposition from a vocal minority of opponents. Scientific debate is essential to all environmental matters but the problem is that the anti-fluoridation lobby does not use robust science, relies too heavily on non-peer reviewed documents and resorts to conspiracy theories and scaremongering. The Internet is awash with web sites that claim that fluoridation constitutes a plot to dispose of waste industrial fluoride, and a communist plot to weaken the moral and physical fibre of developed nations (epitomized by the movie *Dr Strangelove*). A recent (July 2003) article included at the top of its list of reasons for not fluoridating water, a statement declaring that '. . . fluoride is the active ingredient in most pesticides . . .'. This is, of course, nonsense but it is this kind of falsehood that may confuse the reader who has little knowledge of chemistry or pesticides, so our final aim is to provide background information to allow readers to make more informed judgements about environmental questions relating to fluoride.

Acknowledgements

Books are rarely written without lots of help, and this book is no exception. For example, whenever we had a need for the latest information or to review the chapter on naturally occurring organofluorine compounds in plants, we turned to our guru, Dr Laurie Twigg of the Western Australia Department of Agriculture, whose responses were always immediate and thorough. For questions relating to biochemistry, we thank Dr Gary W. Winston, a Chief Toxicologist with the Ministry of Health in Jerusalem, Israel, and a very close confidant to LHW. We thank Robert E. Feder, Esq. of Cuddy & Feder, White Plains, NY, who is not only a dear friend to both of us, and legal advisor to LHW, but also obtained the transcripts of the actions in the several *Martin* v. *Reynolds* legal actions, and gave generously of his time to clarify the legal lingua. David Doley of the University of Queensland in Brisbane, Australia, our friend and a fine plant physiologist, provided the list of Australian and New Zealand plant species given in Table 6.1. His book complements several aspects of this publication. Answers to several questions relating to effects of fluoride on teeth were generously provided by Dr Gary Whitford of the University of Georgia Medical College, Augusta, GA. When it came to writing the chapter on fluorocarbons, Archie McCulloch (Marbury Consulting) provided masses of information and useful comments on the draft.

Thanks are due to the Boyce Thompson Institute for its generous support. Kathleen Kramer, the Boyce Thompson Institute librarian, went far beyond any reasonable expectations by poring through web sites and libraries to acquire references, reprints and copyright permissions. Joan Curtis and Elaine van Etten, of the Boyce Thompson Institute Computer Services, kept one of us relatively computer literate and Dr Richard C. Staples, Emeritus Plant Pathologist at BTI, gave one of us (LHW) almost daily computer advice.

Bernard Willson (ex-Alcan, UK) and the late Serge Friar (Alcan, Canada) were responsible for the authors' first meeting and the start of a lasting friendship that resulted in this book. Both Bernard and Serge were excellent sources of information about the aluminium industry, emissions and regulation. We must also thank the late James Alcock for his advice and information about fluoride effects on animals. Field work will never be the same without Jim's acerbic comments and constant stories about veterinary life. The chapter on animals would not have been the same without the sterling work of one of AWD's first graduate students, Professor John Cooke, and Professor Mike Johnson and their excellent team of workers. Their contribution is gratefully acknowledged. Thanks are also due to the many others who have helped, including: Peter Jones (ex-Anglesey Aluminium), Andy Lindley (INEOS Fluor Ltd), Professor Keith Gregg (Murdoch University), Dr Darryl Stevens (CSIRO) and Dr David Imrie. Thanks go to Patrick Atkins, Director of Environmental Affairs at Alcoa, Inc., and Michel Lalonde, Director for Environmental Health and Safety at Alcan, Inc., for their generous funding of the colour plates in this volume.

And last, but foremost, we thank our wives Sylvia and Carole whose inordinate patience, good humour, support and love carried us to the last word. Thanks for the many days they have spent waiting patiently under trying conditions while we waded through forests and herbage enthusing about plants and fluorides.

1

The History of Fluorine and Sources of Fluorides in the Environment

Introduction

The fluoride[1] story may begin as long ago as the time of Pliny the Elder, who, it is believed by some, was dispatched by fluoride-containing fumes from a Vesuvian eruption. Whether the story is true or not, fluoride was certainly the agent responsible for the death of sheep after the volcanic eruption described in the Icelandic sagas, and fluoride emissions from volcanoes continue to affect the health of humans and livestock today (Georgsson and Petursson, 1972; Fridriksson, 1983; Araya et al., 1990; Baxter et al., 1999; Cronin and Sharp, 2002). The study of fluorine, fluorides and their environmental impact began much later, in the 16th century, with the use of fluorspar (fluorite, CaF_2) as a flux in metal smelting.[2] The German physician Georgius Agricola described how miners used rocks called fluores (fluorspar) to aid the smelting of ores. The name fluores is derived from the Latin fluere, meaning to flow, and in smelting the fluores acted as a kind of solvent, reducing the amount of heat required. More than a century later, in 1670, it is said that Heinrich Schwanhard of Nuremberg, a member of a family of glass-cutters, found that when he treated fluorspar with strong acid the lenses of his spectacles were etched. This led him to develop a new method for etching glass and he used it to make high-quality decorative glassware. At about the same time Venetian craftsmen produced semi-transparent glass by adding fluoride to the matrix.

Over the next century or so several chemists noted that, when fluorspar was dissolved in sulphuric acid, glass retorts were etched and even penetrated. This was due to acid fumes dissolving the silicate. Among these was the Swedish chemist Scheele, who proposed that fluorspar contained an acid, which he called fluorspar acid or fluoric acid. When muriatic acid was found to consist of hydrogen and chlorine, it was suggested that fluoric acid might be a compound of hydrogen and an unknown element with properties similar to chlorine. It took many decades to isolate the element, and the fumes given off during experiments affected the health of several chemists, including Davy, Gay-Lussac, Thenard and the Knox brothers. Some died through exposure to the fumes, notably Louyet and Nicklès. Eventually, in 1886 Moissan devised an ingenious electrolytic method and the element fluorine was finally isolated. Because his health was so affected, Moissan is said to have remarked: 'Le fluor aura raccourci ma vie de dix ans'.

For over two centuries the only way of determining the presence of fluorine was by converting it to hydrofluoric acid and etching glass. Ingenious methods were invented to determine the concentrations of fluoride in materials but they all had very low sensitivity. Nevertheless, fluorine was found to be in rock-phosphate deposits in 1790 and from the 1800s onwards it was found in bones, teeth and urine. Despite the shortcomings of the analytical methods, it was even realized that there was usually a higher concentration in fossil than in fresh bones. This was because fluoride is absorbed from the surrounding soil and water over time. Middleton (1845) understood this and suggested that the fluoride content could be used to determine the geological age of fossil bones – a technique that is still in use today (Johnsson, 1997). By the middle of the 19th century fluorides had been detected in blood, plants, hay, coal, rocks and sea water. In 1852, George Wilson made the important statement that animals get their fluorine from their drinking-water and the vegetables they eat, so it was beginning to be recognized that fluorides are found in all parts of our environment and that they are cycled from soil to plants and animals.

Interest in the medical aspects of fluorides started early, one bizarre example being cited by Meiers (2003). In this case, a Doctor Krimer suggested that fluoric acid might be used to dissolve accidentally swallowed pieces of glass. To test the idea he tried drinking a few drops of dilute hydrofluoric acid solution, and, not surprisingly, he became ill! Some of the most far-sighted medical ideas arose in relation to teeth because it seems that there was, by the 1870s, a firmly held view that fluoride had a role to play in protecting teeth from caries. Meiers (2003) cites Doctor Erhardt, who in 1874 recommended potassium fluoride for the preservation of teeth. In doing so he referred to Hunter's fluoride pastilles, which had been introduced in England some years before. The pastilles were recommended especially for 'children during dentition and for women during pregnancy when the teeth so frequently suffer'. Sir James Crichton-Browne (1892) was also convinced of a link between fluoride and dental health. In an address that was published in the *Lancet* he remarked that caries was becoming much more prevalent in England and went on to say:

> It is impossible to believe that the British Empire would have become what it is today if amongst those hardy Norsemen who pushed up their keels upon the shore . . . and entered upon the making of England there had been only one sound set of teeth in every ten.

He thought that one of the reasons for the poor state of the nation's teeth was a lack of fluorine in the diet and so he said, 'I think it is well worthy of consideration whether the reintroduction into our diet, and especially the diet of childbearing women and of children, of a supply of fluorine in some suitable natural form.' In 1907 the German chemist Deninger (Meiers, 2003) recommended calcium fluoride pills for the prevention of tooth decay, but fluoridation of water was not instituted until 1945 in the USA.

Awareness of the deleterious effects of fluorides on the health of industrial workers came principally with the publication of the classic work by Roholm (1937). At about the same time, the first publications appeared that identified the cause of a crippling condition in India as being due to excess fluoride in the drinking-water (Shortt *et al.*, 1937). However, the condition, known as endemic fluorosis, must have occurred ever since humans settled in areas that have high fluoride in the water, so it is an ancient problem. Subsequently, the condition has been identified in at least 30 countries and in some it is a major socio-economic problem that dwarfs the effects of industrial fluoride pollution.

The toxic properties of soluble inorganic fluorides were also known and exploited in the 19th century. One of the most important developments was in the 1880s when fluoride began to be used to control unfavourable microbial activity during malting and fermentation. It was found that microbial metabolism was inhibited and that led to sodium fluoride being used as a food preservative. Inorganic fluorides have also been used as insecticides for over a century, the first known patent being by Higbee in 1896 (Meiers, 2003). Eventually it was realized that the toxic action of inorganic fluoride is due to the fact that it inhibits certain enzymes (e.g. Kastle and Loevenhart, 1900). That property was put to very productive use in the first half of the 20th century by biochemists who used fluoride as a tool to help them determine biochemical pathways (Emden and Lehnartz, 1924; Warburg and Christian, 1942), and that in turn contributed to our understanding of how fluoride affects plants and animals.

Air pollution by fluorides must have started with the use of fluorspar in metal smelting, but there do not appear to be any records of plant or animal damage until 1848, when Stöckhardt and Schröder (cited in Weinstein, 1977) published the first detailed description of fluoride injury to plants. From the mid-19th century onwards injury was described in plants growing near HF- and phosphate-manufacturing plants, brickworks, glass factories and copper smelters (Weinstein, 1977). Meiers (2003) stated that in 1894 workers in Silesia experienced problems from the fluorides evolved from phosphate manufacture and reported etching of glass windows in the neighbourhood. Plant damage due to emissions from aluminium smelters was not described until much later, in the decade from about 1915 to 1925, but it became much more frequent with the huge increase in manufacture of aluminium during and after the Second World War.

Only a few organic fluorine compounds were manufactured in the 19th century but in the 20th century there was an explosive increase in the number of new organofluoride compounds that were synthesized and released into the environment. Awareness of the environmental effects of some of these compounds began with concerns over chlorofluorocarbons (CFCs) and it continues in relation to pesticides, surfactants and a number of other compounds (Key et al., 1997). Organofluorides are discussed in Chapters 8 and 9.

The Properties of Fluorine

Standard inorganic chemistry texts have accounts of the physicochemical properties of fluorine, so we shall give only a brief outline of a few of the features that are of importance in an environmental context.

Fluorine is the most reactive and the most electronegative of all elements, meaning that it has a powerful attraction for electrons and that it is able to attack all other elements, with the exception of oxygen and nitrogen, so it is not found in the free elemental state in nature. Fluorine can form both covalent and electrovalent bonds with other elements and its extreme electronegativity makes the covalent bonds strongly polar. The small size of the covalently bound fluorine atom and its high electronegativity also allow it to engage in hydrogen bonding. Fluorine forms very strong bonds with carbon so they are resistant to chemical and biological attack, a feature that is of great industrial, medical and environmental importance and which is discussed further in Chapters 8 and 9. The small size of the atom (1.36 Å) also confers upon fluoride salts very different properties from those of the other halides (e.g. chlorine, 1.81 Å), which usually have lower boiling- and melting-points. Also, because of its electron configuration, there are no known compounds in which fluorine has a valency greater than one, separating fluorine from the other halogens, which may have variable valencies.

The fluorine atom can substitute for hydrogen atoms and hydroxyl ions in molecules. For example, it can substitute for most of the hydrogen atoms in hydrocarbons, which is one of the main reasons why chemists are able to synthesize so many organofluorine compounds. Substitution alters the properties of the molecules, so fluorination can be used to design molecules with desirable properties. Similarly, when fluorine substitutes for hydroxyl ions in the calcium phosphate that forms the basis of bone, it makes the mineral harder and more resistant to attack by acids.

The naturally occurring form of fluorine is ^{19}F. Other isotopes, prepared by nuclear reactions, have short half-lives, the longest of which is ^{18}F at 1.87 h. Radioactive isotopes play an important part in biological research because they can be used to trace biochemical pathways and movements of ions, but the very short half-life of ^{18}F restricts its use, particularly for work with plants, where processes are slower. However, that same short half-life is put to use in positron emission tomography, a non-invasive imaging technique that is an important tool used in clinical research. The fluorine is incorporated into a suitable molecule and the emission of positrons from ^{18}F allows the quantitative determination of the distribution and metabolism of the molecule in the human body.

Hydrogen fluoride, which is central to much of this book, is used extensively in industry, especially as an intermediate in the manufacture of most fluoride-containing products, many of which are discussed in Chapters 8 and 9. It is a colourless gas or fuming liquid (hydrofluoric acid) and it is strongly irritant and very corrosive. The odour detection limit is around 30–130 μg/m^3. Hydrogen fluoride is produced in industry in the same way that it has been for four centuries, by the treatment of CaF_2 (fluorite) with concentrated sulphuric acid. The product is volatile HF, which is condensed and purified by distillation. In the past, industrial atmospheres often contained several mg/m^3, but they are now generally much lower. Humans are reasonably tolerant of gaseous hydrogen fluoride but it is the most toxic of all air pollutants where plants are concerned and it may also have profound effects on grazing animals if excessive amounts contaminate their forage. Effects on animals and plants are considered in Chapters 3 and 4.

Analytical Determination of Fluorine

Before outlining the sources and concentrations of fluoride in the environment, it is necessary to comment on the reliability of fluoride determinations. When glass etching was the basis of the method it obviously lacked sensitivity and reliability, and there was a great deal of heated debate about the data that were produced. The introduction of colorimetric techniques increased sensitivity and they formed the basis of determination until the 1960s, when the fluoride-specific ion electrode was introduced. It is probably the most widespread method of detection and it can be used routinely to determine background levels of fluoride in most materials. Other techniques currently in use include ion chromatography and the electron microprobe.

However, it is not only the detection method that affects the results; equally important are having a sampling protocol that ensures that the material is representative and having an appropriate method of sample preparation. There are many publications in which sampling protocols are either inadequate or not well defined, and others in which sample preparation was not in accordance with good laboratory practice. What this means is that published data on the fluoride content of air, rocks, soil, water and biological materials have to be judged in relation to the sampling and analytical techniques that were used. All data on fluoride contents have to be interpreted with caution. For example, McClure (1949) carried out an extensive survey of the fluoride content of foods but many of his values appeared to be excessive and, as pointed out by Singer and Ophaug (1983), at that time there had been little documentation of the accuracy and precision of the methods used to analyse fluoride in foods. In an attempt to assess the reliability of methods used routinely for the determination of fluoride in vegetation, Jacobson and McCune (1969, 1972), Jacobson and Heller (1978) and Jacobson et al. (1982) carried out interlaboratory collaborative studies in the USA and other countries. In the first study (Jacobson and McCune, 1969), 31 laboratories analysed five different plant species using their routine analytical methods, and wide differences in results were obtained. The results of duplicate analyses indicated that some laboratories had difficulty in obtaining reproducible results (Table 1.1). They concluded that variations in procedures, techniques and operating conditions were likely causes for the large relative standard deviations between laboratories (12.7% to 2.4%). In a follow-up study with 64 participants (Jacobson and McCune, 1972), effects on the magnitude and variability of analytical results were evaluated. Detailed procedures were submitted by the

Table 1.1. Results reported by 31 laboratories[a] for analyses of samples of leaf tissues of several plant species (mg F/kg dry weight) (from Jacobson and McCune, 1969). (Reprinted from the *Journal of AOAC International*, 1969, 52, 894–899. ©1969, by AOAC International.)

Laboratory number	Lucerne	Citrus	Gladiolus	Pine	Cocksfoot grass
1	39.4	121.3	38.9	61.2	44.3
2	70.0	126.0	55.0	72.0	56.0
3	59.5	131.5	63.5	84.5	66.0
4	51.7	108.0	49.4	68.9	66.0
5	33.7	107.3	39.4	62.5	31.5
6	28.0	122.0	44.5	50.5	37.0
7	137.5[b]	123.7	91.0	80.7	58.4
8	59.5	118.5	40.0	80.0	64.0
9	45.7	48.6[b]	41.5	14.7[b]	145.4[b]
10	56.5	130.5	54.0	79.5	52.0
11	66.0	87.5	74.5	40.8	83.0
12	59.0	147.5	62.5	100.5	44.0
15	67.0	135.9	59.0	83.7	70.5
16	53.0	113.5	58.0	79.5	59.0
17	42.5	90.0	39.0	62.5	42.0
18	56.9	124.5	54.6	85.7	65.7
19	53.2	117.7	47.6	65.0	50.1
20	50.4	115.8	46.4	72.4	48.0
21	77.5	114.7	96.7	87.2	76.0
22	55.0[b]	125.0	56.5	76.0	61.0
23	51.5	109.5	49.5	73.5	60.0
24	55.0	123.5	52.0	78.5	58.0
25	58.5	129.0	57.5	71.5	60.0
26	56.5	121.0	53.5	96.0	66.0
27	40.0	127.0	42.0	59.0	24.0
28	46.0	111.0	44.5	68.5	48.5
29	39.0	89.0	34.5	67.0	52.0
30	65.5	107.0	46.6	77.0	60.8
31	46.0	123.5	55.0	79.0	47.0
32	52.5	86.5	50.0	80.0	52.0
33	60.5	137.0	61.0	88.0	118.0[b]

[a]Numbers 13 and 14 were not assigned. [b]Aberrant results not included in calculation of mean and standard deviation.

participants along with their results. Relative standard deviations were high and ranged from 14% to 27% for samples containing more than 27 mg/kg fluoride. The authors provided detailed statistical analyses of the analytical results, pooled according to the techniques used by the participants. From these two studies (Jacobson and McCune, 1969, 1972), it was clear that a major source of variation was the amount of fluoride in the samples and the type of vegetation analysed. In a later study (Jacobson *et al.*, 1982), it was found that, despite improvement in speed and simplicity of fluoride analysis, agreement between laboratories had not improved because of the variety of methods and techniques used, the inherent differences between methods and poor laboratory quality control. In a recent study, which included 40 laboratories from the USA, Canada, Great Britain, Brazil and Australia (L.H. Weinstein and D.C. McCune, unpublished), one method in common use was found to give very low results for some samples, especially for the US National Institute of Standards and Technology standard grass sample. This level of variability has serious implications for any situation where monitoring, enforcement or litigation are involved but it is still not given sufficiently serious consideration. Clearly, laboratories should use good quality-assurance practices and include standard samples in their routine but the inescapable message is that data on fluoride concentrations have to be treated with caution until there is confirmation by independent sources.

Natural Sources

Minerals

Fluorine is widely distributed throughout the earth's crust as the fluoride ion. Fluorine is reported to be the 13th (Smith and Hodge, 1979) or 17th (Fleischer, 1953; Bell *et al.*, 1970; NAS, 1971) most abundant element in the earth's crust, occurring in igneous and sedimentary rocks at concentrations estimated to be between 0.06% and 0.09% by weight of the upper layers of the lithosphere (Koritnig, 1951). According to Fleischer (1953), fluorides account for 0.032% of the earth's crust. Several hundred minerals are known to contain fluoride but they vary greatly in fluoride content, from as high as 73% in the rare mineral griceite (LiF) to many others with less than 0.2%. However, the major

minerals that are exploited commercially are fluorspar (fluorite) (CaF_2), fluorapatite ($Ca_{10}F_2(PO_4)_6$) and cryolite (Na_3AlF_6). The environmental transfer and cycling of fluoride is depicted in Fig. 1.1.

The term fluorite is used to denote the natural mineral form of calcium fluoride, whereas the term fluorspar often includes not only the natural mineral but also manufactured calcium fluoride. The identified world reserves of fluorite are around 500 million t and annual mine production is in excess of 4.5 million t, the main producing nations being China, Mexico and South Africa (USGS, 2002). By far the main use of fluorspar is as the source of hydrofluoric acid, which is a feed chemical for a multitude of processes that produce thousands of inorganic and organic fluorine compounds, including insecticides, pharmaceuticals and fabric conditioners. In the USA, for example, about 80% is used for hydrofluoric acid production and the rest is used as a flux in a variety of industries. Mining the mineral leaves solid waste and, in places such as the Pennines of England, the extracted waste still has very high concentrations of fluoride. This is reflected in the fact that vegetation growing on such material has the highest concentrations of fluoride recorded (Cooke *et al.*, 1976; Andrews *et al.*, 1989;

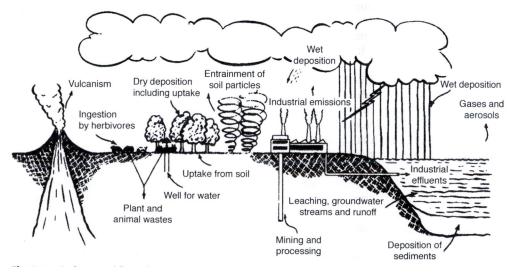

Fig. 1.1. Pathways of fluoride movement and transfer in the environment (redrawn and modified from Weinstein, 1977). (Courtesy of ACOEM.)

Burkinshaw *et al.*, 2001), much higher than are usually found in relation to air pollution. Furthermore, the soils are usually contaminated with lead, zinc and cadmium so plants may experience multiple stresses.

The mineral apatite occurs in a variety of forms but the most economically important deposits are contained in the rock phosphates that are mined for the production of phosphorus and phosphate fertilizers. Rock phosphates are mostly ancient marine sediments, which were first used as fertilizer around the middle of the 19th century. Ever-increasing demands led to huge amounts being mined and, today, about 130 million t are mined each year in over 40 countries (USGS, 2001). The USA is by far the biggest producer, at about 32 million t, but the largest reserves are in Morocco and the western Sahara. About 90% of US production is used for fertilizer and animal feed supplements, so much of the fluoride in them is eventually deposited on to the soil. Current world use of phosphate fertilizer is just above 30 million t/year and the fluoride content ranges from under 1.5 to over 3% (McLaughlin *et al.*, 1996). Phosphate that is destined to be used in animal feeds has to be defluorinated to prevent fluorosis, while processing the raw rock phosphate leads to the release of HF and SiF_4 into water and the air. Consequently, phosphate plants and fertilizers are potentially important sources of fluoride contamination.

Roholm (1937) summarized the basic information on cryolite. It is rare in nature, being found mainly in Greenland and in much smaller quantities in the Urals and Pikes Peak, Colorado. The Greenland deposit was the only commercially useful source until it was exhausted. The mineral source at Ivigtut, Greenland, consisted of a huge deposit of pure cryolite with impurities of quartz, siderite, zinc blende, galena, chalcopyrite and iron pyrite. There were also sporadic occurrences of several other fluorine-containing minerals: pachnolite ($AlF_3.CaF_2.NaF.H_2O$), hagemannite (blends of themsenolite, iron compounds, etc.), cryolithionite ($2AlF_3.3NaF.3LiF$), chiolite ($3AlF_3.5NaF$) and fluorite (CaF_2). Cryolite has had many industrial uses over more than 100 years. It was used to produce crystal soda between 1859 and 1870, by mixing the cryolite with lime and applying great heat. By about 1870, it was being used in the manufacture of opal glass. As the molten glass cools, fluoride separates out into tiny crystals of sodium fluoride, providing the glass's opalescence. Somewhat later it was used in the enamel industry. Beginning about 1890, a unique use of cryolite emerged in the smelting of certain metallic raw materials, the most important of which was aluminium metal by the Hall–Héroult reduction process. In this process, alumina (Al_2O_3) is reduced electrolytically to aluminium metal in a bath of molten cryolite, the cryolite acting as both a flux and an electrolyte. During the 25-year period before 1937, the aluminium industry consumed about 60% of the cryolite produced, with 27% going to the enamel industry, 10% to the glass industry and about 3% for all other purposes. Although the large natural sources of cryolite in Greenland have been depleted, the enormous quantities used in aluminium smelting throughout the world are now produced synthetically. For a period of time from the 1930s onwards, cryolite was a popular insecticide; then its use declined, but recently it has reappeared, especially in the USA (Chapters 2 and 3). The attraction of cryolite for this purpose is that it is very effective against some target species, it has low solubility so it is not dispersed too quickly and it is relatively environmentally safe. Because it is a natural product it has been classified as suitable for organic crop culture.

Rocks and soil

The main reservoirs of fluoride in the biosphere are surface rocks and deposits, soil and the oceans. Estimates of the fluoride content of rocks vary from < 100 to > 1000 mg F/kg. Fleischer and Robinson (1963) quoted average fluoride concentrations for various rocks in the USA as: sandstone, 180; andesite, 210; limestone, 220; dolomite, 260; basalt, 360; and shale,

800 p.p.m. (= mg/kg). There are similar data for many other countries, such as Tibet, where Zhang *et al.* (2002) found averages from 389 to 609 mg F/kg, including sandstone, igneous rocks, shale and limestone.

Soil fluoride is derived from the parent materials by weathering and the actions of microorganisms, plants and animals. Some is also derived by deposition on the surface from the atmosphere or by flooding, but in non-industrial areas this is a very small part of the total soil pool. Reports of the fluoride content of soils vary from under 20 to several thousand mg/kg (Davison, 1983). The higher concentrations are mostly found in areas with sedimentary phosphate or fluorite deposits but in their absence, fluoride concentrations range up to about 700 mg F/kg (NAS, 1971; McLaughlin *et al.*, 1996; Cronin *et al.*, 2000). Most of the fluoride is associated with the clay fraction, so heavier soils tend to have substantially higher concentrations than do sandy soils. Gemmell (1946), for example, reported New Zealand soils to range from 68 mg F/kg in a coarse sand to 540 in a clay, while Nommik (1953) reported Swedish sandy soils to range from 43 to 198 mg F/kg and clay soils from 248 to 657 mg F/kg. Where the underlying parent rocks vary in fluoride content the soil concentration can vary over short distances, as shown, for example, in work by Geeson *et al.* (1998).

Natural waters

The concentration of fluoride in groundwater depends on the geology, chemistry, physical characteristics and climate of the area. Generally, spring and well waters tend to contain higher concentrations of fluoride than surface waters from lakes and streams. Data are available on the concentrations of fluoride in water in different countries (WHO, 1970, 2002a) but they give only a general indication because few surveys have been designed to give robust descriptive statistics. In general, the literature suggests that, if water is not in contact with high-fluoride minerals, the range of

concentrations is from about 0.01 to 0.4 mg/l (WHO, 2002a), but even where most water is within this range there may be local areas where fluorite or phosphate deposits lead to much higher concentrations. There may be considerable variation over short distances in relation to the underlying geology. Lalonde (1976), for example, found that the concentration went from < 0.1 to > 0.8 mg/l over a distance of about 8 km, and he considered that fluoride concentration is a useful indicator of mineral deposits. Even more extreme variation was reported by Banks *et al.* (1998) in Norway, where the fluoride ranged from < 0.05 to 8.3 mg/l over short distances in relation to geology.

The highest fluoride concentrations tend to occur in arid regions, where evaporation concentrates the fluoride. Wang and Cheng (2001) described an arid area of northwest China that covers at least 300,000 km^2. Some of the rocks have minerals containing 2–5% fluorine and groundwater contains from < 0.5 to 5 mg/l. Moving from the mountains to the arid basins, the fluoride content of the water in this region increases due to leaching, changes in water chemistry and evaporative concentration (Fig. 1.2).

The Rift Valley of East Africa has the highest concentrations on record because up to 300 mg/l and 700 mg/l have been reported for parts of Ethiopia and Kenya, respectively (Chernet *et al.*, 2001). Dissanayake and Chandrajith (1999) cite data of Aswathanarayana *et al.* indicating that rivers, ponds, thermal lakes and soda lakes of Tanzania have up to 26, 65, 63 and 690 mg F/l, respectively. In the lakes region of Ethiopia, Chernet *et al.* (2001) have described how evaporation in the arid environment leads to concentration of fluoride. The solubility of fluoride is controlled by equilibrium with CaF_2, but concentration leads to precipitation of calcite ($CaCO_3$) and that increases the solubility of fluoride: hence the very high concentrations.

Fluorosis (Chapter 3) is a potentially serious health problem for humans and livestock that occurs when the fluoride content of drinking-water is above *c.* 1.5 mg/l, so an indication of the global extent of

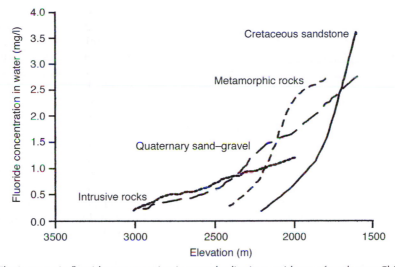

Fig. 1.2. The increase in fluoride concentration in water bodies in an arid area of north-west China with decreasing altitude. The effect, which is due to leaching, changes in water chemistry and evaporative concentration, is most pronounced in areas with Cretaceous sandstone. (Graph redrawn from Wang and Cheng (2001). Wang, G.X. and Cheng, G.D. (2001) Fluoride distribution in water and the governing factors of environment in arid north-west China. *Journal of Arid Environments* 49, 601–614.)

high-fluoride water can be gained from the fact that the United Nations Children's Fund (UNICEF) has identified 27 countries where there is sufficient fluoride in water-supplies to cause fluorosis in humans and livestock. The list is incomplete because areas with > 1.5 mg/l are known in many other countries, including Canada, the USA, Ghana, England and Norway. The countries listed by UNICEF are:

Algeria	Iraq	Palestine
Argentina	Japan	Senegal
Australia	Jordan	Sri Lanka
Bangladesh	Kenya	Syria
China	Libya	Tanzania
Egypt	Mexico	Thailand
Ethiopia	Morocco	Turkey
India	New Zealand	Uganda
Iran	Pakistan	United Arab Emirates

The fluoride content of fresh water is increased by several human activities, such as fluoridation and disposal of sewage and industrial effluents. When drinking-water is fluoridated, the target concentration is 1 mg/l. Contaminants in sewage and industrial waste water also increase the fluoride concentration of fresh water, but potential environmental effects have to be interpreted in the context of other contaminants, which may be more important. This is illustrated by a survey of users of recycled water who were asked about concerns relating to its quality (Higgins *et al.*, 2002). About 79% of respondents had concerns, but fluoride was rated near the bottom of a list of 67 components, which included at the top: pathogens, salinity and phosphorus.

When fluoride is in solution, it is necessary to know not only the total amount present but also the chemical species, because they differ in bioavailability and toxicity (Chapters 2 and 3). In fresh waters the speciation depends on the pH and the presence of complexing agents. Aluminium complexes predominate in acid waters (e.g. Bi *et al.*, 2001; Environment Canada, 2001), but as the pH increases anionic F^- prevails (Table 1.2).

In estuaries the fluoride concentration varies greatly because of the mixing of river and sea water. Chamblee *et al.* (1984) reported concentrations at various positions in the Pamlico estuary as ranging from 0.09

Table 1.2. The chemical forms of fluoride present in water containing 1 mg F/l (data from Jackson *et al.*, 2002). (From Jackson, P.J., Harvey, P.W. and Young, W.F. (2002) *Chemistry and Bioavailability Aspects of Fluoride in Drinking Water.* Report No. 5037. Wrc-NSF Ltd, Henley Road, Medmenham, Marlow, Bucks SL7 2HD, UK.)

			Per cent of total dissolved fluoride		
pH	F^-	AlF^{2+}	AlF_2^+	AlF_3^0	AlF_4^-
6	21.35	5.36	60.4	12.82	0.08
7	97.46	0.02	1.26	1.22	0.03
8	100	0	0	0	0
9	100	0	0	0	0

to 2.45 mg total F/l, but the highest levels were associated with a local industrial source. The same authors found that the fluoride concentration increased with salinity, which is not surprising, because the fluoride concentration in sea water is higher than in most surface fresh waters, averaging about 1.3 mg/l (Carpenter, 1969; Environment Canada, 2001). Although most reported concentrations for sea water are between 1.2 and 1.4 mg/l (Carpenter, 1969), there are also many anomalously high and low concentrations in the literature (Smith and Hodge, 1979). High concentrations were reported from sea water collected in deep waters off the French coast and unusually low values have been reported from the Mediterranean Sea, the Red Sea, the Baltic Sea and near the Azores (Smith and Hodge, 1979). It has been suggested that the high values may be related to nearby industrial effluents into the sea, although it is also possible that anomalous values were the result of analytical deficiencies.

On the basis of his estimate that sea water has an average of 1.3 mg F/l, Carpenter (1969) estimated that, with a total ocean volume of 1.37×10^{21} l, the total oceanic reservoir of fluoride is 1.781×10^{12} t. Fluoride is constantly entering the oceans in river water and there is an input from undersea volcanoes and vents. There is also an input by deposition of gases and particulate materials from the atmosphere, but it is small in comparison with the former sources. Industrial discharges direct to the sea provide local increases in fluoride but in most situations these are rapidly mixed and diluted, so the effects are local. As the concentration in the sea is constant, the inputs must be balanced by processes such as incorporation into the calcium carbonate- and phosphate-containing hard tissues of marine organisms (Carpenter, 1969), sedimentation of undissolved materials and formation of fluorapatite in the sediments. In addition, about 20 kt of fluoride have been estimated to be released from the ocean surfaces into the atmosphere as aerosols (Cadle, 1980, cited by Symonds *et al.*, 1988). The average residence time for oceanic fluoride was estimated to be $2-3 \times 10^6$ years (Carpenter, 1969). In the sea, the speciation of the fluoride is different from that in fresh waters because of the dominance of magnesium in sea water (Table 1.3). About 50% is present as F^- and 47% as MgF^+.

The natural background concentration of fluoride in the air: volcanoes, forest fires and the oceans

As indicated in the opening paragraphs of this chapter, volcanoes and the crevices, vents and hydrothermal systems around them are a source of atmospheric fluorides (Symonds *et al.*, 1988; Halmer *et al.*, 2002). Burning timber in forest fires also makes a contribution and some originates from ocean surfaces (Cadle, 1980, cited by Symonds *et al.*, 1988), but it is the 600 known, active volcanoes that provide most of the background fluoride that is in the air. As the magma rises to the surface, the decrease in pressure causes dissolved gases to form bubbles and they are released when

Table 1.3. The chemical forms of fluoride present in sea water (cited in Environment Canada, 2001). (Data from Stumm and Morgan (1981) and Miller and Kester (1989), cited in *Environment Canada*, 2001.)

Stumm and Morgan (1981)			Miller and Kester (1989)	
Form of fluoride	% of total	Concentration (mg/l)	Form of fluoride	% of total
F^-	51	0.78	F^-	50
MgF^+	47	0.7	MgF^+	47
CaF^+	2	0.03	CaF^+	2.1
			NaF^0	1.1

the melt rises to the surface and as it erupts. Halmer *et al.* (2002) have estimated the annual global emission of HF from volcanoes as being from 7000 to 8600 kt. However, concentrations decrease rapidly with distance from the point of emission so the environmental effects are restricted to the vicinity of the source and to periods just after major events. At distances more than a few kilometres from these natural sources, the background fluoride concentration in the atmosphere is very low. Data cited by the World Health Organization (WHO, 2002b) indicate that the total atmospheric inorganic fluoride (gaseous + particulates) is usually well under 0.1 and mostly < 0.05 μg/m³.

Fridriksson (1983) published a comprehensive review of volcanic eruptions in Iceland and evidence of the toxicity of the emissions (Chapter 5). There are more recent assessments, such as Baxter *et al.* (1999) for an island in the Azores and Cronin and Sharp (2002) for islands in the Vanuatu archipelago. Volcanic eruptions cause damage to biota through the explosive shock or heat, choking ash, mud flows, collapse of buildings, asphyxia by CO_2, poisoning by sulphurous gases and hydrochloric acid and deposition of acid rain (Baxter *et al.*, 1999; Cronin and Sharp, 2002). In human-health terms they are much more important than fluorides. The main form of fluoride emitted is HF but emissions from Vesuvius and Vulcano in Italy have been shown to contain hydrogen fluoride, ammonium fluoride (NH_4F), silicon tetrafluoride (SiF_4), ammonium fluorosilicate ($(NH_4)_2SiF_6$), sodium fluorosilicate ($NaSiF_6$), potassium fluorosilicate (K_2SiF_6) and potassium fluoroborate

(KBF_4). In addition, Roholm (1937) also reported the presence of sodium fluoride (NaF), potassium fluoride (KF), magnesium fluoride (MgF_2) and calcium fluoride (CaF_2). Volcanoes are also an important source of organofluorides, including some CFCs. They are discussed further in Chapters 8 and 9 along with the other naturally occurring organofluorides.

Industrial Uses and Atmospheric Sources

Many types of industrial activities result in the emission of gaseous and particulate fluorides into the atmosphere. It is difficult to distinguish between those activities in which fluoride is an intrinsic part of the product and those in which fluoride is employed as a flux or catalyst. The basic processes associated with these emissions are drying, grinding and calcining of fluoride-containing minerals, their reaction with acids, smelting and electrochemical reduction of metals using fluoride-containing fluxes or electrolytes, firing of brick or ceramic materials, high-temperature melting of raw materials in glass manufacture and the use of fluoride-containing chemicals for cleaning, electroplating and etching in various processes. In addition to these industrial processes, the combustion of coal is a major source of potential fluoride emissions, and the introduction of fluoride in waters, toothpaste, fabrics, pharmaceuticals, wood preservatives and agrochemicals disseminates fluoride into the environment. Here we focus on eight of the main sources of atmospheric emissions because they have been or still are the source of most out-of-plant environmental problems.

Phosphate fertilizers and elemental phosphorus

Phosphoric acid (H_3PO_4) is produced by two methods: the thermal or furnace process and the wet process. The major emission from the thermal process is phosphoric acid mist in fine particulate form, but the wet process is a potential source of hydrogen fluoride and silicon tetrafluoride. Wet-process H_3PO_4 is mostly used in fertilizer production and in the USA it has been the source of most of the new phosphate production over the past two decades. In this process, sulphuric, nitric or hydrochloric acids are reacted with the rock phosphate. Nitric acid is no longer used in the USA and hydrochloric acid is not competitive for fertilizer production. Therefore, sulphuric is the major acid used today. Ground phosphate rock is digested by sulphuric acid in a stoichiometric ratio based upon the calcium oxide equivalents in the rock. The end-products are mostly phosphoric acid and calcium sulphate (gypsum), with the latter being pumped and stored in large ponds (or 'stacks').

The reaction of sulphuric acid with phosphate rock is a multistaged process (Slack, 1969). The first step is the reaction of tricalcium phosphate in the rock with sulphuric acid to form monocalcium phosphate, which in turn reacts with sulphuric acid to yield phosphoric acid and gypsum. Hydrogen fluoride is also formed, as follows:

$$Ca_{10}(PO_4)_6F_2CaCO_3 + 11H_2SO_4 =$$
fluorapatite   sulphuric acid

$$6H_3PO_4 + 11CaSO_4.nH_2O +$$
phosphoric   gypsum
acid

$$2HF + CO_2 + H_2O$$
hydrogen   carbon   water
fluoride   diioxide

The processing of rock phosphate also results in the emission of large amounts of silicon tetrafluoride because, in addition to apatite, the rock contains silica in the form of SiO_2:

$$SiO_2 + 4HF = SiF_4 + 2H_2O$$
silicon   hydrogen   silicon   water
dioxide   fluoride   tetrafluoride

During the evaporation step, much of the fluoride in the rock is volatilized and the fluorides produced are wet-scrubbed, often resulting in the production of commercial products, such as fluorosilicic acid, fluoro-silicates, cryolite and aluminium fluoride. Where wet-process phosphoric acid is concentrated to form superphosphoric acid, most of the fluoride is volatilized, leaving only 0.2–0.3% fluoride in the final product. Addition of reactive silica during evaporation increases SiF_4 volatilization and reduces the fluoride content to 0.1% or less. The evaporators operate under vacuum and the volatilized fluorides are scrubbed in a large volume of process water, resulting in low emissions. In Florida, average fluoride emission factors in the production of normal superphosphate are reported to be 0.10 kg/t in the mixer and den, and 1.90 kg/t from the curing building. The latter emissions are scrubbed with recycled water (Mann, 1992a). Fluoride remaining in wet-process phosphoric acid for use in animal-feed products is defluorinated by converting the remaining fluoride into silicon tetrafluoride. In many countries there are regulations governing the amount of residual fluoride allowed in animal feeds.

In older plants, the drying and grinding of phosphate rock results in the emission of a considerable amount of particulate matter. The emission is largely controlled by the use of various filtering devices, especially bag filters, although dust generated by the mechanical handling of rock is also removed by the wet scrubber that removes gaseous fluorides. Fugitive particulate emissions have little effect on plant life because of their low solubility, and the greatest effect in the event of large emissions might be a physical one on plant foliage or an increased suscep-tibility of certain plants to pest or pathogen infestation (Chapters 3 and 4). In 1969, the efficiency of the various types of cleaning equipment ranged from 95% to 99.9% (USHEW, 1969).

In the wet process, enormous amounts of slurries and other wastes are produced that contain gypsum ($CaSO_4.2H_2O$), HF, fluorosilicic acid and sodium and potassium fluorosilicates. Facilities in some coastal areas dispose of these wastes into the sea. In other areas, enormous settling ponds (usually called gypsum or phosphogypsum ponds or 'stacks') are constructed into which the wastes are directed. The gypsum stacks serve to stack, or accumulate, the gypsum by-product. The gypsum stacks are usually accompanied by process-water ponds that provide surge volume for wet and dry weather conditions and avoid water treatment and water discharge, and to conserve the make-up of fresh water. The concentrations of soluble fluoride in the ponds range from 4000 to 14,000 mg/l, with values greater than 10,000 mg/l being common (Cross and Ross, 1969; Crane et al., 1970; Tatera, 1970; Schiff et al., 1981). Emission of volatile fluoride into the atmosphere occurs both from the pond-water surfaces and from the gypsum stacks, but estimates based on fluoride vapour pressure are difficult because of variations in pond-water composition and temperature. Nevertheless, estimates have been published. Cross and Ross (1969) calculated that the emission rate was 26 lb/day (11.8 kg/day) of gaseous fluoride released from a 160-acre (65 ha) pond in Florida. In addition, a 100-acre (40 ha) cooling pond was estimated to add 16 lb/day (7.3 kg/day) to give a total of 42 lb/day (19.1 kg/day). Crane et al. (1970), using a similar technique, showed that emissions were temperature-dependent, and found that they ranged from 0.56 kg/ha/day at 50°C to about 2.8 kg/ha at 80°C. They suggested that these values were probably on the low side. Tatera (1970) made measurements of pond water transported to the laboratory and he concluded that emission rates could range from 0.36 to 20.5 kg/ha/day.

Linero and Baker (1978) reviewed the available literature up to 1978 and concluded that none of the previous studies provided a sufficient basis for defining an emission factor for fluoride. In a theoretical study based upon the vapour pressures of HF–water solutions, they estimated fluxes from 122 to 195 kg HF/day for a 450 t P_2O_5 plant, with 60% or more of the total release coming from the ponds (Moore, 1987). In 1999, LaCosse et al. made measurements over the surface of cooling ponds in Florida during both the summer and the winter, using a Fourier transform infrared (FTIR) open-path spectrophotometer coupled with a meteorological sensing station. The concentration of SiF_4 was very low and emissions from the ponds were primarily as HF. From these data, the emission rate, using a time-weighted chemical model, was vastly lower than earlier estimates at 0.018 lb/acre/day (0.02 kg/ha/day) and, using a time-weighted analytical model, it was 0.092 lb/acre/day (0.103 kg/ha/day). However, the degree of vegetation injury observed by one of the current authors (LHW) near ponds suggests that the values of LaCosse et al. (1999) may be an underestimate, at least for some ponds. Ball et al. (1999) also examined fluoride concentrations and speciation above a phosphogypsum storage area, using FTIR and a tunable diode laser. Their data showed that HF was mostly emitted from the aqueous areas and SiF_4 from the drier surfaces. Emissions of both were related to temperature and, based on the data they presented, concentrations of HF and SiF_4 at 2–3 m above the storage areas ranged from 0 to 45 and 0 to 450 µg/m^3, respectively. Concentrations as high as these would be expected to affect vegetation growing in the vicinity, so, despite the variability in the estimates of emissions, it is concluded that at least some ponds and stacks are significant sources of fluoride emissions to the environment.

Diammonium phosphate ($2NH_3.H_3PO_4$) is an important fertilizer that provides two important plant nutrients, nitrogen and phosphorus, in a relatively condensed form. It is manufactured by neutralizing phosphoric acid with ammonia and removing excess water by evaporation. Because the reaction is with phosphoric acid, not fluorapatite, fluoride emissions are low.

Triple superphosphate is produced by reacting phosphate rock with phosphoric acid or a mixture of predominantly phosphoric acid and a lesser amount of sulphuric acid. In the USA in 1990, there were only six fertilizer plants capable of producing triple superphosphate: one in North Carolina, one in Idaho and four in Florida. In 1989, about 3.5 million tons (3.2 million t) of triple superphosphate were produced (Mann, 1992b). Silicon tetrafluoride and HF emissions occur in the acidulation process from the reactors and from the den, granulator and drier. According to Mann (1992b), there is little fluoride evolved from the curing building. Scrubbing of fluoride-containing gases from the reactor, den and granulator is accomplished by cyclonic scrubbers, using recycled pond water. Emissions from the drier, screens, mills, product-transfer buildings and storage buildings pass first to a cyclone separator to remove particulate materials and then to wet scrubbers to remove fluorides. Reported efficiencies for fluoride control range from less than 90% to over 99% in different systems (Mann, 1992b). US federal regulations in 1975 limited fluoride emissions from reactors, granulators, driers, screens and mills to 0.1 kg/t of P_2O_5 fed in the process.

Elemental phosphorus is produced by an electric-furnace process. In the process, ground phosphate rock is passed through a heated rotary kiln to produce particles of suitable size, then mixed with silica and coke and finally heated in a carbon-lined furnace with carbon electrodes. In the process, carbon monoxide and phosphorus are evolved as gases, together with SiF_4 formed through a reaction of silica and fluorapatite. The phosphorus is then condensed in a water-spray tower, and the carbon monoxide is used as a fuel or burned as a waste product. Silicon tetrafluoride may escape from the furnace through seals around the electrodes or through the supply chutes for feed materials. Silicon tetrafluoride and HF are largely removed by spray towers, wet cyclone scrubbers or venturi scrubbers in modern plants, but older plants still emit a substantial amount of gaseous fluorides to the atmosphere. The total amount of fluoride emitted to the atmosphere in elemental phosphorus production in the USA in 1968 was estimated to be 5500 US tons (5000 t).

Primary aluminium smelting

The process by which alumina (Al_2O_3) is reduced to aluminium on a commercial scale was discovered independently in 1886 by Charles M. Hall of Oberlin, Ohio, and Paul Louis Toussaint Héroult of France when they were both 22 years old. They discovered that alumina dissolved in cryolite (Na_3AlF_6) will produce aluminium after undergoing electrolysis. The first commercial production of aluminium took place in 1888 in Pittsburgh, Pennsylvania (Wei, 1992).

Aluminium production is carried out in large electrolytic cells called pots with an input of direct current of up to 280,000 A and about 5 V. The pots may be up to 40 m^2 and they are lined with a refractory insulation on which are placed carbon blocks to form the cathode. The anode also consists of carbon and it is immersed in the molten cryolite to complete the electrical circuit. There are two main technologies used in aluminium production: the prebake process and the Söderberg process. Most smelters constructed in the last 30 years are of the prebake type. Söderberg cells are less costly because they avoid prebaking the anodes, but they are less efficient with respect of current use and are subject to much greater emissions of fluoride as well as many hydrocarbons, including polycyclic aromatic hydrocarbons ('pitch volatiles'), the latter being mostly consumed in the prebake process.

The average pot emits about 20–35 kg of gaseous and particulate fluorides per tonne of aluminium produced (Wei, 1992). In 1990, world production of aluminium was 17,832,000 t; thus, emissions of gaseous and particulate fluorides into the atmosphere immediately above the pots can be estimated to be between 356,640 and 624,120 t. Fortunately, due to effective containment and highly efficient scrubbers, most of this is captured and is not released into the outside environment. Each pot is enclosed by a hood in order to collect off-gases from the reduction process and the fumes are drawn into an exhaust duct. Hooding efficiency (the percentage of fumes captured) should be at least 95% for a prebaked pot but is less for Söderberg pots. Several operating factors can affect the efficiency of pollutant collection. For example, when a spent anode is removed from a pot and placed in the pot-room aisle or out of doors, there is a considerable amount of uncontrolled fluoride emissions because the hot anode continues to burn and evolve fluorides from the adhering bath material.

For many years the emissions collected in the hooding system were scrubbed using water, but in the 1970s a brilliant and extremely effective system was introduced for controlling emissions. It utilizes alumina (Al_2O_3), the ore used to produce the aluminium metal, to scrub the off-gases and then the fluoride is reclaimed. The fine alumina particles have a high capacity to chemisorb hydrogen fluoride and they can be filtered out of the airstream. The fluoride-containing alumina, or reacted ore, is then returned to the pots, giving a highly efficient system, not only for controlling fluoride emissions but in economic terms as well. There are basically two types of system in use today. Both direct the off-gases into contact with alumina. One system utilizes a fluidized bed of alumina through which the off-gases are passed and the reacted ore is collected in a baghouse. In the second system, the hot gases drawn from the pots are first passed through an evaporative cooler. Alumina is then injected into the duct work. Although much of the reaction is completed in the duct work, the final steps are completed by passing the gases and alumina to a contact chamber and a baghouse. Often, a portion of the alumina is recycled from the baghouse to increase efficiency. In both cases, the reacted ore is transported back to the pots.

In prebake smelters, the collection efficiency is 95–98% and the scrubbing efficiency is 99.99%. Thus, most of the fluoride released to the environment is gas and fine particles that are not collected by the hooding and which pass out from the roof and vents. In Söderberg smelters, collection efficiency is 80–85% but the scrubber efficiency is about 99.99%. Again, the great bulk of emitted fluoride is from the uncollected portion. Routine operations, such as replacing anodes, siphoning metal and so on, are done on a cyclical basis so emissions from individuals pots are not constant. However, smelters usually have several hundred pots and the process is a continuous one so the emission rate tends to be steady over time. Most smelters have backup systems for scrubbing equipment to allow for breakdowns and routine maintenance, but emissions may be higher during these events.

HF alkylation in petroleum refining

HF alkylation is a key process in the production of petrol blending components. Alkylate is valuable due to its high octane and low vapour pressure. The process reacts light olefins (i.e. C_3–C_5 olefins such as propylene, butylenes and amylenes) with isobutane in the presence of HF as a catalyst, to produce branched, seven- to eight-carbon paraffins, collectively known as alkylates. The HF alkylate capacity worldwide in 2000 was about 850,000 barrels per day. The concentrated HF is recycled with a net consumption of 0.3 lb (0.14 kg) per barrel of alkylates produced (US EPA, 1995c). HF alkylation units accounted for 47% of the alkylate produced in the USA in 1990 (US EPA, 1992). Because of its toxicity, Amoco, Mobil, Allied Chemical and DuPont sponsored a series of HF spill tests in the Nevada desert to determine the distance that the air plume would travel. HF concentrations were found to be ten to 100 times greater than was predicted by HF dispersion models current at the time of the tests (1986). The results suggested that refineries with HF units can pose more serious air-pollution problems than had previously been believed.

Glass and fibreglass manufacture

Fluoride is used in the glass industry for etching, cleaning and making opal glass but it is also added to certain types of glass to produce desirable properties. For example, it alters the bandwidth and dispersion of glass used for lenses, while heavy-metal fluoride glass is an extremely transparent material being developed for use in optical fibres that transmit infrared radiation. During the processing of the glass at high temperatures, a proportion of the fluoride is emitted, so manufacture can be a significant source of HF. Writing in 1977, Bulcraig mentioned that a glass-fibre factory in Wales used glass with 0.6% fluoride and that maximum emissions from each furnace were 9.1 kg/h. Chimneys were designed to keep the ground-level concentrations to less than 1.0 $\mu g/m^3$, which is high by modern standards. Now, in the developed countries, emissions are controlled using venturi and packed-bed scrubbers (Teller and Hsieh, 1992a) and glass production is a much less important source of environmental fluoride. The major types of glass produced are container glass and flat glass, the two comprising about 75% of total glass production. Pressed and blown glasses, which include lead and borosilicate glass and frit for ceramic coating, are produced in lesser quantities. Emissions are primarily from the melting furnaces in the production of borosilicate glass and frits.

Fibreglass production includes continuous filament fibreglass, often used for textile products, and fibreglass wool, used for insulation (Teller and Hsieh, 1992b). The insulation product is used for thermal and acoustic insulation because the small cells of air entrained in the wool prevent the movement of air and sound waves. Fluoride emissions as HF and SiF_4 occur primarily from the melting–refining furnace during textile fibreglass manufacture. The two gases combined are present in concentrations ranging from 20 to 160 p.p.m. (16 to 131 mg/m^3) in the emissions. Emissions from insulation fibreglass operations contain no fluoride. According to Schorr (1977), fibreglass manufacturing with controls emits about 6 lb of fluoride/ton (2.5 kg/t) of raw material processed. Emissions are controlled with one of two types of wet scrubbers: a venturi packed-bed scrubber or a nucleation cross-flow scrubber. Another type of scrubber used is a quench reactor–dry venturi baghouse (Teller and Hsieh, 1992b).

Brick, tile, pottery and cement manufacture

The clays used in the manufacture of bricks, tiles, pottery and cement contain fluoride in the form of hydrated micas, such as muscovite and illite. For brick, tile and pottery production, the formed clays are fired in kilns at gradually increasing temperatures up to about 1100°C. At this elevated temperature, gaseous HF and SiF_4 are evolved. In brick making, 30–95% of the original fluoride in the clay is released (Semrau, 1957; Bohne, 1964). Hydrogen fluoride, silicon tetrafluoride and fluoride-containing dusts are often emitted directly to the atmosphere, especially from brickworks in Europe, South America and China. Emissions have caused serious environmental problems in the past, especially in areas with a high density of high-capacity facilities such as the Bedfordshire area of England (Burns and Allcroft, 1964; Chapter 3). In western, developed countries, the problem of emissions from brick and tile making has diminished but, in China, fluoride emissions from brick manufacturing are still a very serious problem, which affects human health and has endangered the silkworm industry in some areas (Chapter 3). The problem is caused by the fact that small, local kilns are used and the fluoride content of the clay may be extremely high. Chen *et al.* (1993) cite mud used for making tiles as having a fluoride content of 10,000 mg/kg.

Portland cement is manufactured by combining clay and limestone and gradually heating the mixture to about 1450°C in a rotary kiln, forming clinker, which is combined with a small amount of gypsum and ground to a fine powder. During firing, the fluoride in the clay combines with lime to form calcium fluoride. This is mainly incorporated into the clinker and the final product. Therefore fluoride emissions to the atmosphere are low, especially if fugitive dusts are collected efficiently by cyclones or glass-fabric filters. In making 'white cements', cryolite and fluorspar are often added as fluxing agents, so the dusts from this process contain more fluoride than Portland cement.

Iron and steel manufacture

Two types of steel-making technology are in use currently: the basic oxygen furnace and the electric-arc furnace. Fluorspar is incorporated as a flux to increase the fluidity of the slags and improve the removal of phosphorus and sulphur impurities from the melts (McGannon, 1964; NAS, 1971). For iron and steel facilities, a total of about 230,000 lb (104 t)/year of HF was released to the atmosphere in 1993 (US EPA, 1995b).

Other industrial uses of HF

HF has a number of other industrial uses, such as: metal cleaning and removing oxides in electroplating; wet chemical etching of semiconductor wafers in the electronics industry; an ingredient in corrosive flux used in welding; an ingredient in the production of organofluorides, such as CFCs; a reactant in converting uranium to uranium hexafluoride for enrichment; and a chemical derivative in other manufacturing processes. In the USA in the 1990s, from semiconductor manufacturing facilities alone there was a total release into the environment of over 70,000 lb (32 t)/year, of which about 60,000 lb (27 t) was emitted to the atmosphere (US EPA, 1995a).

Coal combustion

Although often ignored as a trace component of coal, fluoride is a normal constituent and is released as HF during the combustion process. About 50 years ago, it was observed that combustion of some coals led to corrosion of ceramic and metal surfaces, etching of glass and porcelain, and an increase in the fluoride content of barley grains dried over coal (Swaine, 1990). Francis (1954), writing in a British textbook, stated that 'The fluorine is largely volatilised during combustion . . . this is responsible for much corrosion of gas scrubbers due to hydrofluoric acid, hydrofluosilicic acid or ammonium fluoride.' The source of the fluoride in coal has not been settled. Various investigators have reported its presence as inclusions of fluorapatite, fluorite, kaolinites, montmorillonites, illite, micas, amphiboles, tourmaline and topaz in different coals (Swaine, 1990). The most common fluoride-containing inclusion is fluorapatite, but kaolinite, montmorillonite and illite have been identified where F^- has replaced OH^-. Often there is a statistical correlation between the amount of fluoride and phosphorus, which points to fluorapatite as being an important source. Crossley's (1944) data, for example, showed such a relationship: $[F] = 9.6 \times [PO_4] - 140$, $n = 23$, $r^2 = 0.868$. However, the other minerals are also possible sources of fluoride (Godbeer and Swaine, 1987; Godbeer et al., 1994).

The fluoride content of coal ranges from a trace (< 20 mg/kg) to several thousand mg/kg (Table 1.4). In most countries there is a wide variation but in Europe, North America and Australia the range is about the same, from about 20 to a few hundred mg/kg. In China there are areas with a similar range but there are also regions where the mean is much higher (Zheng et al., 1999; Ando et al., 2001; Finkelman et al., 2002). If coal has any shale mixed in with it, the fluoride content may be increased because shales tend to have a high fluoride content (Table 1.4).

Six nations have 80% of the estimated world reserves of 1×10^{12} t of coal: the USA, the former Soviet Union, China, Australia, India and Germany (AEI, 1999). Production was estimated in 1997 as being about 4100 million t. With an average fluoride content of 100 mg/kg, consumption could potentially release about 410,000 t of fluoride per

Table 1.4. Examples of reported fluoride contents (mg F/kg) of coal and accompanying shale.

Area	Range of concentrations	Source
Western USA	19–140	Gluskoter, 1977; Valković, 1983
Eastern USA	50–150	
Illinois basin, USA	29–140	Zubovic et al., 1979
Western Canada, Yukon	31–930	Godbeer et al., 1994
Latrobe Valley, Australia	4–79	Swaine, 1990; Valković, 1983
Collie, Australia	16–55	
Britain – coal	< 0–170	Crossley, 1944
Britain – coal-associated shale	25–55	
	120–440	
North-west China, steam coal	48–149	Luo et al., 2002
China – non-polluted area	Mean 152, SD 19	Ando et al., 2001
China, medium pollution	Mean 212, SD 29	
China, high pollution	Mean 656, SD 37	
China, data for 337 samples	15–2350, mean 268	Zheng et al., 1999

year, but the actual amount will be less because of incomplete volatilization and scrubbing. For comparison, Luo *et al.* (2002) estimated that in China combustion of steam coal releases about 66,398 t of fluoride into the atmosphere each year. In the USA, an estimated total of 19,500 tons (17,700 t) of HF was emitted in 1990, and the projected emissions for 2010 are estimated to be 25,600 tons (23,200 t) (US EPA, 1997a). However, unlike China, most of the HF arising from combustion in the USA and other countries is removed by scrubbers.

The environmental effects of fluoride released from coal depend on the way it is used. In power generators between 75 and 97% of the fluoride is driven off, depending on the temperature of combustion (Luo *et al.*, 2002). The fly ash, which is sometimes used for by-products, has a very low fluoride content. In many developed countries, scrubbers are used to remove SO_2 and this is also very efficient at removing the fluoride, which ends up in the solid by-product, gypsum. Emission from high chimneys dilutes the fluoride so that concentrations are low at ground level and the result is that there are few cases reported where fluoride from power generation causes environmental problems. It is the use of coal for domestic purposes that causes the important environmental problems. In Europe 30–40 years ago, fluoride emissions from domestic coal used for heating and cooking raised

concentrations outside the homes to the level where it injured sensitive plants (Davison *et al.*, 1973), but there is no indication that it affected human health. However, in China, not only does the coal have a high fluoride content but it is also used to dry food, which causes direct contamination by fluoride and other toxic elements, such as arsenic and selenium. The size of the area where coal causes human fluorosis in China is shown in Fig. 1.3. In China about 400 million people depend primarily on coal for their home energy needs (Zheng *et al.*, 1999; Ando *et al.*, 2001; Finkelman *et al.*, 2002).

Accidental Releases of Hydrogen Fluoride

Periods of higher than normal emissions occur in most industries due to routine maintenance or failure of scrubbing equipment, but the scale of the increase is usually relatively small. Serious releases tend to occur not in those industries that use solid fluoride as a flux but in those that handle large quantities of hydrofluoric acid. Accidental releases that are sufficient to cause severe environmental damage are fortunately rare but there are a few instructive cases that have been documented in the USA. Such releases undoubtedly occur in other countries but the information is not

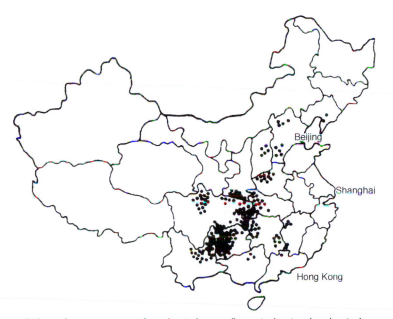

Fig. 1.3. Map of China showing areas with endemic human fluorosis that is related to indoor combustion of coal. (Redrawn from Zheng, B., Ding, Z., Huang, R., Zhu, J., Yu, X., Wang, A., Zhou, D., Mao, D. and Su, H. (1999) Issues of health and disease relating to coal use in southwestern China. *International Journal of Coal Geology* 40, 119–132.)

as accessible. The US Environmental Protection Agency (US EPA, 1992) compiled information on several accidents associated with HF releases. Unfortunately, the report does not include effects on plant life, perhaps the most sensitive biological receptor. However, in two of the cases discussed, one of the current authors (LHW) inspected the areas at least once during the post-exposure period. The examples show the causes and the scale of such releases.

Marathon Oil Company

Perhaps the most serious accident occurred at the Marathon Oil Company refinery in Texas City, USA, on 30 October 1987. The incident resulted in the largest recorded accidental release of HF in the USA and involved an error in procedures for lifting equipment by a contractor. In this case, the contractor attempted to lift a heater convection unit over, rather than around, an HF storage vessel at the refinery alkylation unit. The load was too great for the crane and the operator released the heater convection unit on to the HF vessel, rupturing two lines attached to the top of the vessel and resulting in a gigantic vapour release. Of the 290,000 lb (131,544 kg) of anhydrous HF originally in the vessel, about 53,200 lb (24,131 kg) was released, mostly during the first few minutes, but it did not stop completely for the next 44 h (US EPA, 1992). The fumes migrated through a residential area adjacent to the refinery. Eighty-five square blocks with 5800 residents were evacuated; 1037 residents were treated for skin burns and irritation to the eyes, nose, throat and lungs. Injury to vegetation was severe from the acute insult, and aerial photographs attested to a spreading cone of necrotic foliar tissues from the point of release for about 1 km downwind (Plate 1). Many coniferous trees, especially pines, were killed and most deciduous trees and shrubs were defoliated but, by the following spring, aerial photographs showed that the deciduous plants were green and recovering. An inspection of the vegetation in 1990 of areas affected by the release showed that there was mild fluoride-induced plant injury on several broad-leaved species, but it appeared to be chronic and to have been induced during that year by normal operations (L.H. Weinstein, personal observation, 1990).

Exxon–Mobil Oil Corporation

In 1987, there was a 100 lb (45 kg) release of HF from the Mobil Oil Company refinery in Torrance, California (US EPA, 1992). The release occurred following an undetected flow of HF to the alkylation unit's propane heater. The propane heater used potassium hydroxide to neutralize trace amounts of HF in alkylation by-products. The HF reacted violently with the caustic material, generating a significant amount of heat and pressure, causing an overpressurization in the treater vessel and resulting in a violent explosion and fire, which burned for 2 days. There were ten injuries to workers and passers-by but no reports of effects on vegetation.

Kerr–McGee uranium-processing plant

On 4 January 1986, there was a large release of HF from the Kerr–McGee uranium-processing plant in Gore, Oklahoma (US EPA, 1992). A uranium hexafluoride (UF_6) cylinder was heated excessively while liquefying the contents. The cylinder ruptured and released 29,500 lb (13,381 kg) of UF_6, generating a large cloud of HF and uranyl fluoride. The amount of HF released from the reaction of UF_6 with atmospheric moisture was estimated to have been 3350 lb (1520 kg). There

was one death and at least 35 injuries. The cloud was dispersed by favourable winds. No mention was made of vegetation injury.

Great Lakes Chemical Corporation

The Great Lakes Chemical Corporation in El Dorado, Arizona, released 1320 lb (599 kg) of HF on 27 June 1989. A diaphragm associated with an HF storage tank failed due to corrosion. There were no reported injuries and no mention was made of injury to vegetation.

Citgo Petroleum Corporation

On 12 May 1997, an explosion and fire occurred at the alkylation unit of the Citgo refinery in Corpus Christi, Texas. There were claims that as much as 17,000 lb (7711 kg) of anhydrous HF was released over an 8.5 h period, but the actual amount was probably less. Some plant injury was seen close to the release point but little was seen beyond the first few hundred metres and most of it was found on plant species considered to be of high or intermediate sensitivity to fluoride. Further details are given in Chapter 6 of the biomonitor that was used to establish the area that was potentially affected.

Notes

[1]   'Fluorine' and 'fluoride' are often used interchangeably, but in this book 'fluoride' is used as a general term in all instances, unless there is a need to refer to the element (F_2), in which case 'fluorine' is used.

[2]   Much of the historical account is based on Weeks and Leicester (1968) and Meiers (2003).

2

Uptake, Transport and Accumulation of Inorganic Fluorides by Plants and Animals

Introduction

The fluoride content of the air, soil, water, plants and animals is immensely variable so the aim of this chapter is to provide the scientific basis for interpreting this variation in an environmental context and to demonstrate how fluorides move in the biosphere. It begins with an outline of the dispersion of inorganic fluorides in air, deposition on soil and solubility in soil and water and then it progresses to uptake by plants and animals. Organic compounds are discussed in Chapters 8 and 9.

Fluoride in the Air

Dispersion and dilution

All gases and particles that are emitted into the atmosphere, whether they are from a volcano or a factory stack, are diluted as they are carried by the wind and mixed by turbulence. The pattern of dispersion and the concentration gradients at ground level depend on the height of emission, the temperature of the plume, wind direction, wind speed, turbulence, precipitation, topography and characteristics of the vegetation. However, in general, emissions from a point source, such as a single factory stack, will show patterns similar to those in Fig. 2.1, depending on the weather. When the plume

is hotter than the surrounding air, it is less dense, so it rises vertically, increasing the effective height of the stack. As it is carried away from the stack by the wind, the plume is diluted by diffusion and turbulent mixing. It expands and eventually reaches ground level. Usually this leads to low concentrations of pollutant at ground level near the source and maximum concentrations downwind where the centre line of the plume reaches the ground.

Generally, the higher the height of emission and the wind speed, the further the plume will be carried and diluted before reaching the ground. Under sunny conditions when the wind speed is low, the plume may loop (Fig. 2.1a) as it is carried by large eddies, leading to plants being exposed to high concentrations of pollutants for short periods of time. On the other hand, when it is cloudy and the boundary layer is well mixed, the turbulent eddies are small and the plume takes on a cone shape (Fig. 2.1b). When the boundary layer is stable, typically under clear night skies, there is little vertical rise and the plume spreads out so that, from above, it resembles a fan (Fig. 2.1c). Often this condition is the precursor to a temperature inversion – the normal decrease in temperature with height is reversed so the plume is more dense than the surrounding air and it does not rise (Fig. 2.1d). The pollutants are trapped below the inversion layer and the concentration continues to build up while the inversion lasts. Prolonged

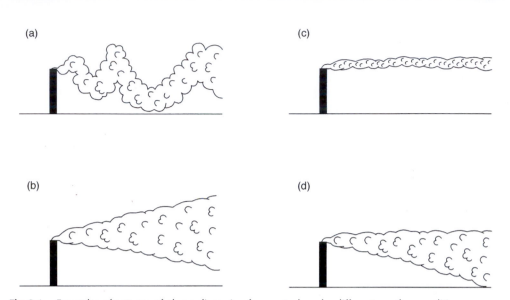

Fig. 2.1.   Examples of patterns of plume dispersion from a stack under different weather conditions.
(a) Sunny conditions, wind speed is low and the plume is being carried by large eddies. (b) Cloudy, with
well-mixed boundary layer producing small eddies. (c) Stable boundary layer, typically occurring with clear
night skies, producing little vertical rise so the plume spreads out. (d) Temperature inversion. The plume is
more dense than the surrounding air and pollutants are trapped below the inversion layer.

temperature inversions have been vital components of many of the worst pollution episodes, notably the London smogs of the 1950s, so modern factories are built with the tops of the stacks above the normal inversion height for the area. If the terrain rises in the downwind direction, the maximum concentration may be higher than on flat land, and the converse is true if the terrain falls away. However, the effects of topography are most obvious when the pollution source is built in a narrow valley. The air moves along the valley, hemmed in by the hillsides, so the pollutant concentration may remain high over long distances. There are several good examples of this in Norwegian valleys (AMS, 1994).

It is rare for fluoride to be emitted solely from a single point at the source because, even in modern factories, some escapes through vents, doors and windows. In some industries, such as phosphate-fertilizer manufacture, there may be secondary sources, such as gypsum piles or open ponds, so there may be several points and heights of emission, rather than just one stack. Emissions from low-level sources

such as these lead to the highest concentration immediately outside the buildings or next to the ponds; then the concentration falls off in a non-linear way.

Concentrations of airborne fluoride near pollution sources

There are relatively few data in the scientific literature that can be used to illustrate the actual concentrations of airborne fluorides at different distances from sources. This is largely due to the lack of availability of real-time analysers that have sufficient sensitivity to detect typical ambient levels and that can distinguish between gaseous and particulate forms. In recent years, spectrophotometric methods (e.g. Fourier transform infrared (FTIR)) have been developed that give real-time data in industrial environments but they still do not have the sensitivity required for use at the very low concentrations that are typical of the external environment. Most current methods still rely on drawing air through an alkali medium for periods of hours to days, so averaging times are at least 30 min. However, enough data are available to indicate the typical range of concentrations and the variability.

A good example of the decrease in concentration with distance is provided by Sidhu's (1979) data from a phosphorus plant in Newfoundland (Fig. 2.2). At distances of 0.7 and 12 km from the plant the weekly mean gaseous fluoride concentrations were estimated (using alkali papers (Chapter 6)) to be 5.14 µg/m³ and 0.15 µg/m³, respectively. The non-linearity is typical and this is usually reflected in the pattern of effects on plants and animals.

Even when emission rates are relatively constant, the wind and weather result in considerable fluctuations in the concentration at ground level. For example, Taylor et al. (1981) monitored a uranium-enrichment facility where there was little variation in the rate of emission of HF. The highest 7-day mean HF concentrations were immediately downwind and the frequency distribution consisted predominantly of lower concentrations, with infrequent peaks. Over a 6-year period, 87% of the observations were ≤ 1 µg HF/m³ and 13% were > 1 µg/m³. McCune et al. (1976) produced similar data for HF concentrations at different distances from an aluminium smelter (Fig. 2.3). Their data show very clearly that concentrations vary much more near the source, where concentrations are highest, than further away. This is important in relation to vegetation because the exposure dynamics – the sequence and range of concentrations – play a part in determining plant response. However, there is a complication because the range of concentrations that is measured at any sampling point depends on the averaging time: the longer the averaging time, the lower the range of concentrations. Bulcraig's (1977) data for a glass factory show this. Over a 5-year period, the 2-week mean concentration fluctuated between 0.2 and 2 µg/m³, a range of ten times, but the annual mean concentration was from 0.31 to 0.51 µg/m³, a range of less than two times.

It is important to know the concentration of both gaseous and particulate forms because of the difference in toxicity, but there are few published reports of concentrations of the latter. Furthermore, the chemical nature, size and proportion of particulate fluorides vary with the process and these have rarely been determined at different distances from sources. The only techniques that are available for determining particulate fluoride are similar to that proposed by Weinstein and Mandl (1971). Air is drawn through a citric-acid-impregnated prefilter that retains the particulate fluoride but allows the acidic HF to pass. The HF is then retained on an alkali medium. Davison et al. (1973), Blakemore (1978), Craggs et al. (1985) and Craggs and Davison (1987a,b) used this method for monitoring fluorides in a town where the fluoride was from coal burning and near an aluminium smelter. In the town, a high-volume sampler was used to obtain 1 h mean concentrations at six locations (Davison et al., 1973). The mean concentrations of gaseous and particulate fluorides at these locations were 0.21 and 0.12 µg/m³, respectively. At a site very close to an aluminium smelter, the weekly mean

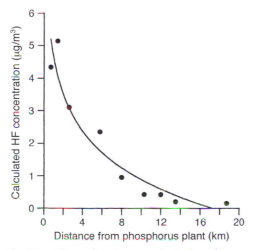

Fig. 2.2. The gradient in atmospheric fluoride concentration with distance from a phosphorus plant in Newfoundland. [HF] = 4.62 − 3.78 × $\log_{10}$ (distance, km), r^2 = 0.91. Note that concentrations were estimated from deposition on alkaline papers. (Drawn from data in Sidhu, 1979.) (Data used from the *Journal of Air Pollution Control Association* 29, 1069–1072, 1979, with permission of the Air and Waste Management Association.)

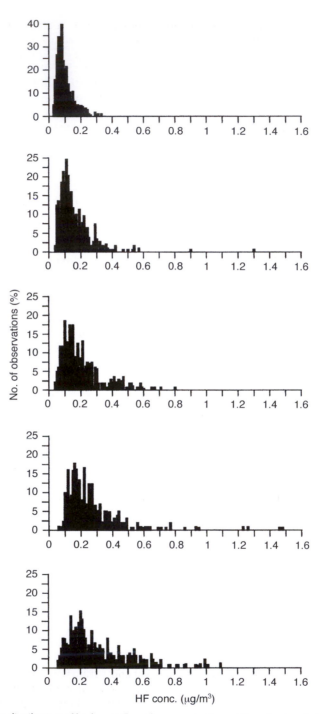

Fig. 2.3. Frequency distribution of hydrogen fluoride concentrations at five sites located around an aluminium smelter. (Redrawn from McCune, D.C., MacLean, D.C. and Schneider, R.E. (1976) Experimental approaches to the effects of airborne fluoride on plants. In: Mansfield, T.A. (ed.) *Effects of Air Pollutants on Plants*. Cambridge University Press, Cambridge, pp. 31–43.)

concentrations of gaseous and particulate fluoride over a 2-year period were 0.358 and 0.338, respectively. Aluminium smelters release a complex mixture of particles that depends on the process and the operating conditions (Less *et al.*, 1975), so the ratio is specific to the particular operating conditions and the location of the sampling site. In this case and at this particular site, the ratio of gaseous to particulate forms was close to 1, but the ratio increases with distance from the source because of the greater rate of deposition of the larger particles.

Deposition of pollutants from the atmosphere

The concentration of pollutants in a plume decreases with distance from the source, not only because of diffusion and turbulence, but also because of wet and dry deposition. Wet deposition is the removal of pollutants by rain, mist and snow. It is very effective and can cause a significant reduction in pollutant concentration during precipitation events. Wind-driven cloud is particularly efficient at removing and depositing pollutants on rough structures, such as the edges of forests or hedges – they act rather like sieves. Dry deposition occurs when the pollutant impinges on a surface or is taken up into leaves through the stomatal pores. Not surprisingly, surfaces such as the cuticle of a leaf, soil or a lake differ considerably in their capacity to adsorb gaseous and particulate fluorides. In general, deposition is greater to a wet surface than to a dry one, which means greater dry deposition when leaves are wet from rain or dew.

Measuring deposition of any pollutant is very difficult, and unfortunately one of the best methods is not usable with fluorides because it depends on having sensitive, fast-response analysers. Most available information about the rates of deposition comes from the use of simple, open collecting bowls or standard rain-gauges. Although these give a measure of the total amount of

wet-deposited fluoride, they do not give a good estimate of the dry deposition because they do not measure the gaseous fluoride that is taken up by leaves. Furthermore, the rate of deposition is affected by the geometry and roughness of the surfaces, so a bowl or funnel does not give an estimate of deposition to a hedge or a forest. Murray (1982), for example, reported total deposition in the open as 10 kg F/ha/year but 32 kg/ha/year in the throughfall of an adjacent *Eucalyptus* forest. Throughfall is the material collected immediately under the canopy of the tree.

Because deposition bowls collect both wet and dry deposition, most estimates of the fluoride content of precipitation are overestimates, especially older ones (Barnard and Nordstrom, 1982). For example, Garber (1970) used a 30 cm funnel to collect samples over monthly periods and he reported a 'control' site as having 0.16 mg F/l and values up to 1.6 mg/l in industrial areas. The magnitude of the error depends on the relative importance of wet and dry deposition in the locality. Sidhu's (1982) data provide a more reliable estimate than Garber's because he collected rain after each event. Barnard and Nordstrom (1982) used automated wet/dry collectors that minimized dry deposition and they found concentrations of fluoride in precipitation ranging from < 0.002 to 0.024 mg F/l at two sites in the USA. The sites were judged to have no local fluoride pollution sources. As there was no correlation with sodium, the authors concluded that the fluoride content was not being enriched at the sea surface. After constructing a mass balance, they concluded that most of the fluoride was of anthropogenic origin. However, the origins of fluoride in precipitation in non-industrial areas are still debatable because Mahadevan *et al.* (1986) found that fluoride in precipitation collected at sites around the Indian subcontinent was significantly correlated with sodium content, suggesting a marine influence. They concluded that the background fluoride content of precipitation is 0.001–0.012 mg/l, with a mean around 0.005.

Rates of fluoride deposition from the air to soil

In order to assess the potential for soil contamination by atmospheric fluorides, it is necessary to calculate or measure the rate of deposition. Although it is difficult to produce precise estimates, an indication can be obtained using the deposition velocity, which is the ratio of the rate of deposition to the concentration in air (Chamberlain, 1966). Therefore, the rate of deposition to a surface is the product of the concentration in the air and the deposition velocity:

rate of deposition ($\mu g/m^2/s$) = concentration in air ($\mu g/m^3$) × deposition velocity (m/s)

The deposition velocity, V_g, varies with the form of the fluoride (gas or particulate), particle size, wind speed and the nature of the plant canopy, particularly its architecture and wetness. Table 2.1 shows examples of the deposition velocity of inert particles to a short-grass sward derived from experiments in a wind-tunnel (Chamberlain, 1966). There are no comparable wind-tunnel studies using fluoride-containing particles but there have been a few studies in the field and using fumigation chambers (summarized in Davison, 1982, 1983) and these produce estimates of V_g that are broadly comparable to Chamberlain's (1966) data. Table 2.1 demonstrates the effects of wind speed and the rapid increase in V_g that occurs with increase in particle size. The table also shows that over the size range 1–20 μm the rate of deposition increases by a factor of > 100 times. This is why the concentration of larger particles decreases rapidly with distance from a source. The World Health Organization (WHO, 2002a) cites work by Sloof et al. (1989) stating that the deposition velocity of particulate fluorides varies by less than 10%. The authors have not seen the report but it is difficult to reconcile the statement with the extensive literature on the deposition velocity of particulate materials, unless the authors were dealing with a very limited range of particle sizes.

The deposition velocity of the gas HF can be calculated from the data of Hill (1971), Israel (1974c), Davison and Blakemore (1976) and Schwela (1979). These data were mostly from the field so the estimates are variable because of the uncontrolled environmental influences. Nevertheless, the data are reasonably consistent, giving a range of V_g from 0.002 to 0.008 m/s (Davison, 1983). WHO (2002) cited Sloof et al. (1989) stating that the average large-scale deposition velocity for 'soluble' fluorides was calculated to be 0.014 m/s. The discrepancy is probably because of a difference in the reference height. Using the range of 0.002 to 0.008, a concentration of HF in the air of 1 μg/m³ is predicted to produce a rate of deposition from 0.63 to 2.52 kg F/ha/year. This is reasonably close to Schwela's (1979) estimate that an average concentration of 1 μg/m³ resulted in the deposition of 750 kg F/km²/year which is equal to 7.5 kg F/ha/year.

Soil Fluoride

Retention and movement of soil fluoride

Water, charged with CO_2, inorganic ions and organic compounds, leaches elements from the upper horizons of soils so they are gradually depleted. There are a few reports that might indicate significant leaching, such as that of Robinson and Edgington (1946), who recorded surface fluoride concentrations of 50 to 590 mg/kg in soils that had subsoil concentrations in excess of 1000 mg/kg. Similarly, the soils surveyed by Omueti and Jones (1980) approximately doubled in concentration from around 200–300 to about 500–600 mg/kg between the surface and 1 m depth. However, such gradients are not commonly reported because fluoride is so strongly adsorbed by soil that leaching is slow and in many soils there is little or no detectable gradient with depth. Also, if the soil is ploughed or if there is a significant input to the surface from plant remains, atmospheric deposition or fertilizer, leaching losses are obscured (Fig. 2.4).

Investigation of the adsorption of fluoride by soil was stimulated in the 1930s by

Table 2.1. Deposition velocity of particles to a short-grass sward in a wind-tunnel at two wind speeds (Chamberlain, 1966) and calculated rate of deposition from an atmospheric concentration of 1 µg particulate fluoride/m³.

Particle size (µm)	Deposition velocity (m/s)		Rate of deposition (kg/ha/year)	
	Wind speed 1 m/s	Wind speed 4 m/s	Wind speed 1 m/s	Wind speed 4 m/s
1	0.00019	0.0006	0.06	0.19
5	0.0018	0.012	0.57	3.78
10	0.0105	0.034	3.31	10.72
20	0.0255	0.063	8.04	19.87

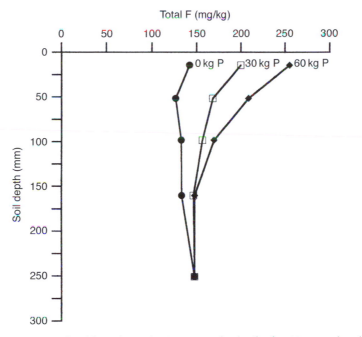

Fig. 2.4. The increase in total soil fluoride (mg/kg) in New Zealand soils after 10 years of application of superphosphate at three rates: 0, 30 and 60 kg P/ha/year. (Redrawn from Loganathan, P., Hedley, G.C., Wallace, G.C. and Roberts, A.H.C. (2001) Fluoride accumulation in pasture forages and soils following long-term application of phosphorus fertilisers. *Environmental Pollution* 115, 275–282, with permission from Elsevier.)

concerns about the fate of fluoride that was being added to soil as inorganic pesticides and as a contaminant of phosphate fertilizers and as slag. It was important to know if the contamination would lead to greater uptake by plants or leaching into water-supplies. As a result of these studies, principally by MacIntire and his colleagues (MacIntire *et al.*, 1942, 1948, 1951a,b, 1955a,b, 1958; MacIntire, 1957; Specht and MacIntire, 1961) it was realized that fluoride is strongly held by soil and, when more is added, the

soil has a capacity to bind it. MacIntire's data showed that even at exceptional rates of application the percentage of the fluoride that is fixed and therefore held by the soil is very high. In one case, 98% of the fluoride added as 4932 lb of CaF_2/acre (about 5308 kg/ha) to a silt loam was retained over a 10-year period (MacIntire *et al.*, 1948). Later, Omueti and Jones (1977a) examined soil samples collected from field plots over a 67-year period and found that much of the fluoride that had been added as a contaminant between 1904 and 1924 was still retained in 1955. They calculated that the average fluoride loss between 1922 and 1944 was only 2.5 mg/kg/year.

In their earlier papers, MacIntire and his colleagues (MacIntire et al., 1942) considered that fixation was due to reaction with calcium because it was 'the dominant exchangeable base and main precipitative ion'. This idea was supported by many studies in which lime was applied. However, it was also realized that calcium is not the sole agent responsible for fixation, because MacIntire (1957) noted that retention of fluoride added to acid soils was proportional to their Al_2O_3 content. Later, Omueti and Jones (1977b) showed that the adsorption capacity of Illinois soils was related to pH, clay, organic matter and amorphous aluminium compounds. They suggested that adsorption was due primarily to the presence of amorphous aluminium oxyhydroxides, which are common weathering products in such soils. Similarly, Mizota and Wada (1980) concluded from a study of Japanese volcanic soils that added fluoride reacted with any 'active' aluminium that might be bound to humus or allophane (non-crystalline, colloidal $Al_2O_3.2SiO_2H_2O$), allophane-like constituents or poorly crystalline layer silicates.

Much less is known about fluoride chemistry and adsorption in organic horizons and peaty soils. A large part of the world's land surface is covered by organic matter – litter, humus and peat – but there are relatively few reports on the fluoride content or chemistry of these materials. In the absence of air pollution, the two main sources of fluoride in organic materials are atmospheric deposits and traces of minerals, so concentrations would be expected to be low in non-industrial areas. This is borne out by Omueti and Jones (1980), who reported that the fluoride associated with the organic matter in Illinois silt loams was only 9 mg/kg. Thompson et al. (1979) reported similar concentrations in humus samples collected more than 14 km from a phosphate plant. There are, however, two reports of much higher concentration (Mun et al., 1966; Perel'man, 1977), so more

data are needed to explain the sources and chemistry of these materials.

Saline soils present a special case because the pH is high and the dominant exchangeable cation is sodium rather than calcium. As a result, the water-soluble fluoride fraction of saline soils is higher than that of non-saline soils, and the fraction increases with the percentage of the cation exchange sites that are occupied by sodium (Lavado and Reinaudi, 1979). In the La Pampa region of Argentina, Lavado and Reinaudi (1979) found that the total soil fluoride was the same as in non-saline soils (24 to 1200 mg/kg) but, because of the dominance of sodium and the low calcium, the water-soluble fluoride was much higher. Much of the interest in the fluoride chemistry of these soils is generated by the use of by-product gypsum for desalination because the gypsum obtained from phosphoric acid manufacture contains 0.2–2.4% fluoride (Kitchen and Skinner, 1969, 1971; Singh et al., 1979). Studies have shown that, when it is added to saline soils, fluoride is largely immobilized in the first 8 days (Chhabra et al., 1980).

Sources of soil contamination by fluoride and effects on soil concentrations

Atmospheric deposition has already been discussed as a source of soil contamination and it was shown that it may lead to addition of anything from a trace to several kilograms per hectare per year. Other important sources of soil contamination are inorganic pesticides, phosphate fertilizers, slag and reclaimed gypsum. It is important to assess the potential input to soil from these sources.

The effect of adding fluoride to soil on the concentration and its distribution through the profile depend on how the soil is managed and its properties. If the soil is undisturbed (e.g. natural vegetation, permanent pasture) the deposited fluoride remains in the surface few centimetres (McClenahan and Weidensaul, 1977; McLaughlin et al., 1996, 2001; Loganathan et al., 2001), but if the soil is ploughed it will be mixed to a depth of 30–50 cm. Mineral soil has a bulk density of about 1.4 g/cm^3, so a 1 ha layer of soil 5 cm deep weighs about 700 t. If fluoride is retained in this 5 cm layer, addition of 1–10 kg F/ha/year will lead to an increase in concentration of about 1.4–14 mg/kg. On the other hand, if the soil is ploughed to a depth of 30 cm, the mixing will result in an average concentration only one-sixth of these values. Another important variable that determines the concentration is the organic content, because it has a bulk density that is as little as 20% of that of mineral soil (Davison, 1983). Therefore, if added fluoride is incorporated into a humus layer 5 cm deep, when the concentration is expressed on a weight basis, the fluoride concentration will be five times greater than that of a mineral soil.

Several authors have reported significant increases in soil fluoride near pollution sources. For example, McClenahan and Weidensaul (1977) reported an increase of about 40 mg/kg in the top 5 cm of a silt loam around a source that started emission about 18 years before the measurements were made. This represents a net annual gain of 2.2 mg F/kg. In a similar, but more extensive study, van Hook (1974) found an average of 873 mg F/kg in soil within a mile (1.5 km) of a 23-year-old aluminium smelter and 444 mg/kg at sites over 6 km distant. Assuming that this difference indicates an addition of about 429 mg/kg, the net annual rate of fluoride input would be 18.6 mg/kg. The distant site may not have been representative of the area but the fact that the rate was higher than in the McClenahan and Weidensaul (1977) case could be explained by the smelter having less efficient control of emissions and the sites being much closer to the source. These are probably the same reasons why the net annual rate of input found by Polomski *et al.* (1980) for soil 0.5 km from a 72-year-old source was as high as 25 mg/kg. In contrast, Sidhu (1979) reported a rate that appears to be considerably higher than those in previous reports. At a site 0.7 km from an 8-year-old fertilizer factory, where the airborne fluoride was 4–5 µg F/m^3, the increase in the fluoride content of the humus

layer was in the region of 1600 mg/kg. This resolves to a net annual figure of about 44 mg/kg per 1 µg F/m^3, but in making a comparison it is important to allow for the difference in bulk density between organic matter and mineral soil, as mentioned earlier. If this allowance is made, the rate of input becomes comparable to both the gross rates of input calculated from deposition velocities and the other estimates of rates of net increase for mineral soils. Nevertheless, the high concentration is a further indication that more information is needed on the behaviour of fluoride in organic soils.

In the last decade or so, there has been a resurgence in the use of cryolite (Na$_3$AlF$_6$) in the USA for the control of pests such as vine weevil. Gianessi and Marcelli (2000) state that about 2.6 million lb (1.2 million kg) were applied in the USA in 1997, almost all of which was used in California and in the Great Lakes and north-eastern states. The rate of application of a dust containing 50% cryolite is typically from about 7 to 60 lb/acre (1.3 to 11 kg/ha) depending on the target pest and the crop (Gowan Company, Prokil Cryolite 50 Dust data sheet). Some crops receive multiple applications but there is usually a prescribed annual total, ranging from about 15–300 lb/acre (2.7 to 55 kg/ha) per season. There do not appear to have been any measurements of how much of this reaches the soil but, if 50% of an annual application of a 50% dust at 30 kg/ha/year reached the soil, it would amount to about 4 kg F/ha/year, a rate comparable to deposition from 1 µg/m^3 of atmospheric fluoride.

Probably the single most widespread source of soil contamination arises from the use of phosphate fertilizers. They are applied in large quantities in all the intensive agricultural regions of the world. Oelschläger (1971) estimated that, for a range of crops fertilized with 400–1000 kg phosphate/ha, the amount of fluoride added was from 8 to 20 kg/ha/year. Harvesting was estimated to remove less than 0.05 kg/ha and a maximum of 0.1–0.4% of the amount added in the fertilizer. Losses by groundwater were also small, at 0.02–0.4 kg/ha/year, so phosphate clearly leads to a significant net gain in soil fluoride content. Oelschläger (1971) estimated this as being about 1–4% of the total per year. McLaughlin et al. (1996) did a similar calculation for Australian soils (Table 2.2). With a fluoride input of 4–16 kg/ha/year, and an initial soil concentration of 300 mg/kg, they estimated that it would take between 25 and 100 years to double the fluoride concentration in the 0–10 cm layer of the soil. This is comparable to Oelschläger's estimate, but ploughing of arable soil would increase these times considerably. Permanent pasture is not ploughed and this is reflected in the finding of Loganathan et al. (2001) that one-third to two-thirds of the fluoride applied in fertilizer is retained in the top 75 cm of New Zealand soils. Consequently, in soils with a long-term history of fertilizer application, the fluoride content of the upper layers increased significantly (Fig. 2.4, p. 27). The strong retention of fluoride in the surface layers means that addition of fluoride does not increase the concentration in drainage water, though there is an increased risk of fluorosis in grazing livestock (Chapter 3).

Fluoride in the soil solution

Roots take up fluoride from the solution immediately around them, so it is important to have a knowledge of the concentration in the solution and to understand what controls that concentration. Many researchers have tried to measure the soluble fluoride fraction, using a variety of extracting solutions. Most earlier studies employed the technique of Shaw (1954) or Brewer (1965) and used a 1 : 1 soil : water extract to measure the 'water-soluble' fluoride. Some,

however, have used a saturated paste extract (Cooke et al., 1976; Somani, 1977) and many have included an electrolyte (Larsen and Widdowson, 1971; Gilpin and Johnson, 1980; Arnesen, 1998; Loganathan et al., 2001; McLaughlin et al., 2001) or buffering and complexing agents (Gisiger, 1968; Sidhu, 1979). Larsen and Widdowson (1971) adopted a different approach, using an anion-exchange resin to extract the 'labile' fluoride. If these techniques were a good measure of the fluoride that is available to roots, there would be a statistical correlation between the soluble fluoride and plant fluoride, but the correlations found by different authors vary enormously; many are rather low. Recently, Stevens et al. (1997) commented that published correlations give r^2 values ranging from 0 to 0.78. There are four main reasons why this variation is to be expected.

First, as Larsen and Widdowson (1971) pointed out, the amount available to roots depends not only on the concentration in solution at any one time, but also on the fraction that is readily desorbed as the soil solution is depleted by root absorption. This is why they devised the resin extraction technique, but it has not been used to any extent (Braen and Weinstein, 1985).

Secondly, when the 1:1 extraction technique was devised (Shaw, 1954), it was not examined for the effects of electrolytes in the soil solution, equilibration time or soil–solution ratio. All of these affect the amount of fluoride extracted from a soil. Figure 2.5 shows the effect of increasing the soil : solution ratio and using different electrolytes on the fluoride extracted from two of Larsen and Widdowson's soils (1971). Using $CaCl_2$ or KCl with an acid soil, the concentration of fluoride in solution fell to the same extent

Table 2.2. Balance of gain and loss of fluorides from soils fertilized with superphosphate (data from McLaughlin et al., 1996). (©CSIRO. Reproduced from Australian Journal of Soil Research 34, 1–54, 1996, with permission of CSIRO Publishing.)

	Input (kg F/ha/year)	Crop removal (kg F/ha/year)	Net addition (kg F/ha/year)	Background soil concentration (mg/kg)	Years to double soil F in the 0–10 cm layer
Wheat	4	0.003	3.99	300	100
Potato	16	0.01	15.99	300	25

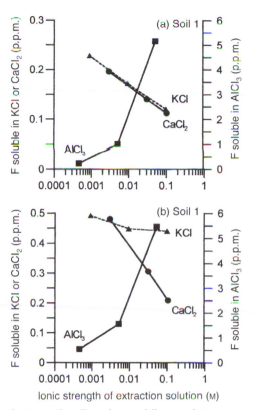

Fig. 2.5. The effect of using different solutions and soil : solution ratios on the fluoride extracted from two soils. (Redrawn from Larsen, S. and Widdowson, A. (1971) Soil fluorine. *Journal of Soil Science* 22, 210–221.)

with increasing ionic strength of both extractants. In the case of a calcareous soil, the concentration of fluoride decreased much more with increasing ionic strength of $CaCl_2$ than with KCl. This was probably due to the formation of insoluble CaF_2. Using $AlCl_3$, the fluoride increased in concentration with ionic strength because the aluminium formed soluble complexes with the fluoride.

The third reason why there is often a poor correlation between extractable and plant fluoride is because its solubility shows a distinctly non-linear relationship with pH. Several authors have shown that fluoride is strongly adsorbed to soils between pH 5.5 and 6.5 and that the solubility increases below 5.5 and above 6.5 (see Stevens *et al.*, 2000). Figure 2.6 demonstrates this using

data of Wenzel and Blum (1992). This means that if the soils being tested all lie in the range of lowest solubility, roughly pH 5–7, the amounts of fluoride in solution will be so low that normal sampling and analytical errors will tend to obscure correlations. On the other hand, if the range spans acid to neutral, then the gradient in solubility would tend to dominate the relationship.

The final reason is variation in the chemical form of the fluoride, the speciation, in the soil solution. Most early publications did not consider this and measurements were only made of the total fluoride in solution. Interest in fluoride speciation began to increase in the 1980s, partly triggered by concerns over the toxicity of aluminium in acid soils and waters (e.g. Cameron *et al.*, 1986). Häni (1978) added fluoride to an acid soil and recorded uptake by plants. Maize plants grown in treated soil showed an increase in both fluoride and aluminium in their leaves and it paralleled the concentrations of the elements in the soil solution, suggesting a link between the two elements. Davison (1983) suggested that, as the solubility of aluminium increases steeply with decreasing pH, it is possible that in acid soil most of the available fluoride exists as soluble aluminium–fluoride complexes. Boron is another element that forms soluble complexes with fluoride, but this interaction has been much less researched (Stevens *et al.*, 1998b). Over 30 years ago it was shown to be associated with large increases in fluoride uptake when both elements were present as fluoroborate in fertilizer (Bolay *et al.*, 1971a). In solution culture, it increased fluoride accumulation even when the boron content of the plant was not increased (Collet, 1969). In the last few years great progress has been made in identifying and quantifying the fluoride complexes that occur in soil solution, most notably by an Australian group (McLaughlin *et al.*, 1996, 2001; Stevens *et al.*, 1997, 1998a,b, 2000). As a result of their studies, it is now known that in soils with a slightly acid pH (5.5–6.5) most of the fluoride is adsorbed to the soil (Stevens *et al.*, 1998a) and the soluble fraction is mostly present as the anion, F^-. In acid soils, theoretically,

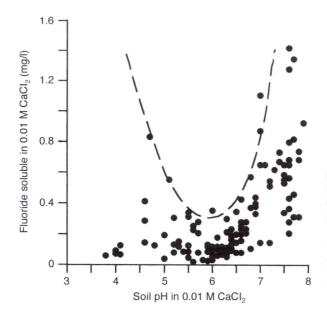

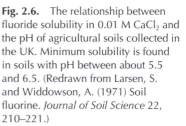

Fig. 2.6. The relationship between fluoride solubility in 0.01 M CaCl$_2$ and the pH of agricultural soils collected in the UK. Minimum solubility is found in soils with pH between about 5.5 and 6.5. (Redrawn from Larsen, S. and Widdowson, A. (1971) Soil fluorine. *Journal of Soil Science* 22, 210–221.)

fluoride can exist as the free F ion (F$^-$) or form soluble complexes, such as SiF$_6^{2-}$, AlF^{2+}, AlF$_2^+$, AlF$_3^0$, AlF$_4^-$, BF$_4^-$ and HF (Stevens *et al.*, 1998b). In alkaline soils the anion F$^-$ predominates. The significance of this variation in the ionic form of fluoride is that plant uptake depends more on the chemical species that are present in the soil solution than on the total amount. Furthermore, the different chemical species vary in their toxicity in terms of both fluoride and aluminium (see Chapters 3 and 4).

Fluoride in Plants

Uptake from the soil

The fluoride concentration in plant leaves varies from less than 1 to several thousand mg/kg. In order to understand the reasons for this large range and to interpret such data in an environmental context, it is necessary to examine the factors that control uptake from the soil and the air. Starting with uptake from the soil, the most important factor is the low solubility of soil fluoride. Plants growing in soils that contain up to about 600–800 mg/kg fluoride usually have from < 2 to about 20 mg/kg in the

leaves (Weinstein, 1977). Where the soil fluoride content is much higher, such as in fluorspar mine waste, leaf concentrations may reach several thousand mg/kg. For example, Cooke *et al.* (1976) reported concentrations from 280 to over 4000 mg/kg in a range of grass and legume species. Some were natural colonists and others were commercial varieties sown for reclamation purposes (Table 2.3).

The controlling influence of the concentration of fluoride in the soil solution has been confirmed many times by growing plants in aqueous culture media. Under these conditions, the fluoride content of the leaves increases approximately in proportion to the concentration bathing the roots (Fig. 2.7). In general, the concentration increases with time and is highest in the roots and is progressively lower in younger leaves that are more distant from the root. The concentrations differ from species to species.

The chemical species of the fluoride in solution is also a major factor controlling fluoride uptake. As Stevens *et al.* (2000) remarked, the species of fluoride that are most readily taken up by plants exist at the pH values where it is most soluble. The converse is that the species that is least readily taken up, the negatively charged anion F$^-$,

Table 2.3. Fluoride concentrations in plants growing on fluorspar mine waste in England (data from Cooke *et al.*, 1976). (Reprinted from *Environmental Pollution* 11, 9–23, 1976, with permission from Elsevier.)

Family	Commercial varieties sown for reclamation and natural vegetation	Fluoride concentration (p.p.m.)
Poaceae (grasses)	*Festuca rubra* var. S59	2125
	Festuca rubra, natural vegetation	2279
	Lolium perenne var. S23	2745
Polygonaceae (docks)	*Rumex obtusifolius*, natural vegetation	320
Campanulaceae (harebells)	*Trifolium repens* var. Kent	4486
	Trifolium repens, natural vegetation	4308
Fabaceae (clovers)	*Campanula rotundifolia*, natural vegetation	548

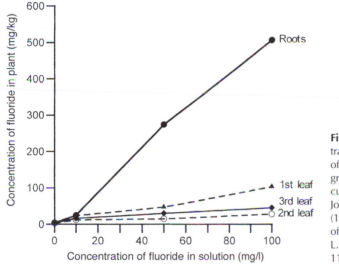

Fig. 2.7. The fluoride concentration in the roots and leaves of sunflower (*Helianthus annuus*) grown for 5 weeks in solution culture. (From Cooke, J.A., Johnson, M.S. and Davison, A.W. (1978) Uptake and translocation of fluoride in *Helianthus annuus* L. grown in sand culture. *Fluoride* 11, 76–88. Courtesy of *Fluoride*.)

predominates in soils with a pH from around 6 upwards. Much of the most productive agricultural land falls within this range.

At least 95% of the fluoride present in roots can be readily extracted by washing with water (Venkateswarlu *et al.*, 1965; Cooke *et al.*, 1978; Garrec and Letourneur, 1981; Takmaz-Nisancioglu and Davison, 1988), which is in marked contrast to another halogen, chloride. This is interpreted as evidence that most of the fluoride in the root and the transport across the root is in the cell walls and intercellular spaces (the apoplast) rather than through the cell membranes and the endodermis (the symplastic route). It is thought that the endodermis acts as a barrier to entry into the conducting system, which limits transport to the shoot. One reason why fluoride is mostly confined to the apoplast is that cell walls have fixed negative charges so they promote exclusion of negatively charged F^- ions, limiting the amount that is held at sites where it can pass the endodermal barrier (Takmaz-Nisancioglu and Davison, 1988). Furthermore, the permeability of cell membranes to F^- is very low, which limits diffusion into cells. The permeability of HF is about six orders of magnitude higher than that of F^- (Gutknecht and Walter, 1981), so it is taken up much more readily than F^-.

Kronberger (1988a,b) proposed a mechanism for the passive uptake of fluoride, based on the action of proton pumps in the cell membranes. He suggested that proton pumps cause a locally lower pH near the membrane surfaces and that this promotes the formation of HF:

$$F^- + H^+ \rightleftarrows HF$$

The HF diffuses through the membranes and then, at the higher pH inside the cell, F^- is formed:

$$HF \rightleftarrows H^+ + F^-$$

However, the fact that most of the fluoride can be removed from roots by washing suggests that this is not a very significant mechanism. Because the cell membranes and the endodermis provide such a barrier to the uptake and transport of F^-, Davison et al. (1985) proposed that most of the fluoride reaching the shoot leaks past the barrier at certain points. Steudle et al. (1993) showed that the Casparian band of maize endodermis is discontinuous at the root tips and at sites of developing lateral roots, the cross-sectional area of this 'bypass' being about 0.08% of the total area of the endodermis. Pitman (1982) estimated that this non-selective bypass pathway carries about 2% of the water flow and this is supported by observations of Perry and Greenway (1973), who found that molecules that did not enter the symplast were transported into the xylem at a rate equivalent to 2–3% of the water flow. A simple calculation based on the concentration of fluoride in the soil solution and the rate of water use suggests that this pathway could account for the normal background concentrations that are found in leaves (Davison et al., 1985). It also explains the increase in shoot fluoride when the concentration is increased in culture media (Fig. 2.7). If this pathway does operate, it suggests that the background fluoride content of leaves is related to water use and that it may be higher in species with a highly branched root system.

Interest in fluoride–aluminium complexes began in the 1970s because of concern about aluminium toxicity and also because of so-called fluoride accumulators. These are species that contain much higher concentrations of fluoride than other species, even when they are growing on soils with 'normal' levels of fluoride. A number of species that are common in the USA are known accumulators, such as hickory (Carya sp.) and flowering dogwood (Cornus florida) (Weinstein and Alscher-Herman, 1982), but the highest concentrations are found in members of the family Theaceae. This includes the horticultural camellia (Camellia japonica) and tea (Camellia sinensis). Zimmerman et al. (1957) found that various brands of tea contained 72–115 mg/kg and Chinese tea 131–178 mg/kg of fluoride. One sample of greenhouse-grown tea contained 1530 mg/kg. Cholak (1959) reported that the concentration in tea ranged from 8 to 400 mg/kg, and Davison (1983) reported that four varieties of tea contained between 139 and 233 mg/kg.

Tea also accumulates aluminium (Chenery, 1955), so Davison (1983) analysed a number of species that were known to be aluminium accumulators (Chenery, 1948a,b) to see if they also accumulated fluoride. Almost all of them contained high concentrations of fluoride, ranging from about 30 to 1600 mg/kg (Table 2.4). The link between the two elements was strengthened when it was found that there was a quantitative

Table 2.4. Fluoride concentrations in a selection of fluoride-accumulating species (from Davison, 1983). (Reprinted with permission of the author.)

Family	Geographical range	Species	Fluoride content (p.p.m.)
Theaceae	Native to South-East Asia	Camellia sinensis, Earl Grey tea	214
	Asia	Eurya emarginata	1655
Symplocaceae	Asia	Symplocos paniculata	56
Diapensiaceae	Eastern USA	Pyxidanthera barbulata	280
	Arctic, montane	Diapensia lapponica	361
Melastomataceae	Tropical South America	Tibouchina organensis	65
Cyatheaceae (ferns)	Australia	Cyathea australis	396

relationship between the concentrations in leaves of cultivated *Camellia* varieties (Davison *et al.*, 1985). Takmaz-Nisancioglu and Davison (1988) explored the relationship further by exposing bean plants to equal concentrations of fluoride as NaF or AlF$_3$ in solution culture. The AlF$_3$ promoted much greater uptake and transport of fluoride to the leaves. Even a pretreatment of AlCl$_3$ before giving NaF increased fluoride transport, leading the authors to conclude that aluminium–fluoride complexes are more readily taken up and transported. They suggested that the effect was due to the difference in charge between F$^-$ and the aluminium complexes. The experiments of Stevens *et al.* (1997), which were more sophisticated that those of Takmaz-Nisancioglu and Davison (1988), confirmed that aluminium–fluoride complexes increase fluoride uptake. Stevens *et al.* (2000) have also made significant progress in quantifying the effects of pH and different species of fluoride on plant uptake. They have produced a model that gives, for the first time, a graphic picture of the relationships between total fluoride in solution, pH and the fluoride content of shoots. Figure 2.8 shows that, with a total fluoride content in solution < 0.14 mM (2.65 mg F/l), which is relatively high, uptake is predicted to be very low on soils with a pH > 5.5–6.0. At lower pHs the fluoride content is predicted to be > 20 mg/kg and at those pH values the shoot concentration rises steeply with the amount in solution. The authors used the model to determine whether the addition of fluoride added to soil in phosphate fertilizer could lead to plant- or animal-toxic concentrations in leaves. They concluded that the risk was very low in soils of neutral pH but that there would be a risk in acidic or alkaline soils (Stevens *et al.*, 2000). However, it is important to be aware of the fact that there is still much to be learned about soil fluoride. For example, the model (Stevens *et al.*, 2000) does not take into account the presence of organic compounds that might compete with cations in forming complexes. Also, there is evidence that fungi might play a part in determining fluoride availability by attacking insoluble fluorides (Wainwright and Supharungsun, 1984). The importance of this kind of fungal interaction in the rhizosphere is unknown.

The shoots and leaves of pasture species may have their fluoride content increased by adhesion of soil from rain splash or worm casts or from the hooves of livestock, particularly in low-growing swards. This external fluoride is not a hazard to the plant but it does contribute to the diet of livestock

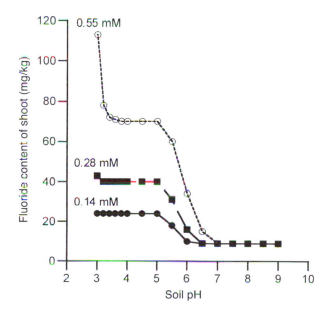

Fig. 2.8. Predicted shoot fluoride concentrations (mg F/kg/dry wt) in relation to soil pH and different concentrations of fluoride in solution: 0.14, 0.28 and 0.55 mM (2.65, 5.27 and 10.54 mg F/l, respectively). (Drawn using data supplied by Stevens from a model described in Stevens, D.P., McLaughlin, M.J., Randall, P.J. and Keerthisinghe, G. (2000) Effect of fluoride supply on fluoride concentrations in five pasture species: levels required to reach phytotoxic or potentially zootoxic concentrations in plant tissue. *Plant and Soil* 227, 223–233. With kind permission of Kluwer Academic Publishers.)

(Healy, 1973; Loganathan et al., 2001). The environmental importance of soil contamination depends on the amount of soil deposited and on the fluoride content of the soil. McClenahan and Weidensaul (1977) incorporated up to 10% (by weight) of soil into dried forage to determine the effect, but the increase in fluoride was only 7 mg/kg. On the other hand, Healy's (1973) data suggest a much greater degree of soil contamination in some seasons. Long-term application of phosphate fertilizer may increase the risk because it increases the fluoride in the surface layers (Loganathan et al., 2001; McLaughlin et al., 2001). There are reports from New Zealand of chronic fluorosis in livestock caused by ingestion of high-fluoride soil (cited by Loganathan et al., 2001), and Cronin et al. (2000) suggested that soils with a total fluoride concentration in the surface (75 mm) layer of about 300–1400 mg/kg have the potential to cause chronic fluorosis in livestock.

Fluoride speciation in leaves

Most of our knowledge of the chemical state of fluoride in plants, whether it originates from the soil or air, is by inference rather than as a result of direct chemical or physical characterization of the compounds or complexes. It is important to know the chemical forms of fluoride because the reactions with calcium, magnesium, aluminium and a few other elements determine the toxicity (Chapter 4). A few species are known to metabolize inorganic fluoride to produce organofluorine compounds and these are discussed in Chapter 8. However, in the vast majority of plants the fluoride remains in the inorganic form and most, if not all, can be extracted by aqueous solvents (Jacobson et al., 1966). It can be detected in the extraction solution by means of a specific ion electrode, which responds only to F^-, provided a metal-complexing agent is added to free the fluoride. This leads to the conclusion that most of the fluoride in leaves is present as free anions or is associated with calcium, magnesium or aluminium (Weinstein and Alscher-Herman, 1982). In some cases, notably plants that grow on fluorspar waste, some of the fluoride-containing complexes are less soluble and the fluoride is only released by treatment with a mineral acid (Cooke et al., 1976). In others, such as the US National Institute of Standards and Technology (NIST) sample referred to in Chapter 1, a proportion of the fluoride can only be released by oxidation and fusion in strong alkali. The predominant association may be with calcium and this is supported by investigations using ^{18}F and the electron microprobe (Bligny et al., 1973b; Garrec et al., 1974, 1977b; Bonte and Garrec, 1980; Garrec and Chopin, 1982; Garrec, 1983). Crystals of calcium fluoride have been detected in the alga Chara fragilis (Lévy and Strauss, 1973). Much less is known about aluminium–fluoride complexes in leaves. Takmaz-Nisancioglu and Davison (1988) analysed fluoride in the xylem sap of tomato plants that had been treated with NaF or AlF_3. In plants treated with NaF 95.3% of the fluoride was present as F^- at the pH used, but in the AlF_3 plants only 56% was F^-, the remaining 44% being complexed, presumably with aluminium. This transport of fluoride as aluminium complexes was confirmed by Liang et al. (1996), using tea plants. They found that 92.2% of the total aluminium content of the xylem sap consisted of fluoro-aluminium complexes, which included AlF^{2+} and AlF^{3+}. However, analysis of leaves indicated that, once in the leaf, the aluminium was complexed with organic acids and the fluoride was released. This was also found by Nagata et al. (2002), who found that most of the aluminium in tea leaves was complexed with catechin and that there was evidence of an aluminium–fluoride complex in only one sample.

Uptake, deposition and translocation of airborne fluoride

The uptake of fluoride by leaves from the air is determined by the boundary layer, the nature of the surfaces and the stomatal apertures. Freely moving, turbulent air carries gases and particles rapidly from place to place but a plant canopy or even a single leaf reduces wind speed, producing a layer of still air that surrounds the canopy and leaves (Fig. 2.9). Movement through this boundary layer is by molecular diffusion, so it is much slower. The thickness of the boundary layer decreases with wind speed but increases with leaf size, so the combination of still air, a dense canopy and large leaves produces a much lower rate of gas conductance than turbulent air, an open canopy and needle-like leaves.

Gases and particles are deposited on leaf surfaces, where they may be retained or lost over a period of time. Leaf cuticles are designed by natural selection to have low permeability to water and other molecules, but low-molecular-weight solutes (e.g. sugars) are known to permeate cuticles via hydrophilic pores. They are lined with fixed negative charges, so cations can permeate but not anions, such as F^- (Marschner, 1995). However, if the leaf is covered by an acidic water film, theoretically HF could permeate the layer and enter the cell walls. There has been only one study of fluoride penetration of cuticles and, in it, Chamel and Garrec

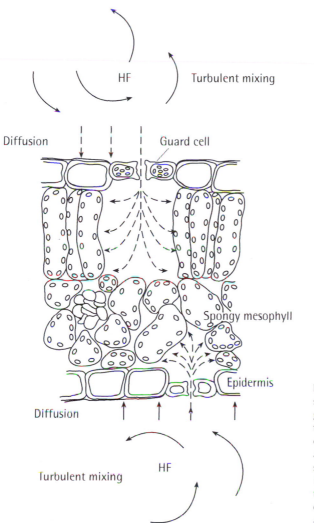

Fig. 2.9. Transverse section of a leaf showing the pathway of uptake of gaseous fluoride. The gas is carried to leaves by turbulence and then it diffuses through the boundary layer and stomata into the intercellular spaces. It impinges on cell walls and is carried by the transpiration stream towards the sites of maximum evapotranspiration, the margins and tip.

(1977) showed that the permeability of pear-leaf cuticle to fluoride is low. However, there is evidence that surface fluoride does penetrate cuticles, probably after weathering or insect damage (Davison, 1982). Aqueous sprays of NaF applied to leaves can cause necrosis identical to that caused by HF, while several workers have shown that heavy particulate deposits may also cause visible injury, especially when leaf surfaces are wet (McCune et al., 1965, 1977). Furthermore, Ares et al. (1980) have shown that cuticular uptake is important in the case of plants that live in very dry habitats.

Hydrogen fluoride gas diffuses into leaves through the boundary layer and then through the stomata (Fig. 2.9), so uptake is determined by their diurnal, environmentally controlled movements. Stomatal apertures are usually greatest from early morning to about noon and they are usually, but not always, closed at night. Water stress in particular reduces stomatal apertures, so uptake is controlled by a large number of variables. Hydrogen fluoride is very soluble (18 mol/l; Doley, 1986) so it readily dissolves in the cell-wall water and dissociates to form F^- (Kronberger, 1988a,b). Dissociation is assumed because the pH of the wall water is neutral to slightly acid, favouring the reaction to the right: $HF \rightleftharpoons H^+ + F^-$. The F^- anions are then carried away from the site of solution by the flow of water, which is driven by evaporation through the stomata. As there is greater water loss from the margins and apex of leaves, the fluoride accumulates in those regions. Some of the fluoride reacts with calcium and some diffuses through the cell membranes but most is swept to the apex and margins. The gradient in concentration has been demonstrated many times by analysing small segments of leaf. Doley (1986), for example, found 270 mg F/kg in the edge 5 mm of a Chardonnay grape leaf but only 84 mg/kg in the rest of the leaf. The smaller the segments that can be analysed, the steeper the gradient, which is illustrated by Garrec et al. (1972), who used a microprobe that allowed them to determine the fluoride in millimetre-wide sections of a fir (Abies alba) needle. The tip 1 mm had almost 60,000 mg/kg but only 3 mm from the tip it

was less than 100 mg/kg. Their work also showed that the fluoride disturbed the normal pattern of calcium distribution in the needle, increasing the concentration in the regions where the fluoride accumulated (Garrec et al., 1974). This transpiration-driven concentration of fluoride in the tip and margins is one of the reasons for its high toxicity to some species.

The process of uptake, from the turbulent air, through the boundary layer and stomata and then into solution and transport to the sites of accumulation takes a significant length of time – tens of minutes, depending on the rate of water movement and the dimensions of the leaf. This buffers the leaf against the changes in fluoride concentration that occur in the air, evening out the concentrations. The result is that the plant does not respond to short-term changes in HF concentration but to the accumulated fluoride in the tip and margins. This is important for monitoring because it means that the shortest time-averaging period that is needed is hours or days rather than minutes.

A variety of other techniques, including ^{18}F, microscopy, solvent extraction and tissue fractionation, have been used to identify pathways of movement and localization of fluoride in tissues. Fractionation procedures used by Ledbetter et al. (1960) showed that, in tomato, fluoride was in its highest concentration in cell walls, followed in order by the chloroplasts, soluble proteins, mitochondria and microsomes. The high concentration in the walls is consistent with the pathway of translocation of water and the high concentration of calcium in that fraction (Clarkson and Hanson, 1980). Concentration in the chloroplasts was also found by Chang and Thompson (1966), and this is in accord with the pH-dependent mechanism of uptake and transport suggested by Kronberger (1988a,b).

It has been suggested that a small proportion of the fluoride may be translocated out of the leaf to the stem or root (Benedict et al., 1964; Keller, 1974; Garrec and Vavasseur, 1978; Kronberger and Halbwachs, 1978; Kronberger et al., 1978). Doley (1986) produced a convincing theoretical argument that fluoride might enter the xylem during the night and therefore be carried to all parts of a leaf, but the possibility of significant transport to remote parts of the plant is much less feasible. Some of the experimental data are equivocal (Weinstein and Alscher-Herman, 1982), so, at present, retranslocation is usually considered to be environmentally unimportant.

Deposition of particulates on vegetation

The rate of deposition of particulate matter to individual leaves can be estimated using the deposition velocity (Davison, 1982). This gives the deposition in terms of the rate per area, but it is conventional to express concentrations on a weight basis, so it is necessary to convert the data using the specific leaf weight (SLW) (g/m^2). This is an important conversion because the SLW varies by at least 150 times from species to species. In the case of a thin, unlignified structure, such as the thallus of a liverwort, the SLW is usually under 1 g/m^2, while in a typical ryegrass leaf it is 10 and in a whole shoot in the region of 20. Many deciduous trees have values between 25 and 50, while in most leathery evergreens (e.g. holly) and conifer needles they are between 80 and 130 g/m^2. Consequently, if the same amount of fluoride is deposited on the same area of leaf of two species that differ in SLW, when the concentration is expressed on a weight basis, it will differ in proportion to the difference in SLW. Table 2.5 gives some hypothetical data for representative values of SLW to demonstrate the effect of this factor and of particle size on the rate of accumulation from an air concentration of 1 µg F/m^3 (Davison, 1982). The data demonstrate that, with the same concentration in the air and the same rate of deposition, a ryegrass leaf (SLW 10 g/m^2) would appear to accumulate fluoride at ten times the rate of a typical evergreen species (SLW 100 g/m^2). The rate of accumulation from the smaller particles is low but for particles greater than about 5 µm diameter it is substantial, especially in thin-leaved species.

Table 2.5. Calculated daily rate of accumulation of particulate fluoride (mg/kg/day) from an atmospheric concentration of 1 µg/m^3, wind speed 4 m/s. Data are presented for four specific leaf weights.

Particle diameter (µm)	Daily rate of accumulation (mg/kg/day) Specific leaf weight (g/m²)			
	10	20	40	100
0.1	1.6	0.8	0.4	0.2
1	1.6	0.8	0.4	0.2
5	34	17	8.6	3.4
10	100	50	25	10

However, the greater rate of deposition of larger particles means that their environmental importance tends to be restricted to areas close to the source.

Variation in the SLW was advanced by Davison (1982, 1983) as one of the main causes of differences in the fluoride content of species found in polluted areas. In addition, very low values of specific weight may well be the reason for the apparently high rates of accumulation by the tassels and styles of maize (Kronberger and Halbwachs, 1978) and the stigmatic surface of strawberry (Bonte and Garrec, 1980).

Uptake of HF by leaves

The first attempts to examine the quantitative relationship between the exposure to HF and the fluoride content of vegetation were by Hitchcock et al. (1971) and McCune and Hitchcock (1971). Using 270 separate fumigation experiments under controlled conditions, the authors showed that the regression of accumulation of fluoride by lucerne and cocksfoot grass on the concentration of hydrogen fluoride and duration of exposure accounted for 70% and 54% of the variation in the data, respectively. As postfumigation harvests showed that the fluoride concentration in lucerne fell with a half-life of about 8 days, a modified equation was produced to take account of this. In later publications (NAS, 1971; Weinstein, 1977), however, the regression model was simplified to produce the dose–rate equation:

$$\Delta F = KCT$$

where ΔF is the change in fluoride concentration (mg/kg), K = an accumulation coefficient, C = the concentration of hydrogen fluoride ($\mu g/m^3$) and T = the duration of exposure (days).

While there is no doubt that accumulation is related to the concentration of HF and the exposure time, the accumulation coefficient, K, is not constant even for one species. For example, K for lucerne has been reported as varying from 0.4 to 7.7; for cocksfoot grass, from 0.4 to 3.8; and, for perennial ryegrass, from 3.0 to 5.1 (NAS, 1971; Weinstein, 1977). Similar variation is apparent in Sidhu's (1979) data for tree species. The reason for this variation is that the model does not allow for differences in SLW within or between species, stomatal apertures, the effects of wind and rain or loss of fluoride. The model has been most successfully used in a comprehensive study of the effects of emissions from Norwegian aluminium smelters (AMS, 1994). One part of this study was an assessment of the relationship between exposure to fluoride and accumulation in leaves, and Horntvedt (1997) showed that there was a significant relationship for

rowan (= mountain ash, *Sorbus aucuparia*) leaves at two smelters. The mean value of K was 1.7 (Fig. 2.10). Pine and spruce had much lower values of K, almost certainly because of the higher SLW and lower stomatal conductance. Clearly, in the case of rowan leaves, the fluoride concentration is dependent on the concentration in the air and the duration of exposure, but, as this is one of the strongest correlations that has been published, it is worth further examination. The first point to note is that, although the relationship was statistically significant, there was a considerable scatter in the points. The 95% confidence interval shows, for example, that an exposure of 200 $\mu g/m^3 \times$ days produced rowan fluoride concentrations ranging from about 230 to 380 mg/kg. This means that, if K were used to predict leaf concentration from atmospheric concentration or vice versa, the estimate would not be very accurate. There are several reasons for the variability, notably errors in measuring the airborne fluorides and variation in the surface deposits (AMS, 1994). On the other hand, the consistency in the data was probably due to the fact that the two smelters were in areas with similar climates and similar day lengths. It is notable that, when two sites at Årdal were compared, there was a greater rate of accumulation in

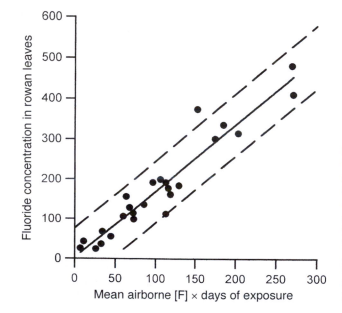

Fig. 2.10. Fluoride concentration in rowan (*Sorbus aucuparia*) leaves growing near two aluminium smelters plotted against fluoride exposure: concentration (μg F/m^3) × duration of exposure (days); 95% confidence intervals shown as dashed lines. (Redrawn from Horntvedt, R. (1997) Accumulation of airborne fluorides in forest trees and vegetation. *European Journal of Forest Research* 27, 73–82.)

tree leaves at a high-elevation site than at a lower-elevation site that had higher atmospheric fluoride concentrations (AMS, 1994). This was thought to be due to a difference in humidity. Even position on the tree affects fluoride accumulation (Vike and Håbjørg, 1995). Day length is important because the duration of exposure is usually expressed in days, but uptake shows a diurnal pattern related to light. Day length varies with season and latitude, so data obtained in more northerly latitudes would not be expected to be compatible with data from more southerly latitudes. A final complication is that there may be differences in the rate of accumulation between individual trees. Vike and Håbjørg (1995) showed that there were clear differences between genotypes of rowan and birch and that the degree of difference varied with the site. This means that values of K are specific to the location and local conditions, so it is not valid to use K to predict accumulation in other regions.

An alternative approach to the study of uptake is by means of a mechanistic model based on an analogy with electrical resistance (Fowler and Unsworth, 1974; Davison, 1982, 1983). Doley (1988) has a comprehensive account of this subject. The resistance to uptake is the reciprocal of the deposition velocity:

$$\text{Flux} (\mu g/m^2/s) = \frac{\text{Difference in concentration between air and sink } (\mu g/m^2)}{\text{Total resistance to uptake } (s/m)}$$

The advantage of this model is that the total resistance to flux by the stomatal pathway is the sum of the component resistances. The total resistance to gas uptake via the stomata can be envisaged as having three components: the boundary layer, stomatal and internal (or mesophyll) resistances:

Total resistance to uptake =
boundary layer resistance + stomatal resistance + internal resistance

In the case of HF, the internal resistance is zero because of the high solubility and because of internal dilution due to water movement. The other two resistances can be measured, so it is possible to use the model to analyse the process of uptake and to determine which of the component factors is most important. The calculation can be made for any particular plant species, provided the stomatal resistance and the SLW are known. Some hypothetical data for 3 days of exposure to 1 μg HF/m³ are shown in Fig. 2.11. Two patterns of stomatal resistance were used, one typical of a fast-growing crop, such as bean (*Phaseolus*), during moist, warm weather (curve a, minimum resistance on the lower surface of 83 s/m, upper surface three times higher) and the other for the same species but during a period when the stomatal resistance was twice as high, e.g. during the onset of drought (curve b). The bean had an SLW of 15 g/m², and it was assumed that there was a constant boundary layer resistance of 100 s/m. After 3 days the fluoride contents were predicted to be 54 and 36 mg/kg for plants in which the stomatal resistance was low and high, respectively. Curve (c) shows the predicted concentration in a plant with the same total resistance as in curve (a) but with an SLW of 100 g/m² (e.g. a leathery-leaved evergreen). The effect of the SLW was to decrease the fluoride from 54 to 8 mg/kg. In practice, species with high SLW usually have a higher stomatal resistance than bean and stomata on only one surface, so curve (d) shows the fluoride content of a species with those characteristics. In that case the fluoride content was predicted to be only 4 mg/kg after 3 days at 1 μg/m³. The difference between curves (a) and (d) is compatible with the difference that Horntvedt (1997) found between rowan and pine. Although this simple model does not include any provision for surface deposition or losses, it does illustrate graphically two of the main reasons why species differ so much in fluoride content, even when they are growing together. The variation between species found by Sidhu (1979), for example, are probably explained largely in terms of differences in resistance and SLW. He found that the fluoride concentrations of balsam fir, black spruce, larch and white birch were 69, 81, 411 and 357 mg/kg,

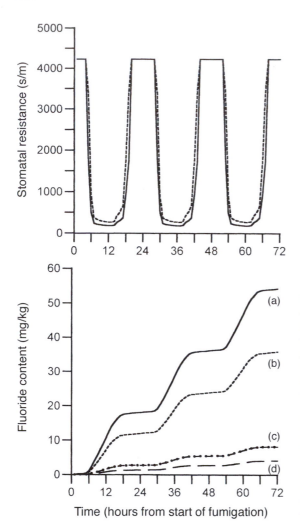

Fig. 2.11. Modelled increase in fluoride concentration in leaves with different specific leaf weight and stomatal resistance exposed for 3 days to 1 μg HF/m³. The upper graph shows the hypothetical stomatal resistance over the 3 days and the lower shows the predicted increase in fluoride content. (a) SLW = 15 g/m² (e.g. bean), minimum resistance on the lower surface = 83 s/m, upper surface three times higher. (b) Same as (a) but during a period when the stomatal resistance was twice as high. (c) Same as (a) but plant with SLW = 100 g/m² (e.g. a leathery-leaved evergreen). (d) Same as (c) but higher stomatal resistance on only one surface and higher resistance typical of a more xeromorphic species.

respectively. Characteristically, the concentrations in evergreen conifer needles were lower than in the deciduous larch and the broad-leaved birch, which have lower resistances and SLW.

To complicate matters further, the SLW is not constant, even within a species. It is affected by light and temperature, and it tends to alter as a leaf ages and senesces. Davison and Blakemore (1976) demonstrated this by analysing the flag leaf of barley as it aged (Fig. 2.12). The fluoride concentration increased steadily over 2 months but this was not due to continued uptake because the content per leaf fluctuated. It was primarily due to a decrease in weight of the leaf as senescence occurred.

A similar model could be used to study deposition of HF on leaf surfaces but the only data on resistances are from short-term exposures to high concentrations of HF (Davison, 1983). Typically these give resistances for a variety of species ranging from 200 to 2000 s/m, which suggests that surface deposition makes a significant contribution to the total leaf fluoride content. This is supported by the fact that it is always possible to wash a significant proportion of the fluoride off leaves (McCune et al., 1965; Jacobson et al., 1969; NAS, 1971; Vike and Håbjørg, 1995). Some of this surface deposit may penetrate the cuticle (Weinstein, 1977; Davison, 1983), but, being superficial it is fully exposed to the effects of weathering

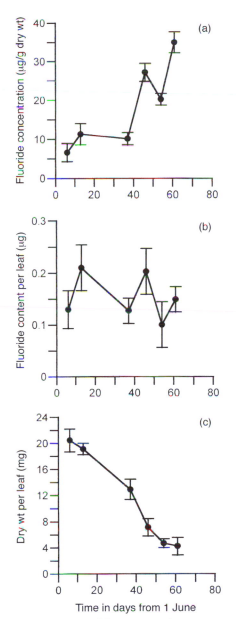

Fig. 2.12. Analysis of the apparent changes in fluoride concentration in the ageing flag leaf of barley growing near an aluminium smelter. (a) The change in concentration over time (mg F/kg dry wt) and 95% confidence intervals. (b) The absolute amount of fluoride per leaf (µg F per leaf). (c) The change in dry weight of the leaf (mg per leaf) as it aged. (Redrawn from Davison, A.W. and Blakemore, J. (1976) Factors determining fluoride accumulation in forage. In: Mansfield, T.A. (ed.) Cambridge University Press, Cambridge, pp. 17–30.)

and it may be this fraction that is partly responsible for the rapid changes in fluoride concentration that have been observed in the field (Davison and Blakemore, 1976; Davison et al., 1979; van der Eerden, 1981). It is also possible that there may be some direct penetration of dissolved fluoride through the stomata, though this is speculative (Eichert et al., 1998). However, leaf surfaces almost certainly have a finite capacity to sorb HF, so uptake is probably non-linear over time and using these short-term resistances will overestimate deposition. There is some evidence that the presence of dusts or other particles may alter the rate of surface deposition because Jacobson et al. (1969) found that washing leaves before HF exposure decreased the postfumigation fluoride content. More research is needed on this subject because surface fluoride may influence disease organisms and the general leaf microflora (Weinstein, 1977).

Fluoride uptake by grass swards

Resistance models are useful for analysing fluoride uptake by a single leaf that lasts over a growing season or two, but a grass sward is too complex for that approach because there is a mixture of leaves of different ages that differ in stomatal resistance and SLW. In tall swards the fluoride concentration varies with leaf position in the canopy (van der Eerden, 1981). Interest in this difficult topic has been stimulated by the need to provide air-quality guidelines that will prevent fluoride accumulation in forage to the level where it will cause fluorosis in livestock and wild herbivores. The most productive approach has been through a combination of field observations and experimental fumigation (see literature cited in van der Eerden, 1991). A few groups have used this approach to produce models that relate grass fluoride to atmospheric fluoride and other environmental factors (Davison and Blakemore, 1976; Davison et al., 1976; de Temmerman et al., 1978; Bunce, 1983; Craggs and Davison, 1987a,b; van der Eerden, 1991).

Most grass fluoride monitoring involves sampling at intervals of a few weeks. As might be expected, this usually shows that concentrations change with time. In some places, the concentration rises substantially in successive months, while in others it falls. Often concentrations are lower in the summer months than in winter (Allcroft *et al.*, 1965; Davison *et al.*, 1976; Suttie, 1977; van der Eerden, 1981, 1991) and this is sometimes ascribed to 'growth dilution', i.e. the grass is increasing dry matter faster than it acquires fluoride. Analysis of such data, however, shows that this seasonality is not due to growth dilution but to differences in pollutant dispersion, atmospheric concentrations and deposition rates (Davison *et al.*, 1976; Craggs *et al.*, 1985; van der Eerden, 1991). Growth dilution is only likely to occur at times when grass is growing very rapidly and the atmospheric fluoride concentration is not much above background levels.

Sampling at shorter time intervals is more instructive because it reveals that in the cool, temperate climate of Europe, pasture fluoride concentrations vary rapidly and that most of the variation is statistically related to airborne fluoride concentrations (Davison and Blakemore, 1976; van der Eerden, 1981, 1991; Davison, 1982, 1983; Craggs *et al.*, 1987a,b). Figure 2.13 shows variation in the fluoride content of pasture grass at two locations, one near an

aluminium smelter in England (Blakemore, 1978) and the other near a brickworks in The Netherlands (van der Eerden, 1981). The most obvious feature is that the fluoride content of the grass changed rapidly over a day or so, in some cases decreasing by over 100 mg/kg in a single day. Both sets of data were collected in the autumn when there was no growth. Weekly sampling also shows large changes over time. In one example (Davison *et al.*, 1979) the mean grass fluoride changed from 508 to 188 mg/kg over a week. The samples were collected from 18 replicate 1 m × 1 m plots, so the differences were quite clearly statistically significant. An analysis of the variation in grass fluoride data that were collected in England, Wales and Canada (Davison *et al.*, 1979) showed that the mean daily change in fluoride (positive or negative) was a function of the mean: that is, the higher the mean fluoride, the greater the absolute daily gains and losses. High concentrations occur near sources, so sites in such areas show the greatest day-to-day gains and losses, whereas the further the site is from the source the lower the variation. This probably explains why researchers find different degrees of day-to-day variation (e.g. van der Eerden, 1991).

In the case of Davison and Blakemore's (1976) data the grass fluoride (Fg) on any one day was related to the content on the previous day (Fg_{t-1}), and to the gaseous

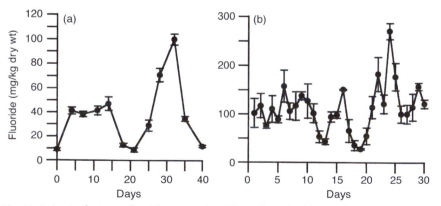

Fig. 2.13. Variation in the mean fluoride content (mg F/kg and standard error) of pasture grass at two locations: (a) near a brickworks in The Netherlands (van der Eerden, 1981) and (b) near an aluminium smelter in England (Blakemore, 1978). Both data sets were collected in the autumn when there was no growth.

and particulate fluoride concentrations (F_{HF}, F_{part}) over the 24 h:

$$Fg = 30 + 0.48\ Fg_{t-1} + 39\ F_{part} + 15\ F_{HF}\ (n = 30,\ r = 0.794)$$

The regression suggested that the influences on the grass fluoride were, in order: $Fg_{t-1} > F_{part} > F_{HF}$. Including the daily rainfall gave a slight improvement to the r^2.

Using data of this kind, sometimes supplemented with data from controlled fumigation, several authors have produced models relating grass fluoride to environmental factors. De Temmerman (1984, cited by van der Eerden, 1991) worked with weekly sampling intervals and produced a model that used the three previous weeks' grass fluoride concentrations (F_{t-1}, F_{t-2}, F_{t-3}) and rainfall. Also using weekly sampling, Blakemore's (1978) regression model accounted for 72% of the total variation over a 2-year period:

$$Fg = 0.67\ Fg_{t-1} + 281\ F_{HF} + 223\ F_{part} - 31$$

Craggs and Davison (1987b) used Blakemore's weekly data to produce the most complex model and it accounted for 84% of the variation, which is remarkably high for field data over a 2-year period. In an effort to produce a simpler model, van der Eerden (1991) used grass grown in standard pots, cut the grass at a standard height to try to remove the effects of soil splash and introduced a seasonal index in the equation:

$$Fg = 90\ Si\ F_A^* \exp(-0.1\ Si^{-1}\ T)$$

where Si = a seasonal index; F_A^* = the highest 24 h average atmospheric F between a day with more than 2 mm of rain and the grass-sampling date; and T is the number of days between the day F_A^* occurred and the sampling date. However, both Craggs and Davison (1987b) and van der Eerden (1991) pointed out that all models of this type are specific to the location and that the accuracy of prediction will be lower in other places. The reason is that the models are based on correlations rather than a mechanistic understanding of the processes involved.

Most of the processes – deposition, uptake and so on – are understood but the major limiting factor is our lack of knowledge of the effects of rain and the mechanisms causing the rapid loss of fluoride.

Rainfall and pasture-grass fluoride

Several older publications comment on the relationship between the fluoride content of precipitation and vegetation. Garber (1970), for example, cited mean rain concentrations at three sites as being 10.0, 5.1 and 1.85 mg F/l, and the concentrations in vegetation at the same sites as 185, 53 and 20 mg/kg, respectively. He concluded that there was a direct correlation between the fluoride in rain and vegetation, which implies that the fluoride content of the grass was determined by wet deposition. In contrast, the studies that were discussed in the preceding paragraph all found a significant inverse correlation between forage fluoride concentrations and the volume of rainfall, the obvious implication being that rain washes fluoride from the grass. This illustrates the fact that precipitation may have conflicting effects on grass fluoride. On the one hand, it scrubs the air, reducing the fluoride concentration while it is raining, but, on the other hand, it may deposit the fluoride on the leaves or leach fluoride from the surfaces. By wetting the surfaces it may also increase the rate of dry deposition (Fowler and Unsworth, 1974). Van der Eerden's (1991) and de Temmerman's (cited in van der Eerden, 1991) models both assume that, overall, rain reduces the fluoride content of vegetation. Intuitively, it would seem logical that rain washes fluoride off leaves, but analysis of field data and artificial rain experiments do not support this idea (Less et al., 1975; Davison and Blakemore, 1976; Blakemore, 1978; Craggs and Davison, 1985, 1987b). Less et al. (1975) found that artificial precipitation (at a rate of 20 mm/h for about 30 min on each of 4 or 5 days during the 10- or 11-day experimental period), rather than decreasing the fluoride concentration in ryegrass foliage, resulted in an increase of about twofold. This was explained by increased absorption of fluoride during the period when the foliage was wet. Less *et al.* (1975) concluded that 'frequent rather than heavy rainfall would lead to greater absorption, but other factors which increase the time during which the grass remains wet include dew, mist, low temperatures and low wind speeds'. Analysis of Blakemore's (1978) data set, in which atmospheric fluorides, grass fluorides, precipitation and temperature were measured in replicated plots, showed that the effect of rainfall was very small and not significant if airborne fluorides were included in the analysis (Blakemore, 1978). Craggs and Davison (1985) used artificial rain to try to resolve the question but there was no consistent effect of the treatment. Blakemore's (1978) data, collected weekly over a 2-year period, show an inverse relationship between the concentration of fluorides in the air and the volume of precipitation (Fig. 2.14). In weeks with less than 10 mm of rain, the HF concentration ranged from 0.02 to 1.1 µg/m³ but, in weeks with > 20–30 mm, the maximum was around 0.4. This suggests that the effect of rain is primarily to reduce the concentration of fluoride in the air.

Loss of fluoride from vegetation

Although it has been known for over 50 years that the fluoride content of vegetation may decrease over short time periods (Zimmerman and Hitchcock, 1946), very little is known about the mechanisms involved. In a review of the subject, Davison (1982) cited 12 reports of substantial reductions in fluoride content in a variety of circumstances and species. For example, Knabe (1970) found that spruce needles lost about 340 mg/kg over 5 months, Georgsson and Petursson (1972) found that grass contaminated with volcanic ash fell from 4300 to 30 mg/kg in 30 days (Chapter 5) and, in fumigation experiments, Weinstein (1961) reported that the fluoride content of tomato decreased after fumigation by 70 mg/kg in 3 days. Davison (1982) listed the known possible mechanisms that might account for these observations as: growth dilution; leaching by rain; leaf death and loss; shedding of surface waxes; guttation; translocation to the stem and root; and volatilization.

Leaching by rain and growth dilution have already been discussed. The latter is not responsible for the commonly observed drop in grass fluoride concentrations during

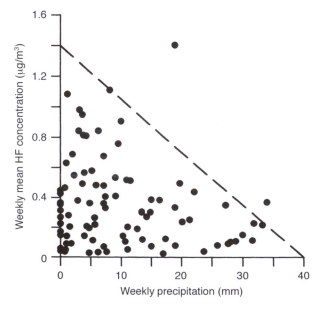

Fig. 2.14. The relationship between the weekly mean concentration of HF (µg/m³) in the air and the total precipitation (mm) per week over a 2-year period near an aluminium smelter. The maximum HF concentration recorded was lower in weeks with higher rainfall. (Drawn from data in Blakemore, 1978.)

the summer and it could not have been responsible for the loss from Knabe's mature spruce leaves (Knabe, 1970) or the daily losses found by Blakemore (Davison and Blakemore, 1976; Blakemore, 1978). Eventually all leaves die, transferring fluoride to the humic layers of the soil. In pastures, where leaf turnover rate is high, this may play a significant part in reducing the amount of fluoride in the sward as a whole, but the evidence suggests that the rate of death is normally too low to offset gains by uptake and deposition unless the atmospheric concentrations are low (Davison, 1983). The wax covering on leaves is continually subjected to weathering by rain, wind, insects and microorganisms, and there is evidence of constant regeneration of the waxes. In the absence of any other plausible explanation, Chamberlain (1970) considered that this was probably responsible for the loss of aerosol that he observed in experiments with radioactive isotopes. Because the rate of loss of surface materials has not been measured, however, the role of this mechanism remains as speculative now as it was in 1970. Volatilization was discussed by Davison (1982, 1983), and evidence was produced that hydrogen fluoride volatilizes rapidly from inert acidic surfaces (Takmaz-Niscancioglu and Davison, 1982). It may therefore be an important mechanism, but it still has not been demonstrated with leaves or in field conditions. Guttation is the expression of droplets of water from the margins and tips of leaves that occurs under certain conditions of temperature, water regime and humidity. It is particularly common in grasses and is responsible for some of the 'dew' that is seen in early morning. Takmaz-Niscancioglu (1983) found that fluoride ions are detectable in guttation droplets on maize, but the rate of expression of the element from the plant was not very great and it probably cannot account for the observed rates of loss. Finally, there is the possibility that a significant proportion of the fluoride in leaves may be translocated to the root. It has been suggested by a few workers, but there is little evidence that it occurs fast enough or on such a scale as to account for a significant proportion of reported losses

(Davison, 1982; Weinstein and Alscher-Herman, 1982). The overall position today remains as it was when the subject was first reviewed (Davison, 1982) and that is that the mechanisms and rates of loss or decrease in fluoride concentration in tissues remain largely unexplored and unexplained.

Fluoride Uptake by Animals

Although most of the information available about fluoride uptake and accumulation is related to humans, livestock and a few small mammal species, it is clear that the fluoride content of animal tissues is determined by four factors: the route of uptake, the presence of complexing agents in the diet, the pH of the digestive system and the occurrence of calcified tissues in the body. The routes of uptake are inhalation, deposition on the outer surfaces and ingestion. The presence of complexing agents and the pH determine the species of fluorides in the digestive system and therefore the bioavailability. Fluoride has a strong affinity for calcium so it accumulates in calcified tissues, whether it is human bone, the shell of a crab or the calciferous gland of an earthworm.

Inhalation is not considered to be an important uptake route for vertebrates because the amount taken in and retained during lung ventilation is small in absolute terms and very small relative to other routes (Largent, 1961; WHO, 1984). Davison (1987) estimated that, in the case of fattening cattle, an individual weighing 500 kg inhales about 144 m^3 of air a day. Assuming that all the fluoride is retained (McIvor, cited in WHO, 2002a), the daily input from air containing a background level of 0.05 µg F/m^3 would be only 7.2 µg (1.44×10^{-5} mg/kg body wt). Food intake is of the order of 10 kg (2% of body weight) and water consumption around 40 l/day. Assuming an average fluoride concentration of 5 mg F/kg in food and 0.1 mg F/l in water, the intake from these sources would be 54 mg/day (0.108 mg/kg body wt), i.e. over 7000 times the amount inhaled. In a polluted environment the relative contributions would be of the same order. With concentrations of 0.3 µg F/m³ in the air and 40 mg F/kg in forage and assuming the water-supply is unpolluted, the amounts inhaled and ingested would be 43 µg and 404 mg, respectively (8.6×10^{-5} and 0.81 mg/kg body wt). Although humans consume different relative amounts of food and water, the end result of the calculation is similar: inhalation is not significant, even in a polluted environment. The only comparable estimate for invertebrates was made by Davies (1989), but it also confirms the very small contribution of inhalation. Using adult butterflies (*Pieris brassicae*) exposed to 7.18 µg F/m³ for 18 days (the normal lifespan of a butterfly), she estimated that respiratory intake accounted for about 0.011% of the total body load (Table 2.6).

There are dozens of reports listing the fluoride content of animal tissues but there have been very few investigations of the rate of deposition of fluoride on the outer surfaces of any animal, whether it is skin, butterfly wings, fur or the shells of terrestrial and marine animals. As Davison (1987) pointed out, surface deposits may be important at some trophic levels because the fluoride will be consumed by predators, whether it is external or internal. In addition, preening by birds, the collection of pollen from the body by bees and coat-licking by cattle would all be expected to transfer deposited fluoride from the surface to the digestive system. In industrial environments, transfer from the hands and lips of workers to the mouth is an important route of entry of fluoride, so personal hygiene plays a vital part in determining uptake by the individual.

Davies (1989) investigated rates of deposition of HF on insect bodies and wings by exposing adult *P. brassicae* under controlled conditions. In the first experiment the butterflies were fumigated for 7 days and then analysed (Table 2.7). The experiment showed that there was significant deposition on the insect surfaces, the wings containing much more fluoride than the bodies when the data were expressed on a weight basis. However, when the data were expressed on an area basis, the pattern was reversed: the rate of deposition on bodies

was six to ten times greater than on wings. This demonstrates once again the importance of expressing data on both a weight and an area basis.

In a second experiment, Davies (1989) sampled over 20 days to assess the rate of HF deposition (Fig. 2.15). As a yardstick for comparison, she also recorded deposition to pieces of alkali-impregnated paper cut to the same size as the wings. An alkali surface acts as a perfect sink, the rate of deposition being controlled only by the HF concentration and the boundary-layer resistance (Davison and Blakemore, 1980). The rate of deposition on the papers was linear over 20 days (Fig. 2.15) and comparable to data reported by Davison and Blakemore (1980) for field experiments. In contrast, the rate of deposition on the wings and bodies was much lower. Further analysis of the data indicated that the rate was not linear with time and that the surfaces had a limited adsorptive capacity. The importance of surface deposition was

Table 2.6. Contribution of different routes of transfer of fluoride to adult large white butterflies (*Pieris brassicae*) exposed to 7.18 µg F/m³ for 18 days (data of Davies, 1989).

Route	% of total body load
Surface deposition of particles	7.3–18.4
Surface deposition of HF	79.8–90.9
Ingestion via nectar	1.82
Respiratory uptake	0.011

Table 2.7. Fluoride concentrations in the wings and bodies of adult *Pieris brassicae* after fumigation with two concentrations of HF for 7 days. Data are expressed on a weight (µg/g) and an area basis (µg/dm³). (Data of Davies, 1989.)

Concentration of HF in air (µg/m³)					
< 0.1		1.56		11.54	
Wings	Body	Wings	Body	Wings	Body
Concentration expressed on weight basis (µg/g dry wt)					
11.6	0.7	87.3	13.5	491.9	49.2
Concentration expressed on area basis (µg/dm³)					
0.27	6.7	2.5	24.8	15.5	88.1

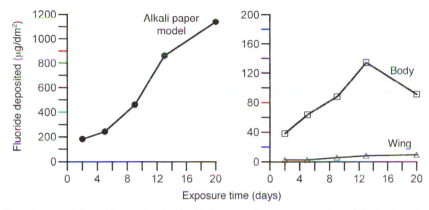

Fig. 2.15. The rate of deposition of fluoride ($\mu g/dm^2$) on alkaline paper models and the body and wings of adult large white butterflies (*Pieris brassicae*) exposed to 1.84 μg HF/m^3 over 20 days (drawn from Davies, 1989).

confirmed by analysis of pine sawfly collected from the field near an aluminium smelter in Wales (Davies *et al.*, 1992). The whole body load was 100 mg/kg and the exuvium (the skin after casting) had 452 mg/kg. The latter constituted 23% of the total body fluoride load.

The loss of deposited fluoride was also investigated (Davies, 1989) by fumigating butterflies, analysing some and then transferring the remainder to clean air for 7 days before analysis (Table 2.8). This showed that surface fluoride is lost from butterflies, the rate being higher from the wings than from the body. This makes interpretation of the fluoride load of field-caught insects very difficult, and it has not usually been taken into account in attempts to determine if there is bioconcentration of fluoride with trophic level.

Surface deposition also occurs in other organisms, particularly those with a calcified shell or exoskeleton, because fluoride has a high affinity for calcium. The calcium occurs as calcium carbonate and calcium phosphate, depending on the taxonomic group. Mollusc (snails, mussels, oysters, etc.) and egg shells are principally calcium carbonate, while the exoskeleton of crustaceans (crabs, shrimps, krill, etc.) is a combination of chitin and calcium phosphate. Exposure of snail or egg shells to atmospheric HF must lead to some surface deposition, but the extent of this has not been

Table 2.8. Postfumigation loss of fluoride. Fluoride concentrations in the wings and bodies of adult *Pieris brassicae* immediately after fumigation with two concentrations of HF for 7 days, and after 7 days in clean air. Data are expressed on a weight basis ($\mu g/g$). (Data of Davies, 1989.)

Air concentration (μg F/m^3)	Organ	μg F/m^3 immediately after fumigation	μg F/m^3 after 7 days in clean air
1.84	Wings	188.2	144.5[a]
	Body	55.6	47.1 NS
14.6	Wings	791.8	557.4[a]
	Body	213.3	242.2 NS

[a]Statistically significant difference after 7 days postfumigation.

examined. In the case of eggs it is probably not significant, as Vikøren and Stuve's (1996a) data seem to indicate that most shell fluoride is obtained from the parent. On the other hand, the fluoride content of sea water is relatively high, at 1.3 mg/l, and it is known that shells and exoskeletons of diverse marine organisms do contain high concentrations of fluoride. Camargo (2003) reviewed the subject and reported concentrations in the whole exoskeletons of marine crustaceans ranging from around 200 to about 6000 mg/kg dry wt. Gregson *et al.* (1979) cite the case of a gastropod, *Archidorus britannica*, as containing 3% fluoride. The processes involved in biomineral formation are not well known but it is thought that much of the fluoride in marine exoskeletons is taken up directly from the water (Camargo, 2003). The Antarctic krill, *Euphausia crystallorophias*, has the highest known concentrations, with the mouth-parts reaching almost 13,000 mg/kg, but the levels vary in different parts of the exoskeleton (Sands *et al.*, 1998). Substitution of fluoride in calcium phosphate makes it harder, so it is thought that fluoride has an important role in hardening the mouth-parts and therefore in the ecology of krill, which are a vital part of the Antarctic food chain. In addition, the cast exoskeletons play an important part in the flux of fluoride in the southern oceans (Nicol and Stolp, 1989). The high fluoride content of krill was also instrumental in limiting the exploitation of krill for aquaculture and animal and human consumption.

Some animals contain silicate-based structures, notably diatoms and sponges, and there are two cases reported of high fluoride concentrations in sponges. One, *Halichondria moorei*, has the exceptionally high fluoride concentration of up to 11.5% fluoride (Gregson *et al.*, 1979). X-ray diffraction showed that the fluoride occurred as potassium fluorosilicate in the spicules. The mechanism of incorporation of the fluoride is unknown.

In most animals, ingestion in food and drink is the main pathway of fluoride uptake. An indication of this has already been given for cattle and a butterfly, but there are many more data available for humans (Hodge and Smith, cited in NAS, 1971; WHO, 1984, 2002a). The daily intake varies greatly in humans depending on age, the fluoride content of water and the diet and the amount consumed. In many circumstances, fluoride in water accounts for half or more of the daily intake, which is much more than in other vertebrates, such as cattle. An adult living in a developed country typically takes in from about 0.2 to 2.0 mg F/day, but obviously it varies with diet and the fluoride content of the water supply. In the USA, an adult residing in a community with non-fluoridated water receives on average 0.22 to 0.76 mg fluoride in their liquid intake and 0.31 to 0.43 mg from the

remainder of the diet (Singer and Ophaug, 1983).

Most vegetables and meat have low fluoride contents but some items are much higher because of the bone content, such as sardines (NAS, 1971; Sherlock, 1984; COMA, 1994). Fluoridated water increases intake significantly because it contains about 1 mg F/l, but in Western countries the increased consumption of carbonated drinks and bottled mineral water is altering the pattern of intake (FSA, 2001). Many bottled mineral waters contain low concentrations of fluoride but others may be well over 1 mg/l and a few are as high as 5–8 mg/l (FSA, 2001). Tea may also add a significant amount to the total load because infusions contain about 0.5–5 mg F/l (Zimmerman *et al.*, 1957; FSA, 2001). Sherlock (1984) estimated that in Britain tea contributed 1.3 mg to the daily intake of 1.8 mg, and Jenkins (1991) showed that it could influence dental caries. On the other hand, excessive consumption of high-fluoride tea makes a significant contribution to fluorosis in China (Cao *et al.*, 1997, 1998, 2000). Fluoridated toothpastes contain 0.1% fluoride, or less if formulated for children (FSA, 2001), and their use about doubles the daily intake by children (FSA, 2001). Humans living in the vicinity of industrial emissions do not usually have a significantly higher than normal fluoride intake (WHO, 1984), because the amount in the diet is not increased, especially the fluoride in drinking-water.

Bioavailability and fluoride retention

The proportion of ingested fluoride that is retained by an animal depends on the acidity of the digestive system, the chemical species in the digestive system and the presence of calcified tissues in the body. Most of the fluoride in the human diet is present in the anionic form, F^-, complexed with other ions, or as insoluble compounds, but the very low pH in the digestive system strongly favours the formation of HF, which is passively absorbed in the stomach and the intestine. The greater the gastric acidity, the greater the amount absorbed, but the range is from about 70 to 90%. Consequently, only a relatively small proportion is eliminated in the faeces, Hodge and Smith (1965) citing it as ranging from about 10 to 30%.

The bioavailability of fluoride is affected by other components of the diet, notably elements that can complex with it (Hodge and Smith, 1965). For example, WHO (2002a) cites a study in which there was almost 100% absorption of sodium fluoride given as tablets but when the same dose was taken with a glass of milk the absorption decreased to 70%. A calcium-rich breakfast reduced availability further. The effect was due to fluoride binding with calcium and other divalent cations. Aluminium has a similar effect in humans and other mammals (Hodge and Smith, 1965; WHO, 2002a) and it has been shown to alleviate fluorosis in cattle (NAS, 1974). The effects of aluminium are illustrated by Wright and Thompson's (1978) experiment, in which they fed rats three forms of fluoride for 7 days and assessed the absorption and route through the animals (Table 2.9). There was a marked difference between sodium fluoride and the two aluminium compounds, both of which are emitted from aluminium smelters.

In adult humans, about half of the absorbed fluoride is excreted in the urine and 99% of the rest is deposited in the calcified tissues, bones and teeth. Normally only small amounts are excreted in sweat but it can be up to 50% of the total under conditions that promote excessive sweating (Hodge and Smith, 1965). Concentrations of fluoride in the soft tissues are usually well under 1 mg/kg (wet-weight basis) but they are higher when the intake is increased (Hodge and Smith, 1965). Some data of Andrews et al. (1989) for field voles illustrate this (Table 2.10). Exposure to fluoride at this site increased soft-tissue fluoride concentrations from about 1.2 to 3 times but those in the hard tissues increased from 100 to 140 times.

Urine concentrations rise rapidly after ingestion of fluoride, but the amount excreted by this route depends on several factors, notably the total fluoride in the diet,

age and urinary pH (FSA, 2001). In general, the rate of urinary clearance is high and it plays an important part in determining the amount of fluoride retained by humans and other omnivores. However, Boulton et al. (1995) have found that the herbivorous vole Microtus agrestis has a comparatively low excretion of urinary fluoride and that consequently it is the initial absorption from the gastrointestinal tract that has the greater influence on retention. The same authors commented that this was in agreement with research on another herbivore, the rabbit, so there may be a fundamental difference

Table 2.9. Effect of fluoride source on availability and route of transport of fluoride in rats given dietary supplements for 1 week (data from Wright and Thompson, 1978). (*British Journal of Nutrition* 40, 139–147, 1978, reprinted with permission.)

Fraction	Sodium fluoride (NaF)	Cryolite (Na$_3$AlF$_6$)	Aluminium fluoride (AlF$_3$.H$_2$O)
% retained	51	55	15
% excreted in urine	48	30	1
% absorbed (= retained + urine)	99	86	16
% excreted in faeces	1	15	84

Table 2.10. The fluoride content of various tissues of field voles (*Microtus agrestis*) collected from a control site and a revegetated fluorspar tailings dam (from Andrews et al., 1989). Note that the data are expressed on a dry-weight basis (mg/kg), but soft tissues are usually on a wet basis, which makes the concentrations more than ten times lower. (Reprinted from *Environmental Pollution* 60, 165–179, 1989, with permission from Elsevier.)

Tissue	Habitat	
	Fluorspar tailings	Control site
Heart	11.0 (2.04)	8.71 (1.23)
Liver	11.3 (2.2)	6.48 (0.94)
Kidney	15.1 (3.72)	8.35 (1.25)
Skeletal muscle	23.8 (8.58)	7.83 (1.71)
Femur	1106 (284)	90.5 (15.2)
Teeth	426 (105)	41.2 (5.19)
Pelvic girdle	1264 (370)	87.7 (19.2)

between herbivores and omnivores in this respect.

There is always some fluoride in the diet, so even in pristine environments bones and teeth contain substantial concentrations, usually in the region of a few hundred mg/kg (Hodge and Smith, 1965; US EPA, 1980; Murray, 1981; WHO, 2002a). Young animals deposit a greater proportion of dietary fluoride in the bones than do adults, so bone fluoride concentrations increase non-linearly with age (Fig. 2.16). The rate of deposition also varies with sex (Pattee *et al.*, 1988; Vikøren and Stuve, 1996a) and, in growing chicks, sexual maturity and factors associated with egg production increase deposition in females (Michel *et al.*, 1984). Fluoride is both gained and lost from bones, so, if the dietary intake decreases over a period of time, there may be a reduction in bone fluoride, although the half-life is long (FSA, 2001; WHO, 2002a).

Cattle absorb a much lower percentage of ingested fluoride than humans or other omnivores – usually in the region of 50%, though it is variable. When forage is supplemented with a soluble form, such as NaF, absorption may be as high as 70–80% but with CaF_2 it is usually < 40% (Shupe *et al.*, 1962). Soil may be a significant source of fluoride in the diet of cattle and sheep and it has an availability from < 10 to 38% (Milhaud *et al.*, 1984; Cronin *et al.*, 2000).

The lower absorption by cattle is largely due to the fact that the pH of the rumen, the main site of fluoride absorption, is around 6, which means that low-permeability F^- predominates. The rumen pH is buffered by huge amounts of alkaline saliva but it varies from as low as 5 to over 6 so absorption of fluoride would be expected to vary accordingly. Just as in humans, about half of the absorbed fluoride is retained, mostly in the bones and teeth, and the concentration increases with age. The concentration of fluoride in cattle faeces is usually higher than in the forage because a greater proportion of the dry matter is absorbed than fluoride. For example, if the forage contains 100 mg F/kg, the digestibility of the forage is 65% and 50% of the fluoride is absorbed, the faecal fluoride will average 143 mg F/kg. The implication is that detritus feeders and decomposers are exposed to higher concentrations than herbivores, but the environmental significance of this has not been examined.

Herbivorous insects are at the opposite end of the scale of fluoride accumulation from humans because they do not have calcified tissues to act as sinks for fluoride and the gut pH is usually alkaline, well over 7. The result is that absorption and body concentrations are extremely low, and the total fluoride load is largely due to surface deposits and the gut contents at the time of

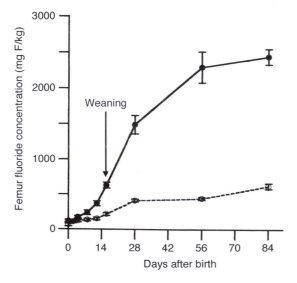

Fig. 2.16. The increase in femur fluoride (mg/kg dry wt) in young voles (*Microtus agrestis*) exposed to diets from a reference site (○) and from a chemical works (●) that emitted fluorides. (Redrawn from Boulton, I.C., Cooke, J.A. and Johnson, M.S. (1994b) Age-accumulation of fluoride in an experimental population of short-tailed field voles (*Microtus agrestis* L.) *Science of the Total Environment* 154, 29–37, with permission from Elsevier.)

collection (Weinstein *et al.*, 1973; Hughes *et al.*, 1985; Davies *et al.*, 1992; Port *et al.*, 1998). Working with Mexican bean beetles, Weinstein *et al.* (1973) found that less than 0.1% of the ingested fluoride was retained, while Hughes *et al.* (1985) showed that cabbage looper larvae retained 9% and 1% of ingested fluoride when it was presented as KF-dosed cabbage and HF-fumigated leaves, respectively. Port *et al.* (1998) reported similar rates for large white butterfly larvae fed on cabbage dosed with HF or AlF_3. Davies *et al.* (1992) used a different approach, collecting pine sawfly larvae in the field near an aluminium smelter and then raising them on unpolluted needles until pupation. A mass balance showed that absorption was less than 2.5%. There was no fluoride detectable in the pupae even though the larvae had fed until the third/fourth instar on needles with up to 170 µg F/g. Mitterböck and Führer (1988) also found lower concentrations in pupae than in larvae of nun moth, but the pupae were not cleared of fluoride to the same extent as in pine sawfly (Davies *et al.*, 1992).

Fluoride uptake, feeding strategy and trophic level

Finally, the amount of fluoride available to animals and the degree of accumulation vary greatly because of differences in feeding strategy, foraging territory, lifespan and, possibly, trophic level. The fluoride content of food plants varies from species to species and in different tissues of the same plant, which is why feeding strategy and the size of the foraging territory play such an important part in determining fluoride loading (Davison *et al.*, 1985). A general grazer that eats whole leaves, such as a butterfly larva, ingests much more fluoride than an adult of the same species feeding on nectar (Davies, 1989). Phloem-feeding aphids are similarly exposed to much lower concentrations, while xylem-feeding froghoppers ingest only the very low concentrations that are being carried from the soil (Davison, 1987; Davies *et al.*, 1998). Small general grazers,

such as voles, may have a lower intake than cattle feeding on the same pasture because they are more selective in what they eat. For example, Davison (1987) observed that, in a pasture where the fluoride concentration of whole grass shoots was in excess of 100 mg/kg, small mammals were feeding on the basal inner sheath tissues, which contained less then 5 mg/kg. However, even when animals are exposed to the same fluoride supply, accumulation may be different, as Boulton *et al.* (1995) demonstrated. Two vole and two mouse species showed a difference in accumulation when they were exposed to NaF in drinking-water, but it was mostly related to differences in the intake of water. Larger mammals, such as deer, and migratory animals, such as birds, have bigger foraging areas than small mammals so they are exposed to an even greater range of concentrations. This may put them at a lesser risk than more sedentary animals but it is more difficult to assess the overall diet and therefore the magnitude of the risk. As young vertebrates accumulate fluoride faster than adults, the fluoride available in the breeding and early feeding areas may be more important than other areas, so the whole foraging territory needs to be assessed. The final factor that determines accumulation is lifespan, especially in animals with calcified tissues. Populations of short-lived species usually show seasonal variations in bone fluoride due to the turnover of individuals (Wright *et al.*, 1978; Andrews *et al.*, 1989), which makes it difficult to estimate accumulation at the population level and to compare species of different longevity.

There have been several reports in which whole-body fluoride concentrations were measured for animals with different feeding strategies, notably by Dewey (1973), Murray (1981), Buse (1986) and Andrews *et al.* (1989). All of these studies (Table 2.11) were solely of invertebrates, which avoids the near-impossible problem of comparing groups that include both invertebrate and vertebrates. As Table 2.11 shows, there were some differences between the groups within individual studies, but few consistent patterns across all of the studies. For example,

Table 2.11.   Data from Dewey (1973), Murray (1981), Buse (1986) and Andrews *et al.* (1989) to illustrate the range of fluoride concentrations found in invertebrates having different feeding strategies. (Reprinted from *Environmental Entomology*, vol. 2, pp. 179–182, 1973; *Science of the Total Environment* 17, 223–241, 1981; *Environmental Pollution* 57, 199–217, 1986; *Environmental Pollution* 60, 165–179, 1989, with permission from Elsevier.)

Feeding strategy	Species	Mean mg F/kg and standard error
Data from Dewey (1973), near aluminium smelter		
Cambial-region feeders	Beetles, borers	20 (8.3)
Predators	Ants, dragonflies	53 (22)
Foliage feeders	Grasshoppers, various larvae, weevils	83 (35)
Pollinators	Bees, moths, butterflies	247 (59)
Data from Murray (1981), near aluminium smelter		
Herbivores	Site 1: crickets, grasshoppers, weevil	20 (7.7)
	Australian Fluorine Chemicals (AFC) site: crickets, moths, beetles, bugs	27.5 (6.4)
Omnivores	Site 2: ants, cockroach	53 (2)
	AFC site: ants, beetles	24 (5.8)
Carnivores	Site 1: spiders, flies, beetles	29 (11)
	Site 2: spiders, flies, beetles	78 (29)
Data from Buse (1986), near aluminium smelter		
Herbivores	Grasshoppers	20
Omnivores	Beetles	50
Decaying and fresh plant	Slugs and snails,	190
material, soil organic matter	earthworms	184
Predators	Centipedes,	48
	spiders,	393
	harvestmen	258
Scavengers	Millipedes, woodlice	1100

Data from Andrews *et al.* (1989), fluorspar-tailings dam

	Mean µg F/g and 95% confidence interval	
	April	October
Herbivores	660 (549)	880 (492)
Carnivores	466 (132)	794 (211)
Detritivores	2162 (1622)	4041 (814)

Buse (1986) found higher concentrations in predators than in herbivores, but Andrews *et al.* (1989) had much the same concentrations in their herbivore and carnivore groups. Similarly, Dewey's (1973) data showed no significant difference between foliage feeders and predators. The one striking pattern is that scavengers and detritivores (Buse, 1986, and Andrews *et al.*, 1989, respectively) had much higher concentrations than all other groups in each study. This may be partly due to ingestion of high-fluoride materials in the surface layers of the soil, but it was probably largely due to the fact that the gut contents were not removed before analysis. Walton (1986) commented that the total fluoride concentration of earthworms was largely due to soil contained in the gut. A further complication in Buse's (1986) data is that the scavengers included woodlice, which have a calcified carapace that acts as a fluoride sink. The most obvious feature of Dewey's (1973) data is the high concentrations in pollinators. As their diet is low in fluoride, this was almost certainly due to dry deposition of fluoride on the surfaces and the fact that the animals have very high surface-to-weight ratios. None of the authors assessed the external deposited fluoride. The final conclusion is that there is no evidence of food-chain biomagnification; fluoride uptake and accumulation are determined much more by factors such as digestive-system pH and the presence of calcified tissues.

3

Effects of Inorganic Fluorides on Animals

Introduction

Not surprisingly, there has been more research on the effects of fluoride on humans than on any other animal, so this chapter begins with an account of the sources of fluoride in the human diet; then it considers the effects of acute and chronic exposures before progressing on to discuss the effects on other mammals and invertebrates.

Humans

In Chapter 2 we saw that, for most animals, ingestion of fluoride in food and drink is the main pathway of uptake. In humans the amount varies greatly from person to person and place to place but in most situations water accounts for half or more of the daily intake. In developed countries and in populations with piped water, the total is typically about 0.2–2.0 mg F/day but fluoridation of public water supplies and consumption of bottled water or carbonated drinks may alter this significantly. In those parts of the world where the fluoride content of drinking-water exceeds 1–2 mg/l, intake is much higher, while in some countries food and tea add a significant burden. Excessive use of fluoridated toothpaste, gels and mouthwashes may also increase intake significantly. For example, dental products meant to be used topically contain fluoride

concentrations ranging from 250 to over 1500 mg/kg (WHO, 2002a). Workers in industry may be exposed to high concentrations in the air and to fluoride-containing dusts, which increases their intake, though industrial exposure is decreasing because of improvements in operating conditions.

Because fluoride occurs naturally in everything we eat and drink, even in pristine environments, it is a normal constituent of human tissues, concentrating in the teeth, bone and any other calcified materials, such as kidney stones. Soft tissues contain around 1 mg F/kg (wet wt basis) and bones and teeth contain a few hundred mg/kg. About 99% of the total body load is retained in the bones and teeth, where it is incorporated into the crystal lattice (WHO, 2002a). Bone is largely composed of a calcium phosphate mineral analogous to crystalline calcium hydroxyapatite ($Ca_{10}(PO_4)_6(OH)_2$). Fluoride ions can substitute in the hydroxyl position, altering the solubility and hardness of the tissue. There appears to be strong evidence that ingestion of fluoride directly stimulates bone and fluorapatite formation *in vivo* (WHO, 1984, 1994; Grynpas, 1990; Mohr and Kragstrup, 1991; Chavassieux *et al.*, 1993; Ohta *et al.*, 1995; Cao *et al.*, 1996; Hou, 1997; Ando *et al.*, 1998, 2001). Contrary to some reports, recent medical and epidemiological evidence suggests that there is no significant relationship between bone strength and fluoride exposure (Sower *et al.*, 1991; Whitford, 1992; Fratzl *et al.*,

1994; Cauley *et al.*, 1995). At low levels fluoride has beneficial effects in preventing dental caries and it has been used as a treatment for osteoporosis for many years. These positive effects have led to an endless debate about whether fluoride is an essential element for humans and livestock (NAS, 1974; WHO, 1984), but it is impossible to test this idea experimentally and the arguments tend to be confounded by emotional concerns about the effects of water fluoridation. It does not seem to be a fruitful topic for discussion here.

Effects of acute exposure

It is difficult to make a clear distinction between acute and chronic effects because the rationale for how an effect is categorized depends upon the concentration of fluoride in the atmosphere, its form (i.e. gaseous or particulate), the length of exposure and sensitivities and tolerances inherent in individual humans. Because workplace exposures in most industries have been shown to be well under allowable threshold limit values (obviously with some exceptions, depending upon the type of industry, the standards promulgated by individual countries and the degree of enforcement), we assume that normal day-to-day exposures in the aluminium, magnesium, phosphate and other common industries fall within the category of chronic effects. Even where daily workday exposures have exceeded present-day standards, we consider these to be chronic exposures because no health effects have been registered after a single exposure.

The intense reaction of elemental fluorine (F_2) with skin produces a thermal burn. Solutions of HF, on the other hand, produce slow-healing chemical burns (Stokinger, 1949; Hodge and Smith, 1977). The primary treatment for HF burns is the application of calcium or magnesium salts, which form insoluble complexes with fluoride, although other traditional burn treatments are also used (Gosselin *et al.*, 1976). Lung tissues are delicate and may be severely or fatally irritated by high concentrations of F_2 or HF (Greendyke and Hodge, 1964). According to Largent (1950) (cited by Hodge and Smith, 1970) the human response to increasing concentrations of gaseous fluoride is as follows:

3 p.p.m. (2.44 mg/m^3) – no local immediate systemic effects observed, although some subjects have complained of transient irritation, reddening and itching of exposed skin;
10 p.p.m. (8.13 mg/m^3) – many persons undergo discomfort;
30 p.p.m. (24.4 mg/m^3) – causes serious complaints and objections;
60 p.p.m. (48.8 mg/m^3) – brief exposures result in definite irritation of the conjunctiva, nasal passages and discomfort of the trachea and pharynx;
120 p.p.m. (97.6 mg/m^3) – the highest concentration tolerated for less than 1 min by two male subjects; smarting of skin also noted.

Most of our knowledge of the effects of acute exposure to fluoride salts comes from accidental or deliberate poisoning (Hodge and Smith, 1965), but a detailed discussion of this aspect is beyond the scope of this book. In many of the cases reported in the USA death was due to ingestion of ant or cockroach poison, but there was one tragic case in which sodium fluoride was unintentionally mixed with scrambled eggs and served to hospital patients: 263 were poisoned and 47 died (Hodge and Smith, 1965). Cases such as this suggest that the acute lethal dose of sodium fluoride for adults is 5–10 g (32–64 mg/kg body wt) and that the minimum acute dose that might lead to adverse health effects is 5 mg/kg body weight (WHO, 2002a). Generation of hydrofluoric acid affects the gastrointestinal system, causing abdominal pain and other symptoms. Other effects, such as cardiac arrest and damage to the central nervous system, are caused by hypocalcaemia and enzyme inhibition. Normally about 50% of the total serum calcium is ionized and it is that form that is biologically active in bone formation, blood coagulation, functioning of the neuromuscular system and other cellular processes. Fluoride complexes with calcium and induces hypocalcaemia.

Dental caries

At low concentrations fluoride decreases the incidence of dental caries. The effect has been attributed to three mechanisms: inhibition of bacterial metabolism; inhibition of demineralization when fluoride is present at the crystal surface during acidification; and enhancing remineralization by forming a low-solubility veneer, similar to fluorapatite (Featherstone, 2000). Many investigators have credited the major benefits of fluoride for anticaries activity to its reactions with tooth mineral, consigning a less important role to its antimicrobial activity. The toothpaste industry promotes the idea that fluoride makes teeth more resistant to acids by issuing protocols for an educational experiment in which eggshells are treated either with water or a fluoride solution and then plunged into vinegar – the fluoride-treated shells are resistant to attack. The mode of action on the human microflora was reviewed recently by Marquis *et al.* (2002). The most direct manner involves the binding of fluoride ions (F^-) or HF to sites on enzymes or other proteins, such as the haem moiety of a number of enzymes, including catalases, peroxidases and cytochromes. In intact microbial cells, enolase, a key enzyme of glycolysis, may be inhibited at very low micromolar concentrations and low pH values. Inhibition of catalase would jeopardize bacteria or mixed bacterial communities in managing with oxidative damage from hydrogen peroxide. Because organic weak acids, food-preservative weak acids and non-steroidal anti-inflammatory agents have a similar anticariogenic effect to that of fluoride, Marquis *et al.* (2002) pointed out that:

> Fluoride has specific effects on biological systems not shared with organic weak acids, mainly anti-enzyme effects due to fluoride binding or to binding of fluoride–metal complexes. Fluoride also has effects shared with organic weak acids, mainly those having to do with enhanced transport of protons across the cell membrane. It is these latter effects that seem to be most pertinent in the antibacterial– anticaries properties of fluoride. Fluoride can even have an anticaries effect when added to sucrose in the diet and can act in concert with other anticaries agents. Moreover, fluoride appears to have important ecological effects on dental plaque in that it acts to reduce acidification and in the long run serves to select for a less acid-tolerant, less cariogenic microbiota.

The authors go on to say that the antimicrobial–anticaries effects of fluoride relate to 'reduction in the acid tolerance of glycolysis by intact, cariogenic bacteria in plaque. As a result, acid production is stopped before the plaque pH drops to values leading to rapid demineralization.'

Chronic exposure

Excessive fluoride intake at chronic levels over long periods of time can lead to dental and skeletal fluorosis. The former is characterized by failure of the enamel covering the teeth to crystallize properly, resulting in flaws that range from barely discernible white inclusions to severe brown stains, surface pitting, brittleness and excessive wear. Skeletal fluorosis is a slow, progressive and crippling affliction. Roholm (1937) produced the first detailed characterization of fluorosis in relation to workers in the cryolite industry. He described how high exposures led to calcification of ligaments, increased mineralization of bone (osteosclerosis) and abnormal outgrowths of new bone (exostoses). At its earliest stage, osteofluorosis is often asymptomatic but can be visualized radiologically as increasing bone density, particularly of the vertebrae and pelvis. This condition appeared in cryolite workers after about 4 years in which daily absorption of fluoride was 20–80 mg. Over the long term, excess fluoride leads to painful joints and greatly restricted mobility. In severe cases the person may be permanently bent and may not be able to walk, but less severe forms can be slowly reversible (Grandjean and Thomsen, 1983).

Osteosclerotic changes have been associated with bone fluoride concentrations of about 5000–6000 mg/kg of dry, fat-free bone (Weidmann and Weatherall, 1970; Zipkin et al., 1970; Smith and Hodge, 1979; WHO, 1984). Pathological changes have been found at bone fluoride concentrations as low as 2000 mg/kg (Baud et al., 1978; Boillat et al., 1979).

Occupational exposure

Since Roholm (1937) published his review there has been a great reduction in industrial exposure to fluorides, especially in the developed countries, but even in the 1970s the US National Institute for Occupational Safety and Hygiene (NIOSH) recognized 92 occupations with potential exposure to fluorides and 57 occupations considered to have potential exposures to hydrogen fluoride (NIOSH, 1976). The available literature mainly relates to workplace exposures by inhalation or dermal contact in aluminium smelting, magnesium processing, phosphate fertilizer and hydrofluoric acid manufacturing and welding operations. The concentration of fluoride in the internal air of workrooms in several industries is shown in Table 3.1. The concentrations vary considerably, from a low value of 0.03 mg/m³ to as high as 16.5 mg/m³. These can be compared with the threshold limit values (TLVs) stipulated by OSHA (2002), which are, for an 8 h day, 5-day week with little or no adverse effects: 0.1 p.p.m. (0.08 mg/m³) for F_2 and 3 p.p.m. (2.44 mg/m³) for HF. The American Conference of Governmental Industrial Hygienists (ACGIH, 1996) proposed a 'ceiling' value for HF of 3 p.p.m. but 1 p.p.m. for F_2. NIOSH (OSHA, undated) recommends an exposure limit for HF of 2.5 p.p.m. as a time-weighted average for up to a 10-hour working day and a 40-hour working week. The NIOSH short-term exposure limit is 6 p.p.m. Clearly, US government agencies are in relatively close agreement on occupational exposures to HF. There are similar standards in force in other countries but it is interesting that the former USSR recommended a much lower threshold limit value of 1.0 mg/m³ expressed as HF (US EPA, 1980). The limits indicate that humans are remarkably tolerant of atmospheric fluorides, much more so than plants, which is one reason why the authors promote the use of plants for surveillance.

Occupational exposure is generally monitored by analysis of urine, measurement of plasma fluoride and radiological examination. Analysis of fluoride in hair has been shown more recently to be a potentially effective method for monitoring (Kokot and Drzewiecki, 2000), but urine analysis is the most widely used technique. About 50% of the fluoride ingested is rapidly excreted in

Table 3.1. Fluoride concentrations in internal workplaces of several industries (data from WHO, 2002a). (Reproduced from *Environmental Health Criteria* 227, Courtesy of WHO, Geneva.)

Industry	Country	Concentration of F (mg/m³)	References cited in WHO (2002a)
Machine-shops and shipyards	The Netherlands	0.03–16.5	Sloof et al., 1989
Aluminium smelting	Sweden	0.90[a]	Sloof et al., 1989
Aluminium smelting	British Columbia, Canada	0.48	Chan-Yeung et al., 1983
Aluminium smelting	Norway	c. 0.5	Søyseth et al., 1995
Aluminium smelting	The Netherlands	c. 0.5[b]	Sorgdrager et al., 1995
Aluminium smelting	Iran	0.93	Akbar-Khanzadeh, 1995
HF manufacture	Mexico	1.78[c] and 0.21[c]	Calderon et al., 1995
Phosphate processing	Poland	Up to 3	Czarnowski and Krechniak, 1990

[a]34% in gaseous form.
[b]Either gaseous or particulate fluoride at two smelters.
[c]During 1987–1988 and 1990–1994, respectively.

the urine and the concentration is closely related to the amount ingested and its bioavailability. Dinman *et al.* (1976) and Hodge and Smith (1977) suggested that no detectable radiological or clinical signs of osteosclerosis will materialize at workplace atmospheric concentrations below 2.5 mg/m[3] and at fluoride concentrations in urine that are at or below 4 mg/l preshift (collected at least 48 h after previous occupational exposure) and 8 mg/l postshift over a long time period. These data were the basis for the US recommendations for the threshold limit values established by NIOSH. Because the amount ingested varies, NIOSH (1975, 1976) recognized that a single urine sample was inadequate to assess the general working environment and proposed that, for workers exposed to inorganic fluorides or HF, end-of-shift urine samples should be collected for 4 or more consecutive days. If the fluoride concentration exceeded 7.0 mg F/l, a preshift sample (an estimate of the worker's skeletal fluoride burden) was to be collected within 2 weeks at the start of a work shift at least 48 h after the previous exposure, and a subsequent postshift sample (an estimate of the exposure conditions during the shift) would be taken at the end of the work week. If the preshift sample exceeded 4.0 mg/l, or the second postshift sample exceeded 7.0 mg/l, the individual's dietary sources, work practices and environmental control were to be evaluated. In the event that the median post-shift urinary fluoride concentrations exceeded 7.0 mg/l, the working environment would undergo an industrial hygiene examination. The efficacy of the 4 mg/l limit is demonstrated by a study by Derryberry *et al.* (1963). The amount of fluoride excreted in urine at the end of the work shift was measured in 74 workers exposed to high concentrations of fluoride. Unexposed workers were used as the control group. An index of exposure was calculated for each worker based upon the percentage of urine specimens containing > 4 mg/l. None of the workers were considered to be disabled, although in 23% of the workers minimal or questionable degrees of increased bone density were found, a condition which the authors state would not have been sufficient to have been recognizable as increased osseous radio-opacity in routine examinations. There were no abnormal findings of gastrointestinal, cardiovascular, metabolic or haematological conditions in the exposed group, although there were more frequent complaints of respiratory conditions.

There have been many epidemiological studies of the health of industrial workers (WHO, 2002a), but, because they are exposed to several potential inciting agents, it is difficult to determine the effects of any single agent unless it causes unique or very specific symptoms. Most reports of skeletal fluorosis have been from aluminium smelters, magnesium foundries, fluorspar processing and phosphate fertilizer-processing plants (Table 3.2), but Leone *et al.* (1970) stated:

> In spite of continual vigilance by industrial health authorities, few cases of human industrial fluorosis have been identified and incidents have mostly been of a trivial character. An exception was in the Danish cryolite industry where cases were identified and investigated by Roholm (1937).

Responses of workers in the aluminium smelting and phosphate fertilizer industries were reviewed by Hodge and Smith (1977) (Table 3.2), but it is important to recall that this was a summary of research that was published from the 1930s to the mid-1970s; it must not be assumed that similar observations would be made today. Dental fluorosis was rarely a symptom because workers were of adult age, but Hodge and Smith (1977) found a number of consistencies within each industry, such as respiratory problems and skeletal pain. Where there were fluoride effects, air concentrations usually exceeded 2.5 mg F/m[3] and urinary fluorides were generally equal to or exceeded 9 mg/l. Not surprisingly, there were also many symptoms reported that were unrelated to accepted fluoride effects, including: gastrointestinal and renal complaints; headache; eye irritation; vascular dysfunction; cardiac enlargement; menstrual irregularities; and inflammation of the uterus, cervix and vagina. These could have been due to several other pollutants or

Table 3.2. Occurrence of industrial osteosclerosis in various industries (from Hodge and Smith, 1977). (Reprinted courtesy of ACOEM.)

Authors cited in Hodge and Smith (1977)	Air concentration (mg/m³)	Urine (mg F/l)	No. of cases/ no. at risk
Aluminium smelting			
Kaltreider *et al.* (1972)	2.4–6	9	76/79
Agate *et al.* (1949)	0.14–3.4	9[a]	'A few'
Tourangeau (1944)	2.5–3.5	–	2/10
Visscher *et al.* (1970)	–	–	9/17
Boillat *et al.* (1975)	0.5–3.7	–	18/20
Franke *et al.* (1972, 1975)	–	–	28/?
Roche *et al.* (1960)	–	–	1/?
Lezovic and Arnost (1969)	–	1.15–6[a]	4/50, 2 slight, 2 mod.
Schlegel (1974)	2–3	–	61/?
Coulon[b]	–	13 postshift	13/631
Roholm (1937); Brun *et al.* (1941)	15–20	Average 16	57/68
Other industries			
Largent *et al.* (1951)	1.2–3.9	> 10[c]	5/16
Peperkorn and Kehling (1944)	–	–	34/47
Dale and McCauley (1948)	–	10.8	24/40
McGarvey and Ernstene (1947)	–	23	1
Henderson (1975)	0.06–8.2	2–6 preshift 4–25 postshift	1/4
Wilkie (1940)	–	15	2
Derryberry *et al.* (1963)	3.4	0.5–44 average 4.7	17/74
Bowler *et al.* (1947)	0.1–0.7	0.5–7.5	1/54
Fritz (1958)	–	–	67/156
Bishop (1936)	–	–	1
Fourrier and Champiex (1956)	–	–	4

[a]mg/24 h urine.
[b]Personal communication to Hodge and Smith.
[c]In 4/5 with severe osteosclerosis.

be unrelated to the industrial environment. A number of epidemiological studies have reported increased rates of various cancers in workers in the aluminium industry but the World Health Organization (WHO, 2002a) stated:

> In general, there has been no consistent pattern and bone cancer was not usually assessed. Although increases in lung cancer were observed in several studies, it is not possible to attribute these increases to fluoride exposure *per se* due to concomitant exposure to other substances. Indeed, in some of these epidemiological studies, the increased morbidity and mortality was attributed to the workers' exposure to aromatic hydrocarbons.

Ambient air near industrial sources

Atmospheric fluoride concentrations outside industrial buildings are usually at least 1000 times lower than inside, so the risk of fluorosis to the population inhabiting the surrounding neighbourhoods would be expected to be minimal. Nevertheless, there have been concerns, especially 30–50 years ago, when emissions were higher and there was less information available. One of the earliest investigations, at Fort William in Scotland, is discussed in Chapter 5 as a case history, but Hodge and Smith (1977) summarized investigations of health near aluminium smelters and phosphate fertilizer factories:

Fluoride from either industry has been found by some investigators to cause mottling of the enamel of children's teeth, while this effect has not been noted by others. Effluents from the aluminium industry have been credited by some investigators with reducing the incidence of dental caries in children, but others have seen no such protective action. In two instances the concentration of fluoride in teeth was increased. Skeletal changes detectable upon x-ray examination were reported in two studies, but were not demonstrable in four others. Urinary excretion of fluoride was also reported unaffected more often than not. On the other hand, hematologic and other blood changes were described in two studies, but were not seen in a third. Afflictions of the respiratory tract were found in three studies. Examining the data on drinking water, airborne fluoride, and urinary excretion reported in the several studies gives the impression of mostly 'normal' values. The origin of the abnormalities described in skeletal, dental, hemotological, pulmonary and other systems is not clear; if fluoride is a factor, other important sources must be unrecognized.

Once again, it is important to remember that this refers to conditions that no longer exist in the developed world and that the difficulty with these studies is that there are so many uncontrolled factors that it is difficult to isolate the effects of any one factor.

An intriguing skin manifestation called 'Chizzola maculae' was reported in the vicinity of an aluminium smelter near the village of Chizzola in Trentino, Italy, in 1932–1933. The maculae appeared as skin lesions, resembling bruises (referred to medically as resembling ecchymosis or erythema nodosum). The lesions continued, albeit at a much reduced rate until 1937. In 1967, there was a recurrence of the condition in Chizzola and the surrounding countryside. An investigation by the Health Commission of the Ministry of Health found that 49% of the children in Chizzola were affected and that 36–52% of control children examined had similar lesions, although they were not exposed to the effluents (Cavagna and Bobbio, 1970, cited in WHO, 1984). Urinary fluoride levels of children living near the plant were no higher than control children

from an uncontaminated area. In 1968, 98 of 100 children presented maculae; during the same year, emission control systems were installed.

Waldbott and Cecilioni (1969) reported Chizzola maculae on the skin of ten of 32 persons living near fertilizer plants in Ontario and Iowa and an iron foundry in Michigan and attributed the maculae to fluoride exposure. This prompted a Royal Commission in Ontario (1968) to conduct an environmental and medical investigation on residents living near the factory and included some of the residents diagnosed by Waldbott and Cecilioni as suffering from fluoride poisoning. The Commission found no evidence of fluoride poisoning in any resident examined. Chizzola maculae have never been reported from areas where fluorosis is endemic because of elevated fluoride in drinking-water or among workers with significant occupational exposure. In 1980, Giovanazzi et al. reported that the prevalence of maculae was significantly higher in children from Chizzola than from a control population of children. For example, they reported that more than 65% of the students in elementary school in Chizzola (64.3 and 70.9% for boys and girls, respectively) presented the typical blue maculae, while only 13.3% were positive in a control school (only girls). In middle school, more than 90% in Chizzola were positive (87.5% and 100% for boys and girls, respectively), and the comparable figure for the control school was about 12% (15.4 and 11.1% for boys and girls, respectively). At the same conference, however, Cristofolini et al. (1980) stated that histologically similar lesions were present in both groups and that the maculae were predominant at trauma sites. They concluded that Chizzola maculae are trauma ecchymoses (black-and-blue marks) not correlated with fluoride pollution, in agreement with Cavagna and Bobbio's emphatic conclusion (1970, cited in WHO, 1984) that the maculae are not fluoride effects. Although unpublished, Norwegian investigators at the conference were of the opinion that the maculae might be related to pitch volatile substances emitted by smelters using

Söderberg technology (Weinstein, personal communication).

There have been other instances of allegations of effects in areas bordering on industry. One is discussed as a case history in Chapter 5, but another example is the elevated urinary fluoride and bone and joint pains in a family living near an ironstone factory, reported by Murray and Wilson (1946). However, X-ray examination did not reveal a condition of osteofluorosis. Following claims of crop, animal and human health effects in the vicinity of a phosphate plant in Ontario and the airing by the Canadian Broadcasting Company (CBC) of a programme entitled 'Air of Death' in 1967, a Royal Commission (1968) was established to investigate the problem. The Royal Commission made a thorough study of the environmental and human health effects, including some residents who were diagnosed by Drs Waldbott and Cecilioni as suffering from fluoride poisoning, based upon a series of symptoms, including Chizzola maculae. As a result of examinations at a university hospital in Toronto, the Commission concluded that atmospheric concentrations were too low to induce osteofluorosis, the food and water-supplies were not contaminated and there was no radiological evidence of exostoses, elevated urinary fluoride or other fluoride-induced maladies. The CBC was excoriated for the broadcast.

Endemic fluorosis: fluoride in water, indoor air and tea

In 1937 Shortt et al. identified the cause of a crippling condition in India as being due to excess fluoride in the drinking-water. Now it is recognized that endemic dental and skeletal fluorosis occurs in over 30 countries worldwide. Although it affects a significant proportion of the population in many of these countries, by far the greatest numbers of sufferers are in India and China. Fluorosis in both of these countries is also caused or exacerbated by consumption of high-fluoride tea and emissions from coal that is used in the home for heating, cooking and drying of foods (Desai et al., 1993; Saralakumari and Ramakrishna, 1993; Gupta et al., 1994; Wang and Huang, 1995; Ando et al., 1998). Often health problems are compounded by other nutritional disorders.

In India, excess fluoride in drinking-water is not a problem in cities with piped water; it occurs in villages where wells contain water that has been in contact with high-fluoride minerals. Figure 3.1 shows the incidence of skeletal fluorosis and the relationship with fluoride in drinking-water in villages in Rajasthan. It is still not clear exactly how widespread endemic fluorosis is in India because it seems to be discovered in a new area every few years, but recently the WHO (n.d.) stated:

> High fluoride concentrations in ground water, beyond the permissible limit of 1.5 p.p.m., has come to stay as a public health hazard affecting a large segment of the rural population to the tune of 25 million . . . The population at risk is estimated around 66 million.

The extent of fluorosis varies from dental lesions, typically seen in children, to severe crippling. Apart from the immediate suffering, the severe cases cause economic hardship because sufferers cannot work or contribute to the family life. Most attempts to provide simple systems for defluorinating water seem to have failed for one reason or another. They range from storing water in aluminium vessels (which sequesters some of the fluoride) to systems using charcoal. The answer lies in the provision of clean water and there are initiatives being promoted by the WHO and the United Nations Children's Fund (UNICEF).

Although fluorosis is an ancient problem in China (Wang and Huang, 1995), it was first recognized in the 1930s and since 1960 there have been systematic surveys of endemic diseases that have quantified the problem. China has large areas with excessive fluoride in the water, which tend to be in the north of the country. They are associated with evaporative concentration of the fluoride in arid areas and concentrations are

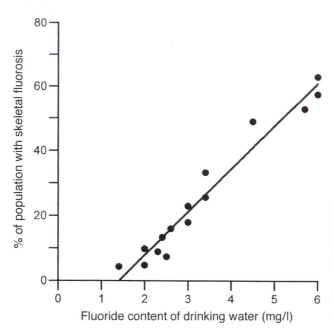

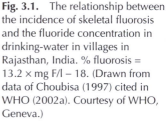

Fig. 3.1. The relationship between the incidence of skeletal fluorosis and the fluoride concentration in drinking-water in villages in Rajasthan, India. % fluorosis = 13.2 × mg F/l − 18. (Drawn from data of Choubisa (1997) cited in WHO (2002a). Courtesy of WHO, Geneva.)

reported up to 20 mg F/l (Wang and Huang, 1995). Tea causes problems where consumption and the fluoride content are high. Wang and Huang (1995) state that in some places a family may consume 10–20 g of tea per day and occasionally up to 40 g. Fluorosis due to coal burning is mostly in southern China (see Fig. 1.3). Li and Tang (1996) reported the fluoride content of coal in Hubei Province to be between 75 and 3858 mg F/kg, with a median of 531 mg/kg. The statistics for the incidence of fluorosis are staggering. In 1997, more than 31 million people may have suffered from indoor airborne fluoride exposure in China (Ando et al., 2001). It has been estimated that there have been 18 million cases of dental fluorosis and 1.46 million cases of skeletal fluorosis caused directly or indirectly from high-fluoride coal emissions (Li and Cao, 1994; Hou, 1997; Li et al., 1999). In the 1990s it was estimated that there were about 42.9 million cases of dental fluorosis and 2.37 million cases of skeletal fluorosis in China (Ji, 1993; Li and Cao, 1994).

When coal or briquettes are burned, both gaseous and aerosol fluorides are emitted and they may contaminate food. Traditionally, food is dried above fires before

storage, so fluoride and other contaminants are deposited directly on the food, leading to high concentrations (Fig. 3.2). Gaseous forms include HF and SiF_4 and, in one study, they were found to contribute 40–84% of total inorganic fluoride emissions, with indoor concentrations ranging from 11 to 155 $\mu g/m^3$ (Zhang and Cao, 1996). The residents were also exposed to a plethora of other hazardous chemicals and chemical compounds, including heavy metals and metalloids (arsenic and selenium), sulphur dioxide and polycyclic aromatic hydrocarbons. An et al. (1997) found that, in villages in China where coal high in arsenic and fluoride was burned indoors, 30% of the villagers developed chronic arsenism. Ando et al. (2001) presented the results of a study of three Chinese villages from Jiangxi Province: one heavily polluted village (severe fluorosis area), one moderately polluted (moderate fluorosis area) and one in a non-polluted area (non-fluorosis/control area). As shown in Table 3.3, the fluoride content of coal used by the residents of the heavily polluted area was high (656 mg F/kg), with concentrations in the moderately polluted and non-polluted areas being much lower (212 mg F/kg and 152 mg F/kg,

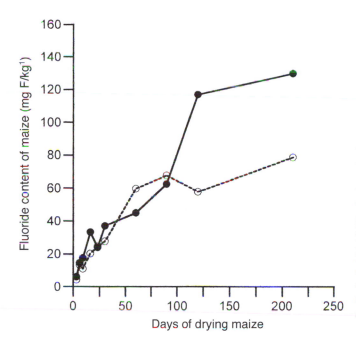

Fig. 3.2. The fluoride content (mg F/kg) of maize dried at two locations in Zhijin County, China. Fresh maize cobs were hung 1 m above and 1.5 m to the left of domestic stoves and the fluoride content measured periodically. (Drawn from data in Zheng, B., Ding, Z., Huang, R., Zhu, J., Yu, X., Wang, A., Zhou, D., Mao, D. and Su, H. (1999) Issues of health and disease relating to coal use in southwestern China. *International Journal of Coal Geology* 40, 119–132.)

Table 3.3. Concentrations of fluoride in coal and soil and main sources of fluoride intake from air, water and food for families in heavily polluted, moderately polluted and unpolluted areas in China (Ando *et al.*, 2001). (Reprinted from *Science of the Total Environment* with permission from Elsevier.)

	Coal (mg/kg)	Soil (mg/kg)	Air (μg/m^3)	Water (mg/l)	Crop (mg/kg)	Chili, vegetables (mg/kg)	Total F intake (mg/person/day)
Heavily polluted area							
Mean	656	620	74	0.19	70.2	195	43.2
SD	133	37	13	0.08	26.5	230	43.2
Moderately polluted area							
Mean	212	4807	76	0.10	3.58	129	14.4
SD	29	828	34	0.02	2.25	71	6.2
Non-polluted area							
Mean	152	285	4.8	0.57	1.26	1.68	2.99
SD	19	35	1.3	0.01	0.37	0.23	1.64

respectively). In the study, airborne gaseous fluoride was very high but the contribution to the total intake per person was small (2%). However, it contaminated the food being dried for storage along the ceiling above the coal stove. Food products, such as maize, chillies and potatoes, were much more contaminated than rice, wheat and 'vegetables'. Drinking-waters from these villages were not fluoridated and therefore were relatively low in fluoride. The concentrations were

0.19 mg/l, 0.10 mg/l and 0.57 mg/l, respectively, for heavily, moderately and non-polluted areas. The non-polluted drinking supply represented only a small proportion of the fluoride ingested in the heavily and moderately polluted areas.

The concentrations of urinary fluoride were in proportion to the total amount of fluoride ingested and were about the same for male and female pupils 10–15 years old. In the heavily polluted area, 99.5% of the pupils examined had some degree of dental fluorosis (211 of 212 pupils). The one pupil with normal teeth came from a family that burned wood for cooking and heating. Nearly three-quarters of the pupils in this group had severe dental fluorosis. In the moderate area, 81.9% of the pupils exhibited dental fluorosis, with 28.5% in the severe category. Although most of the pupils attending school in the heavily and moderately polluted areas exhibited dental fluorosis, no pupils were observed with dental caries. Of course, the lack of caries but the presence of dental fluorosis would not be desirable partners. As might have been expected, the degree of fluorosis was more severe in children than in their parents because of their developing teeth. No typical dental fluorosis was observed in the control area. Skeletal fluorosis was a common malady among the residents of the moderately and severely polluted villages, with 56.9% and 70.0%, respectively, of the residents surveyed having high bone densities in forearms, lower limbs, vertebrae and pelvis. In these patients, osteosclerosis was severe and, in the severely polluted area, osteoporosis was also a common feature of the syndrome. Of the afflicted patients diagnosed with skeletal fluorosis, 88% in the heavily polluted area and 51% of those in the moderately polluted area were classified as severe. Similar results were found by Li *et al.* (1999), who also showed that a high proportion of the residents of Minzhu Town had a high incidence of osteosclerosis and osteoporosis and, as might be intuited, the incidence and severity increased with age. There were significantly fewer cases of skeletal fluorosis in females than in males. An *et al.* (1997) also found that a high proportion of children (in this case 100%) from villages burning high-fluoride coal developed dental fluorosis. About 50% of adults in these villages had skeletal fluorosis. Their data also suggested that, in the presence of arsenic, a lower level of fluoride was required to induce skeletal fluorosis.

Of all the common foodstuffs, tea has one of the highest potentials for increasing the daily fluoride intake because the tea bush, *Camellia sinensis*, accumulates high concentrations of fluoride from normal soils even in clean air (Chapter 2). Tea is an evergreen and fluoride continues to accumulate in leaves over their 3–5-year lifespan, so older leaves have significantly higher concentrations than the young ones. The best quality traditional black and green teas are generally prepared from young leaves but some are made from older leaves and they have correspondingly higher concentrations. The fluoride content of brewed tea depends on the amount of soluble fluoride in the leaves, the length of the brewing period and the concentration of fluoride in the water used for preparation (Kumpulainen and Koivistoinen, 1977; Smid and Kruger, 1985; Schamschula *et al.*, 1988). An analysis of 21 brands of standard tea, 11 brands of decaffeinated tea and 12 brands of herbal tea available in the USA and brewed (with water containing less than 0.02 mg F/l) for 5–120 min resulted in the following fluoride concentrations: 1.5–2.4, 3.2–4.2 and 0.05–0.1 mg F/l (Chan and Koh, 1996). The wide range of fluoride concentrations found in tea is also shown in Table 3.4. The infusions were made using 2.6 g of tea (the average weight used in a UK tea bag) and 250 ml of distilled water or Newcastle tap water, which is fluoridated. It is noticeable that the highest concentrations tended to be in the cheaper, mass-market blends, possibly because of the use of older leaves, but it is important to note that the fluoride content varies considerably between different samples of the same brand. Some brick teas have high fluoride concentrations, notably Bianxiao, which is cited in FSA (2001) as having 441 mg/kg. In brick-teas the leaves are compressed into dense bricks to facilitate storage and transport over long distances, but the reason for the high fluoride content is that it is made from old leaves. Cao *et al.* (2000) reported studies in which it was stated that 20% of brick-tea was twigs, with 26–135 mg F/kg, and the rest was old leaves, with 430–822 mg F/kg. Consumption of brick tea is a major cause of fluorosis in parts of China. Over 90% of it is used in China and it is an important part of the diet of population groups in Tibet, Mongolia and Uygur (Cao *et al.*, 1998). Not only is the

Bianxiao brick-tea used for brewing, but it is also mixed with milk (milk tea), butter (butter tea) or flour (zanba), so it is an important food. Thus, fluoride appears as a major contaminant in both their drinking-waters and their food. Cao *et al.* (1998) cite soluble fluoride concentrations ranging from 352 to 576 mg/kg. The incidence of fluorosis in districts where Bianxiao tea is an important part of the diet is very high, as illustrated by some data of Cao *et al.* (2000). The authors assessed the total dietary intake of fluoride and the incidence of fluorosis in three districts of Tibet. Table 3.5 shows that butter tea and zanba made the largest contribution to the total fluoride intake of children and that the average intake was from 7.97 to 9.42 mg F/day, several times higher than the recommended maximum. The incidence of dental fluorosis was very high and it paralleled the fluoride intake. There is a great awareness of fluorosis in China and there have been control programmes in operation since 1980 (Wang and Huang, 1995).

Table 3.4. The fluoride content of a selection of tea leaves and infusions that are readily available in the UK. The infusions were made using 2.6 g of tea (the average weight used in a UK tea bag) and 250 ml of boiling distilled or Newcastle (fluoridated) tap water. The brewing time was 5 min. (Haley and Davison, unpublished.)

Type/source of tea	Fluoride content		
	Air dry leaves (mg/kg)	Infusion in distilled water (mg/l)	Infusion in fluoridated tap water (mg/l)
Popular blend, supplier no. 1, loose leaf	333	3.5	4.6
Popular blend, supplier no. 2, bags	320	3.5	4.6
Japanese Sencha	238	2.0	3.1
English Breakfast, supplier no. 3	237	2.6	3.6
Pu'er	237	2.2	3.2
Popular blend, supplier no. 3, loose leaf	221	1.9	3.0
Gunpowder	194	1.6	2.5
Darjeeling	131	1.1	1.9
Assam BOP	130	1.3	2.2
Chinese Fujian Oolong	118	0.9	1.9
Assam, supplier no. 3	108	1.2	2.1
Lapsang Souchong	75	0.7	1.6

Table 3.5. The sources of fluoride in the diet in three districts in Tibet and the incidence of fluorosis in children. (Reprinted from Cao *et al.*, *Journal of Fluorine Chemistry* 106, 93–97, 2000, with permission from Elsevier.)

Source	Urban: Lhasa City	Semi-agricultural– pastoral: Dingri County	Pastoral: Naqu County
	Fluoride (mg F/day)		
Butter tea	6.92	7.68	8.03
Zanba	0.95	1.24	1.36
Flour	0.063	0.009	0.02
Meat	0.008	0.006	0.005
Vegetable	0.033	0.021	
Total daily intake (mg)	7.97	9	9.42
Prevalence of dental fluorosis (%)	52.9	75.9	82.7
Dental caries index	1.67	3.11	3.66
% teeth graded severe	21.2	46.7	44.8
Urine fluoride (mg/l)	1.25	1.92	2.26

Fluoridated water

The link between fluoride and dental caries was made in the 19th century and as early as the late 1800s fluoride pastilles were advertised for preservation of teeth (Chapter 1). However, the addition of fluoride to municipal drinking-water as a public-health measure evolved from studies by H. Trendley Dean, head of the Dental Hygiene Unit of the US National Institute of Health. Dean recalled that earlier studies by two dental researchers, Frederick McKay and G. Vardiman Black, in the early 1900s, had found that naturally occurring dental fluorosis was common in Colorado Springs, Colorado, but the permanently stained teeth (Colorado brown stain) were resistant to dental decay. If a lower level of fluoride would not cause mottling, he theorized that the addition of fluoride at a cosmetically safe level would be also effective in reducing tooth decay. Pioneering studies by Dean and his co-workers (Dean, 1938; Dean *et al.*, 1942) led to the introduction of fluoridation in entire municipalities. In 1945, the City Commission of Grand Rapids, Michigan, voted to become the first city in the world to fluoridate its drinking-water and this effort was coordinated by the US Public Health Service and, later, by the National Institute for Dental Research.

The rate of tooth decay was monitored for 30,000 schoolchildren. After 11 years, the caries rate among Grand Rapids schoolchildren had been reduced by more than 60% (NIDCR, n.d.). Fluoridation studies were also conducted in four pairs of cities (intervention and control), also beginning in 1945: Grand Rapids and Muskegon, Michigan; Newburgh and Kingston, New York; Evanston and Oak Park, Illinois; and Brantford and Sarnia, Ontario, Canada. After 13–15 years, during which sequential cross-sectional surveys were conducted in the other paired communities, caries was reduced by between 50 and 70% (CDC, 1999). These findings created a public-health sensation that promised to revolutionize dental care. Today an estimated 360 million people in about 60 countries are exposed to fluoridated water (Hausen,

2000). The effectiveness of fluoridation has changed over the years in Western countries because of the increased use of fluoride toothpastes and mouthwashes and the huge increase in consumption of bottled water and other drinks. Overuse of toothpaste may lead to mild dental fluorosis, while consumption of bottled water negates the effects of municipal fluoridation. Recently, the citizens of Cleveland, Ohio, which has a fluoridated water supply, have had an interesting dilemma. In a study by Case Western Reserve University, of the 57 samples of bottled water available, only three had fluoride levels within the range recommended by the Ohio Environmental Protection Agency (0.80–1.3 mg/l). These bottled waters were not meant to substitute for fluoridated city water, but were being widely consumed by people of all ages. All Cleveland water-supplies were not only within the acceptable range, but were very close to the optimum level of 1.0 mg/l. The study concluded that children drinking bottled water should be considered for prescription fluoride supplements. However, that could create a problem with those children already drinking the brands of bottled water containing adequate fluoride, since excessive ingestion during childhood could lead to dental fluorosis (Case Western Reserve University, 2000).

Fluoridation has been a contentious issue for many years because of concerns about the effects of the fluoride on public health. Opposition has come from many sources, sometimes distinguished researchers and policy-makers, as well as an array of fringe organizations and vocal individuals. The internet has dozens of web sites with headings such as 'Act Now to Ban Fluoride in Drinking Water', 'You're Putting What in Our Drinking Water?', 'Fluoride the Aging Factor: How to Recognize and Avoid the Devastating Effects of Fluoride', 'Fluoride, Teeth and the Atomic Bomb' and 'Opposition by EPA's Headquarters Professionals Union'. Over the years the proponents have demonstrated to the satisfaction of professional dental and medical organizations worldwide that fluoridation is effective in reducing caries and is safe. The opposition has claimed that it does not work and that it causes almost every affliction known to humans. Science must always be challenged because that is the way that progress is made, but the challenge must use scientifically sound experiments and robust analysis. In some cases scientific studies have been badly designed (McDonagh et al., 2000), but the anti-fluoridation lobby weakens its own credibility by making claims that are often completely spurious and by resorting to falsehoods, conspiracy theories and scare tactics.

Concerns about the safety of fluoridation have been made and investigated many times over the last 40 years. One of the most recent summary statements on the safety was by the US National Institute of Dental and Craniofacial Research:

> As with other nutrients, fluoride is safe and effective when used and consumed properly. After more than 50 years of research and practical experience – as well as data evaluation by the U.S. government, committees of experts, and national and international health organizations – the verdict remains the same: fluoridating community water supplies, at optimal levels, is an effective and safe method for preventing tooth decay. Moreover, no credible scientific evidence supports an association between fluoridated water and conditions such as cancer, bone fracture, Down's syndrome, or heart disease as claimed by some opponents of water fluoridation.
>
> (NIDCR, n.d.)

Note that the statement concludes with a response to allegations of increased incidence of cancer, bone fractures, Down's syndrome and heart disease. To this list can be added acquired immunodeficiency syndrome, impaired intelligence, Alzheimer's disease, allergic reactions and other health abnormalities (Hodge, 1986; WHO, 2002b).

Several other important questions have arisen in the last few years about the effects of fluoridation, notably in relation to the risk of bone fractures in the elderly, the size of the reduction in caries and the incidence of dental fluorosis. In relation to the first of these, bone fractures, the current evidence seems to indicate that the optimal amount of fluoride in water does not increase the risk of osteoporotic fractures in elderly people (Hausen, 2000; Phipps et al., 2000). Because the evidence linking dental health and fluoridation was undertaken some years ago, the British Department of Health recently commissioned a review (McDonagh et al., 2000). The report and a follow-up by the Medical Research Council (MRC, 2002) illustrate some of the difficulties of interpreting the existing data on the efficacy of fluoridation and the incidence of dental fluorosis. The reviewers (McDonagh et al., 2000) were careful in their selection of outcome measures and their assessment of the scientific papers. They graded the publications into three categories: A (high quality, low risk of bias); B (moderate quality, risk of bias); and C (lowest quality). Of the 214 publications from around 30 countries none fell into category A and many were category C. So with this cautious approach the main conclusions were:

1. The quality of the evidence on water fluoridation is low.
2. Overall, reductions in the incidence of caries were found, but they were smaller than previously reported.
3. The prevalence of fluorosis is highly associated with the concentration of fluoride in drinking-water.
4. An association of water fluoride with other adverse effects was not found.

As a result of the review, the Department of Health requested the MRC to set up a working group to consider what further research is needed to improve knowledge of fluoridation and health. This they did (MRC, 2002), making a large number of recommendations, including the need to study the impact of water fluoridation on caries reduction in children, against a background of widespread use of fluoride toothpaste, and on the bioavailability of fluoride. Significantly, the MRC report was critical of some aspects of McDonagh et al. (2000). According to MRC (2002), McDonagh et al. (2000) omitted some important studies, such as those concerned with long-term dental effects in communities with naturally fluoridated water and a large, important study of the relationship with cancer risk. However,

one of the aspects that illustrates the difficulty of interpretation is in relation to dental
fluorosis. McDonagh *et al.* (2000) stated that
'The evidence of a reduction in caries should
be considered together with the increased
prevalence of dental fluorosis'. The authors
used sophisticated statistical methods to
interpret a very variable data set of 88 studies relating fluorosis to the fluoride content
of water (their Fig. 4). When they excluded
data points above 1.5 mg F/l, the prevalence
of aesthetically important fluorosis was
predicted to be 10% and 6% in fluoridated
areas (1 mg/l) and non-fluoridated areas
(< 0.4 mg/l), respectively. This is an important conclusion but it should be leavened by
the fact that all but one of the 88 papers was,
by the authors' own assessment, grade C
in quality. Is it safe to make a quantitative
estimate if the quality is so low? Also, the
MRC (2002) pointed out that McDonagh
et al. (2000) had included several publications from countries that have a warmer
climate than Britain so water consumption
would be greater, increasing the risk of
fluorosis. The statistical analysis involved
the concentration in the water, not the more
important total fluoride intake. The MRC
report also commented that there are around
90 causes of enamel defect and three or four
are common. They stated that it is not easy
to diagnose the causes of these defects
and recommended that, in future studies,
photographic techniques should be used
with careful attention to examiner training,
calibration and blinding. Finally, the MRC
(2002) produced data that suggest that
the prevalence of aesthetically important
fluorosis is probably lower than indicated by
McDonagh *et al.* (2000). They cited a study
in which the incidence in fluoridated Newcastle was found to be 3% and in adjoining
non-fluoridated Northumberland it was
0.5%. A European Union (EU) study (Table
3.6) that used photographic techniques
showed similarly lower prevalence rates.
The conclusion of this story is that fluoridation probably does increase the prevalence
of fluorosis but to a lower level than suggested by McDonagh *et al.* (2000). However,
there is no doubt that research, claim and
counter-claim will continue for many years.

Table 3.6.   Prevalence of aesthetically important
dental fluorosis in seven European countries
based on photographic diagnosis (data of
O'Mullane cited by MRC, 2002).

City	Number of children photographed	Prevalence of important fluorosis (%)
Cork, Ireland (fluoridated)	325	4
Knowsley, UK	314	1
Haarlem, The Netherlands	303	4
Athens, Greece	283	0
Alamada, Portugal	210	1
Reykjavik, Iceland	296	1
Oulu, Finland	315	0

Livestock and Wild Mammals

Vertebrate herbivores are much more prone
to develop fluorosis than animals that feed
at other trophic levels (Chapter 2). Because
of the economic importance of farm livestock, there has been a great deal of research
on fluorosis in cattle and sheep although
the effects on horses, goats, hens and other
species are also well known (NAS, 1974;
Suttie, 1977, 1983). The effects on some
wild mammals have also been documented
but there are fewer experimental studies.
Population and ecological effects remain
largely unexplored.

   Acute toxicity has been studied experimentally, using standard laboratory animals, and the symptoms are described in
Hodge and Smith (1965), but acute toxicity
is exceptionally rare in domestic or wild animals. It is most likely to occur as a result of
accidental release of HF or deposition from
volcanoes (Chapter 5). The fluoride concentrations that occur in the environment of the
major emitting sources may cause chronic
fluorosis, which consists of dental lesions,
osteosclerosis, exostoses, stiffness and lameness (NAS, 1974; Suttie, 1977, 1983). Dental
lesions only develop during tooth development, so the state of each individual tooth
reflects the fluoride intake during that
period. For example, the oldest teeth in Plate
2 (fourth and third from the left in the photograph) are normal but the next, younger

tooth (second from the left) has permanent staining and slight roughness on the surface indicating that the fluoride was higher when it was developing. The most recent tooth (first left) has a more advanced degree of fluorosis because it is stained and rough and is showing signs of excess wear. Bones that are badly affected by fluoride (Plate 3) are larger than normal and have chalky, white deposits that make the surfaces rough. Severe stiffness and lameness are thought to be caused by calcification of tendons and the periarticular structures (Burns and Allcroft, 1964; NAS, 1974). Extreme dental wear and lameness lead to reduced milk production, loss of condition and impaired reproduction, with obvious economic consequences, so it is important to monitor the fluoride intake of animals in areas with fluoride exposure. An unusual symptom that has been described in some countries is fracture of the pedal bone (Plate 4), which leads to severe lameness (Burns and Allcroft, 1964). Plate 5 shows an animal with the characteristic cross-legged stance due to bilateral fracture of the inside foot. Burns and Allcroft (1964) noted that in Britain and The Netherlands pedal-bone fracture was associated with hard ground. However, there is evidence linking it to copper levels, but the exact aetiology is not clear.

Many reviewers have discussed the difficulties involved in the diagnosis of chronic fluorosis, and the National Academy of Sciences (NAS, 1974) states:

> It is difficult to define a precise point at which fluoride ingestion becomes harmful to the animal. It can vary from case to case and may be influenced by the following factors: amount of fluoride ingested; duration of ingestion; fluctuation in fluoride intake with time; solubility of fluoride ingested; species of animal involved; age at time of ingestion; general level of nutrition; stress fractures; individual response.

Nevertheless, the dietary tolerances of the main domestic animals are known and widely accepted (Table 3.7) and environmental-quality standards based on them are effective in preventing fluorosis (Chapter 7).

Fluoride emissions from volcanoes are the oldest source of fluorosis in livestock (Chapters 1 and 5), but industrial fluorosis has been recognized for about a century. For example, in 1912 cattle located next to a superphosphate factory in Italy exhibited symptoms resembling osteomalacia (softening of bones) (Bartolucci, 1912, cited by Shupe and Olson, 1983), and it was believed to be related to the fluoride emissions from the factory. NAS (1974) states that in the 1920s the use of feed supplements containing excessive fluoride increased the incidence of toxicosis in animals, and it is clear from the literature of the time (Roholm, 1937) that fluorosis was widely recognized and that there were some severe cases. Burns and Allcroft (1964) referred to veterinary investigations of fluorosis in Britain between 1938 and 1948, mostly in relation to brickworks, calcining of limestone and enamel works. Reports and research publications multiplied from the late 1940s until the 1960s because of the expansion of fluoride-emitting industries. In the 1950s fluorosis was described in sheep by Harvey (1952) and Pierce (1952) and in other domestic animals by Rand and Schmidt (1952) and Neeley and Harbaugh (1954). The peak of the incidence of fluorosis was probably in the 1950s and 1960s in Western countries, but there has been a huge reduction in the incidence over the last 30 years. Reports of fluorosis still appear in the scientific literature (Choubisa, 1999; Patra et al., 2000; Swarup et al., 2001), but knowledge of the dietary tolerance of the common farm animals and adequate surveillance should prevent the development of serious symptoms.

The extent of fluorosis during its peak in the 1950s and 1960s and the strong association with certain industries can best be illustrated by reference to the work of Burns and Allcroft (1964). The first account of fluorosis in Britain was in 1941 in relation to 'a mysterious lameness of cattle' that occurred in the vicinity of brickworks in Bedfordshire, but as the cases increased the Ministry of Agriculture, Fisheries and Food commissioned the Veterinary Investigation Service to survey the incidence of fluorosis in livestock in England and Wales. This was done between 1954 and 1957 by Burns and Allcroft (1964). Figure 3.3 shows a map of

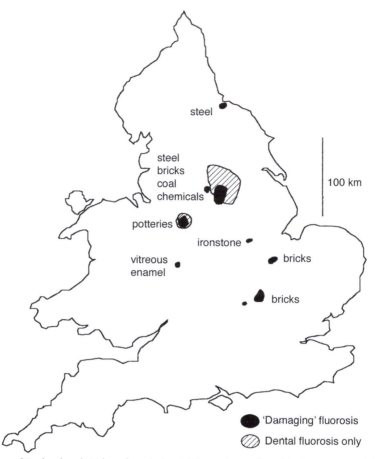

steel

steel
bricks
coal
chemicals

100 km

potteries

ironstone

vitreous
enamel

bricks

bricks

'Damaging' fluorosis

Dental fluorosis only

Fig. 3.3. Areas of England and Wales where industrial fluorosis was found in the 1950s, and the main sources of fluorides. (Redrawn and modified from Burns, K.N., Allcroft, R. (1964) Fluorosis in cattle in England Wales. 1. Occurrence and Effects in Industrial Areas of England and Wales 1954–57. Ministry of Agriculture, Fisheries and Food, Animal Disease Surveys Report No. 2, Part 1. Her Majesty's Stationery Office, London, according to licence number C02W0002815.)

England and Wales with the areas of dental and 'damaging' fluorosis marked. Damaging fluorosis included lameness, loss of milk yield and dental fluorosis that was sufficiently severe to interfere with feeding or reduce the market value of the animals. The centres were associated with the steel, pottery, brick and enamel industries and they covered huge areas. Figure 3.4 shows more detail for the area around Stocksbridge, Rotherham and Sheffield, a centre for steelworks, coke ovens, power stations, small brickworks and some chemical industry. Grazing was abandoned in the worst areas and damaging fluorosis occurred over a large

area. Since 1964, when the report was published, there has been a huge contraction in the industries and a reduction in emissions.

Wild mammals

There have been numerous studies of fluorosis in other mammals, including wild ungulates (deer, elk, bison, moose) and diverse small herbivorous or omnivorous mammals and rodents. A very useful review was published by the WHO, which lists most of the publications (WHO, 2002b). The

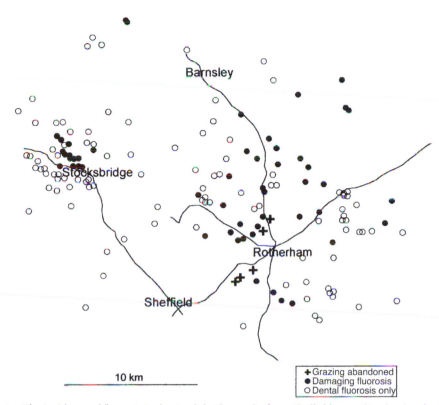

Fig. 3.4. The incidence of fluorosis in the Stocksbridge–Rotherham–Sheffield area of England in the 1950s. Sources of fluoride included open-hearth steelworks, power stations, coke ovens, chemical works and brickworks. (Redrawn and modified from Burns, K.N., Allcroft, R. (1964) Fluorosis in cattle in England Wales. 1. Occurrence and Effects in Industrial Areas of England and Wales 1954–57. Ministry of Agriculture, Fisheries and Food, Animal Disease Surveys Report No. 2, Part 1. Her Majesty's Stationery Office, London, according to licence number C02W0002815.)

first signs of chronic fluorosis in ungulates are mottling, pitting and black discoloration of the teeth, with consequent softening and abnormal wear of the dental surfaces (Kay *et al.*, 1975). The syndrome appears to have been first reported in the literature in the 1950s and was confirmed by Karstad (1967), who found that the mandibular bones of mule deer near an industrial complex contained 4300–7125 mg/kg fluoride/ kg. Teeth exhibited black discoloration, pitting and abnormal wear. Black-tailed deer (*Odocoileus hemionus columbianus*) found near an aluminium smelter in Washington, USA, had dental lesions and fluoride concentrations in ribs that ranged from 2800 to 6800 mg/kg, compared with controls at 160 to 460 mg/kg (Newman and Yu, 1976). Dental fluoride and profound

osteofluorosis were reported from red deer (*Cervus elaphus*) yearlings examined near Norwegian aluminium smelters (Vikøren and Stuve, 1996a). Roe deer (*Capreolus capreolus*) and red deer (*C. elaphus*) in the Ruhr Valley, Germany, and red deer in Germany and Czechoslovakia were found to have developed tooth abnormalities, as well as high concentrations of fluoride in mandibular bones (Kierdorf, 1988; Kierdorf, H. *et al.*, 1996; Kierdorf, U. *et al.*, 1996). These abnormalities were accompanied by structural changes in affected dental enamel (Kierdorf *et al.*, 1993; Kierdorf, H. *et al.*, 1996; Kierdorf, U. *et al.*, 1996), and similar changes have been observed in the enamel of fluorotic teeth of wild boars (*Sus scrofa*) (Kierdorf *et al.*, 2000). The occurrence of severe dental fluorosis and osteofluorosis has been correlated with reduced life expectancy in red deer (Schultz et al., 1998). Kierdorf and Kierdorf (1999) suggested that the presence of dental lesions can be used as a bioindicator in deer to evaluate the presence of fluoride contamination. It was used very effectively in the Saxonian Ore mountains (Kierdorf and Kierdorf, 1999).

Shupe et al. (1984) examined 1125 mule deer (O. hemionus), 105 elk (Cervus canadensis) and 64 American bison (Bison bison). Dental and skeletal lesions were found in all three species exposed to elevated levels of fluorides in waters and vegetation in Utah, Idaho, Wyoming and Montana, USA. Excessive fluoride was provided by naturally high-fluoride waters and associated vegetation and from an aluminium smelter in Montana. Of particular interest are the results found for bison, elk and mule deer inhabiting an area near geothermal springs in Yellowstone National Park, where 29 of 44 bisons examined from one area had severe dental fluorosis. Bones of all species showed the classical bone lesions, ranging in severity from slight to severe, and characterized by irregular, poorly organized and improperly mineralized bone. Osteocytes were clumped abnormally rather than being uniformly distributed throughout the osteons (Shupe et al., 1984).

Walton (1984, 1985, 1986, 1987a,b, 1988) reported significant dental damage and/or tooth wear in field voles (Microtus agrestis), woodmice (Apodemus sylvaticus), moles (Talpa europaea) and shrews (Sorex araneus) at various distances from an aluminium smelter in Wales. Radiographs, however, showed no deformities in skeletons of voles or mice when compared with animals from the control areas. In contrast, cotton rats (Sigmodon hispidus) residing in a petrochemical waste site were found to have symptoms of both dental and bone fluorosis (Paranjpe et al., 1994; Schroder et al., 1999). Approximately 80% of the rats were found to have dental fluorosis.

The most comprehensive studies of small mammals were conducted in Britain by Andrews et al. (1989), Cooke et al. (1990, 1996) and Boulton et al. (1994a,b,c, 1995, 1997, 1999). The strength of their work lies in the fact that it included both field studies and experimental toxicology performed under controlled conditions. They have also investigated different sources of fluoride– fluorspar waste and sites near chemical works and an aluminium smelter. Working at a fluorspar-tailings site, Andrews et al. (1989) showed that the herbivorous vole (M. agrestis) developed dental lesions but the omnivorous shrew, S. araneus, did not, despite the latter having a greater concentration ratio of bone fluoride to food fluoride. The work also showed the high rate of turnover in the populations, which makes it difficult to assess the ecological effects of fluoride. Experimental toxicology in which four species were exposed to 0, 40 and 80 mg F/l in water showed that 40 and 80 mg/l induced premature mortality in M. agrestis and Clethrionymus glareolus (bank vole) but not in woodmouse (Apodemus sylvaticus) or laboratory mouse (Mus musculus). Severe dental lesions were induced in animals that survived the 84-day experiment but 'Overall . . . there were few clear differences in inherent species sensitivity to fluoride, the interspecific variation in metabolism and accumulation rates being attributable mainly to variation in intake.' This is a clear indication that the feeding strategy is vitally important in determining the effects of excess fluoride in the environment. In a later study Boulton et al. (1994) exposed Microtus to diets containing 100–300 mg F/kg based on vegetation taken from the vicinity of an aluminium smelter and a fluorochemical works. The diets reduced live-weight gain and produced 40–100% mortality and marked dental lesions. The most interesting fact was that voles that consumed a diet containing 100 mg F/kg showed no weight loss, no early mortality and only slight dental lesions. The authors concluded that the contrast in severity of the diets was due to differences in the chemical speciation and bioavailability of the fluoride (Chapter 2). This presents an argument for monitoring fluoride, not only by recording the fluoride in the diet but also by examination of the teeth (Boulton et al., 1997, 1999).

There have been very few studies of carnivores, largely because the incidence of fluorosis is so low, but Walton (1984) made some interesting observations on the red fox (*Vulpes vulpes*). Analysis of bones from many parts of Britain showed that, although the skeletal fluoride of small mammals was increased in the vicinity of an aluminium smelter, it was not increased in fox bones, even though they are omnivorous and eat many small rodents. Examination of their faeces showed that skeletal parts of the prey passed through the digestive system largely untouched, which explains the lack of accumulation.

Other vertebrates

Domestic hens are known to be relatively tolerant to fluoride (Table 3.7) and that appears to be true for the few other birds that have been examined. Japanese quail (*Coturnix coturnix japonica*) exhibited no detrimental effects on growth when they were provided with sufficient fluoride to result in tibial fluoride concentrations of 13–2223 mg/kg (ash weight). The 24 h median lethal dose (LD_{50}) for 1-day-old

Table 3.7. The dietary tolerance of domestic animals. Fluoride concentrations assume that the fluoride is in a soluble form, such as NaF, and are at levels that could be fed without clinical interference with normal performance. They do not ensure that there would be some signs of elevated fluoride that do not affect performance. Adapted from NAS (1974).

Species	mg F/kg dry wt
Young beef or dairy heifer	40
Mature beef or dairy heifer	50
Finishing cattle	100
Feeder lamb	150
Breeding ewe	60
Horse	60
Growing dog	100
Finishing pig	150
Breeding sow	150
Growing or broiler chicken	300
Laying or breeding hen	400
Turkey	400

European starling (*Sturnus vulgaris*) chicks was reported to be 50 mg/kg body weight and 17 mg/kg for 16-day-old nestlings. Growth rates were significantly reduced at 13 and 17 mg/kg body weight (Fleming *et al.*, 1987). American kestrels (*Falco sparverius*) and eastern screech owl (*Otus asio*) were found to be resistant to relatively high fluoride concentrations in their diets (Hoffman *et al.*, 1985; Carrière *et al.*, 1987; Pattee *et al.*, 1988). Vikøren and Stuve (1996b) investigated common and herring gulls (*Larus canus* and *Larus argentatus*, respectively) near two aluminium smelters and at three control sites in Norway. Although eggshell and bone fluoride concentrations were elevated in birds taken near the smelters, the levels were not much higher and there were no changes in bone morphology. The same authors (Vikøren and Stuve, 1995) examined bone fluoride in Canada geese (*Branta canadensis*). Again, there was some indication of increased fluoride but there were no gross osteofluorotic lesions.

Fish accumulate fluoride in the bones, just as other vertebrates do, but there do not seem to be any reports of osteosclerosis or any other bony lesions and there are no definitive studies of effects on populations. The toxicity appears to be negatively correlated with water hardness because of complexation with calcium (WHO, 2002a) and it is reduced by the presence of aluminium (Baker, cited in Havas and Jaworski, 1986), illustrating once again the importance of chemical speciation on bioavailability. Eggs of *Catla catla* exposed to an effluent and sodium fluoride showed delayed hatching at concentrations > 3.2 mg F/l (Pillai and Mane, cited in WHO, 2002a), and Neuhold and Sigler (1960) found 20-day $LC_{50}s$[1] of 2.7–4.7 mg F/l for rainbow trout (*Oncorhynchus mykiss*). Camargo (2003) reported a safe concentration for rainbow and brown trout of 5.1 and 7.5 mg F/l, respectively. However, most of the research on fish and other aquatic organisms has involved much higher concentrations (WHO, 2002a). Acute toxicity tests (96 h LC_{50}) give concentrations ranging from 51 in rainbow trout to 460 mg/l in the three-spine

stickleback (*Gasterosteus aculaeatus*). Perhaps the most important point is that the
toxicity levels are all below those that occur
due to fluoridation, so there does not appear
to be any risk from that source.

Fluoride has been known to be toxic to
some invertebrates since the 19th century
and sodium fluoride was patented for use as
a pesticide in 1896. Since then several compounds have been marketed as insecticides,
notably sodium fluoride, cryolite and
fluorosilicates of barium, sodium, ammonium and zinc. Sodium fluoride and sodium
fluorosilicate were widely used until about
30 years ago to control cockroaches, ants and
other domestic pests, but they have been
restricted, banned or removed from the market in many countries because of the dangers
of accidental or deliberate poisoning. In the
1930s cryolite came into prominent use as
an insecticide for the control of crop pests,
such as codling moth (McMullin, 1935;
Dobrosky, 1937, 1943; Marshall *et al.*, 1939;
Dean *et al.*, 1946). Its use declined with the
advent of new, more specific compounds,
but there has been a resurgence in use in the
last decade, particularly in the USA. The
attraction of cryolite is that it is effective,
that insects do not develop tolerance and
that there are minimal problems with
residues in the harvested parts of the crop.
Because it is a natural product, it even quali-
fies as an 'organic' pesticide. Wahlstrom
et al. (1996) described formulations as
'unsurpassed in terms of efficiency and cost/
benefit ratio for the control of serious vine-
yard pests'. It is also very effective against
other important pests, such as the Colorado
potato beetle and vine weevils. The only
case of an unacceptable residue seems to be
elevated fluoride in some California grape
juices and wines. Wahlstrom *et al.* (1996),
for example, showed that concentrations as
low as 1–3 mg F/l affected the fermentation
rates produced by some yeast strains. The
problems are avoided by appropriate timing
of applications.

Many invertebrates possess a calcified
exoskeleton, a shell or calciferous glands.
The calcified tissues attract fluoride so, not
surprisingly those organisms that possess
them usually have relatively high fluoride

concentrations (Chapter 2). In krill, which
have a calcium phosphate-based carapace,
it is considered that fluoride plays an important role in hardening the mouth-parts but
it is not known if there are any positive
benefits in any other groups. Very high
concentrations have been reported in shells
and carapaces (Chapter 2), but there appear
to be no observations of any of the deformities or lesions akin to the osteosclerosis
or exostoses that are seen in vertebrates.
There are no reported studies of effects on
the calciferous glands of earthworms.

Unfortunately, the large-scale use of
fluorides as insecticides has not stimulated
much research into the fundamental mechanism of action or of the basis of differences
in sensitivity between species. Statements
in the manufacturers' promotional literature
simply record that cryolite and similar compounds act as 'stomach poisons'. Fluoride
in the digestive system might interact with
digestive enzymes or interfere with the
gut microflora, but the effects will depend
strongly on the pH, the chemical species of
the fluoride and the presence of other agents,
such as calcium, that can complex with the
fluoride. In addition, Edmunds (1983) suggested that, if fluoride affected parasitoids,
parasites or pathogens, it could result in an
increase or decrease in the growth or fecundity of herbivores. He also pointed out that
fluoride may alter plant metabolism, making
it more or less nutritious, changing the
synthesis of anti-herbivore compounds or
changing colour or surface characteristics of
the plant to make it more or less suitable
as an oviposition or feeding site. Therefore,
the effects of fluoride on invertebrates are
potentially very complex.

Invertebrates

Field studies near fluoride sources

There are dozens of publications that demonstrate elevated fluoride concentrations in
a range of invertebrates living near sources,
but there are comparatively few that have
sought to determine whether the fluoride

had any effect on individuals or populations. Two approaches have been used to study effects: uncontrolled field studies in areas near fluoride sources and laboratory studies exposing animals to HF in the air or fluoride salts in the diet. For two species, silkworm and honey-bees, there are data from both the field and the laboratory. As will become evident, many questions still remain.

Several field studies have shown correlations between fluorides and changes in insects including silkworms (*Bombyx morae*: Colombini *et al.*, 1969; Kuribayashi, 1971, 1972, 1977; Yoshida, 1975), honey-bees (*Apis mellifera*: Kunze, 1929; Maurizio and Staub, 1956; Maurizio, 1957, 1960; Mueller and Worseck, 1970), the European pine-shoot moth (*Rhyacionia buoliana*: Manskova, 1975), bark beetles (Pityokteines: Pfeffer, 1962) and the pine-bud moth (*Exoteleia dodacella*: Hagvar *et al.*, 1976). A detailed account of one of the most intriguing examples is given in Chapter 5. In most instances where fluoride has been implicated in changes in insect populations or behaviour, there were other pollutants present, and that is a fundamental problem with field observations. Pollutants that commonly co-occur with fluorides include sulphur dioxide, nitrogen oxides, ozone, heavy metals, particles and pitch volatiles. Several of these are known to alter plant chemistry and affect feeding herbivores, so, even where a correlation is strong, it does not mean that there is a cause–effect relationship, and it is usually impossible to subtract the effects of other factors or test for interactions.

Forest insects

Carlson and Dewey (1971) concluded that in areas downwind of an aluminium smelter in Montana, USA, an increased infestation of pine-needle scale (*Phenacapsis pinifoliae*) was related to increased fluoride content of the needles, but their data did not support this conclusion. In the text they said, 'Although the correlation is insignificant, the graph does indicate a

trend, and more extensive sampling would likely confirm the relationship.' In fact, the correlation was so low that even the last part of the statement is questionable. Later, Carlson *et al.* (1977) studied a possible relationship between pine-needle sheath miner (*Zellaria haimbachi*), a needle miner (*Ocnerostoma strobivorum*), sugar-pine tortrix (*Choristoneura lambertiana*) and ambient atmospheric concentrations as well as the foliar fluoride concentration in lodgepole pine (*Pinus contorta* var. *latifolia*). They found a significant relationship between *O. strobivorum*, *Z. haimbachi* and foliar fluoride concentrations. In contrast, in a study with a different diaspidid scale insect, *Nuculapsis californica*, resident on ponderosa pine, Edmunds and Allen (1956) found no relationship between fluoride and insect population density.

Using a different approach, but still working in the field, Mitterböck and Führer (1988) supplied nun-moth larvae (*Lymantria monarcha*) with Norway spruce needles containing up to 365 mg F/kg and collected from the vicinity of an aluminium smelter in Austria. Mortality increased up to 75% and development was delayed. Reduced growth rates resulted in pupal weights as low as 45% of the controls. Larvae accumulated very high concentrations of fluoride, with an accumulation factor of 7.3 for larvae that had died in the fifth instar. Pupae, which did not exceed an accumulation factor of 3.8, contained less fluoride. A level of 1400–1500 ± 600 mg F/kg in larvae was lethal, but surviving pupae contained less. The high retention of fluoride in *L. monarcha* is an interesting contrast to other lepidoptera (Hughes *et al.*, 1985; Davies *et al.*, 1992). This approach has much to commend it because it removes some of the confounding factors, but ideally it should be coupled with physiological studies of bioavailability and digestive system function.

Not unexpectedly, there are reports of both no effects and positive effects of fluoride on invertebrates. Beyer *et al.* (1987), for example, collected trap-nesting wasps at different distances from an aluminium smelter to test the theory that fluoride might have changed the communities. In fact, there was no relation between fluoride and relative wasp density and number of cells provisioned with prey. In contrast, Thalenhorst (1974) found a positive correlation between exposure of spruce to fluoride (and other atmospheric pollutants) and the density and survival rate of pseudofundatrices of the aphid *Sacchiphantes abietis*. In addition, increased aphid survival was correlated with a diminished capacity of the trees to produce a specific defensive reaction following penetration of the host by aphid stylets. Although the biochemical basis for this reaction has not been elucidated, secondary compounds, such as phenolics, might be involved. But fluoride exposure is also known to increase the total free amino acids, organic acids and reducing sugars (Weinstein, 1961; Yang and Miller, 1963a,b; Arndt, 1970) and perhaps other compounds that are feeding stimuli for some insects.

Soil- and ground-dwelling invertebrates

Recolonization of areas disturbed by sand mining in the presence of a major source of fluoride pollution has provided an interesting insight into the effects of fluoride (Madden and Fox, 1997). The effect of fluoride on vegetation structure appeared to be reflected in the relative abundance of arthropods. In general, there was an increase in vegetative growth at low fluoride levels, which was beneficial to flies, crickets, mites and overall arthropod biomass. The overall arthropod diversity at the highest-fluoride site was reduced, although there were increases in the populations of mites and cockroaches. Mites were found to benefit in direct response to fluoride. Beetles, spiders and ants declined at the high fluoride levels, i.e. between c. 150–250 mg F/kg, in *Angophora costata* foliage. Wasps were apparently unaffected in high-fluoride areas.

Ground beetles have been studied in Finland. Holopainen *et al.* (1995) investigated the beetle fauna in 16 spring cereal fields around a fluoride-emitting fertilizer plant, using pitfall traps. The 16 sites were divided into three zones with different degrees of fluoride pollution. However, the fluoride concentrations in barley leaves were quite low, the highest site having only about 27 mg F/kg in August. The other two sites were at background level, with about 5–8 mg/kg. Analysis of two of the beetle species paralleled these low levels, because the average concentrations (around 8 mg/kg) were the same as background concentrations reported by other workers. The total catch of beetles was higher at the sites in the background area, but ordination showed that the communities were most strongly related to soil type and the water content of the soil. The authors concluded that there was no direct effect of the fluoride on the beetles but discussed the possibility of indirect effects via the food-chain. However, they had no data or evidence in relation to the food-chain, so their conclusion that fluoride and other pollutants may have an influence on carabid communities in cereal fields is difficult to support. Using their pitfall technique and ordination would probably provide more concrete results in an area with higher fluoride levels.

Although there are several reports of the fluoride content of earthworms (Garrec and Plebin, 1984; Buse, 1986; Walton, 1986; Vogel and Ottow, 1991), there have been few attempts to investigate if it has any effect at either the physiological (e.g. functioning of the calciferous gland) or ecological levels. In one study (Samal and Naik, 1989) the number of individuals and body length of earthworms (*Octochaetona surensis*, *Dichogaster bolaui* and *Perionyx excavatus*) collected from sites downwind of an aluminium smelter in India, including a control site 80 km away, were inversely related to their fluoride accumulation. The population of earthworms per m^2 and the body length was about half at 0.5 km from the smelter as at 9.0 km. Earthworms with gut soil ranged from about 2600 mg F/kg at a distance of 0.5 km to about 140 mg/kg at 9.0 m and, without gut soil, from about 1900 to 80 mg/kg. The question remains whether the reduced performance was due to fluoride or to some other co-occurring factor.

Controlled experiments

There have been a few experiments in which insects were exposed to extremely high concentrations of HF in order to determine the direct toxic and mutagenic effects. The fruit fly, *Drosophila melanogaster*, has most frequently been the subject for the study of the direct effect of HF on arthropods. Mohamed and Kemner (1970), using what appeared to be an extremely high concentration, reported that homozygosity for one of the second chromosomes from a treated male resulted in a reduction in viability, which ranged from subvital to lethal. Gerdes and his co-workers (Gerdes, 1968; Gerdes *et al.*, 1971a,b) found a small increase in the recessive lethal mutation rate following exposure to HF. Exposure of four strains of *Drosophila*, two mutant and two wild type, to 1.3, 2.9, 4.2 and 5.5 mg/m^3 was found to be harmful during the second 12 h exposure (Gerdes *et al.*, 1971a). At 4.2 mg/m^3, there was a clear difference in response of the four strains. The effects appeared to be predominantly linear with respect of concentration, with mutant strains exhibiting the greatest response and the two wild-type strains showing the greatest tolerance. Treatment of oocytes of *Drosophila* resulted in a dose-dependent decrease in fertility and fecundity. Although the data are scientifically interesting, the conclusions are not relevant to a real-life situation because the concentrations were about three orders of magnitude greater than normally found in fluoride-contaminated atmospheres. This is also true of most reports where invertebrates were exposed to fluoride salts in the diet. Even then, they sometimes show surprising results because, when flour beetles (*Tribolium confusum*) were fed with flour containing up to 10,000 mg NaF/kg for 0, 2, 4 and 6 days and then with uncontaminated flour, there was a pronounced stimulation in egg production following the 2- and 4-day treatments (Johansson and Johansson, 1972).

Honey-bees have been used for biomonitoring (Bromenshenk, 1978, 1988, 1992, 1994; Bromenshenk *et al.*, 1985, 1996; see Chapter 6), but there has been persistent concern for over 30 years that fluoride emissions affect honey-bees. For example, Bourbon (1967) reported a high mortality rate among bees in a colony located several hundred metres from an aluminium smelter. However, there are few experimental studies using realistic levels of fluoride, and many questions remain. Bee foraging characteristics and capacity to transport pollen back to their colonies means that they sample particular plant species over a relatively large area. This is often considered to make bees susceptible to pollution and suitable as biomonitors, but the pollen of insect-pollinated plants is usually held within the flower, where it receives minimum contact with air pollutants, and nectar is produced from the phloem, which has near-background concentrations of fluoride (Davies, 1989). In fact, bees are exposed to far lower concentrations than detritus feeders and some other groups (Chapter 2).

The toxicity of fluoride to bees has been assessed by injection and by feeding studies. For bees in a free-flying colony, Maurizio (1960) found the LD$_{50}$ for soluble fluorides to be 5–8 µg per bee, but for cryolite it was 130–160 µg. She also pointed out that the quantity of fluid in which the fluoride was contained had a significant influence on toxicity: that is, if the fluoride was dissolved in a small amount of fluid, the concentration was higher, resulting in a lower LD$_{50}$ value. Trautwein *et al.* (1968) reported that the LD$_{50}$ value for NaF was 11 µg per bee over 24 h but for HF it was 5 µg per bee over a 24 h period. In some instances, bees alleged to have been killed by fluoride contained as little as 0.6 to 2.8 µg per bee and, in 57 samples of normal bees, the fluoride concentration ranged from 0.63 to 4.81 µg per bee. Dreher (1965) reported an LD$_{50}$ of 10 µg per bee and Bourbon (1967) found that in bees with a high mortality rate located several hundred metres from an aluminium smelter workers contained 18 µg per bee, while bees that did not leave the colony contained 10 µg.

In order to determine the risk to bees in the field, it is necessary to know the routes of transfer and the amount in the digestive system. As indicated in Chapter 2, fluoride may reach honey-bee colonies by dry deposition on the wings and body, by intake through spiracles, by mistaking dust for pollen (Tong *et al.*, 1975) or from pollen and nectar. Colonies might also be contaminated by forced air circulation and evaporative cooling employed by the bees to control temperature and humidity in individual colonies. Bromenshenk *et al.* (1996) stated that 'fluoride usually cannot be washed off a bee, except where fluoride levels are very high, and then only a small amount can be removed'. Without disputing these findings, it is also a fact that a large proportion of the fluoride emitted from aluminium smelters and phosphate-rock processing are in the form of fluoride-containing particles of varying solubility, so it is logical to assume that some of the fluoride found associated with bees collected from the field is lodged in the body hairs, legs or other appendages rather than being in the body. In using washing to determine the surface contamination there may be a problem of wettability so perhaps a detergent should be used. However, one simple technique for quantifying surface deposition is that used by Davies (1989) (see Chapter 2). She exposed dead insects to known concentrations of HF and determined the fluoride content of the wings and bodies after 9 days (Table 3.8). The fluoride concentrations were all significantly higher than in controls, which were kept in scrubbed air, demonstrating that the surfaces of bees collect fluoride by dry deposition. The fact that there was no significant difference in the fluoride content of wings exposed to 1.85 and 14.6 µg HF/m³ indicates that they have a finite capacity to hold fluoride. However, the data also support the statement of Mayer *et al.* (1988) that in the

field much of the fluoride in honey-bees is due to surface deposition.

One of the most informative field studies was that of Mayer *et al.* (1988), who used replicated hives in the field to determine the effects of emissions from an aluminium smelter. They established four honey-bee colonies in each of three locations selected to provide high, medium and low exposures to fluoride emissions from a nearby aluminium smelter. A fourth location, a control, was established 200 km to the east. Samples for fluoride analysis were collected from: live adult bees at the hive entrance; dead bees from Todd traps; teneral adults from 50 capped cells; stored pollen from 50 cells per colony; and all the honey from one frame per colony at two different times each year. Live control bees had fluoride contents ranging from 11 to 15 mg/kg and the mean for bees at highly exposed colonies was around 220 mg/kg (Table 3.9). These are similar to data reported by Dewey (1973) and Bromenshenk *et al.* (1985). The mean fluoride per bee at the highest site was 10.5 µg and some individuals had up to 14.3 µg, which is considerably higher than the LD_{50} of 10 µg per bee estimated by Dreher (1965). There was no significant difference in fluoride content of live and dead bees, indicating that mortality was not associated with fluoride (Table 3.9). The authors suspected that most of the fluoride was on the external surfaces of the bees, which is supported by Davies's (1989) experiments on dry deposition of fluoride to insect surfaces. The fluoride content of pollen was surprisingly high, ranging from 16 to 60 mg F/kg at the different sites, but honey was consistently low at 0.4–1.4 mg F/kg. The latter is similar

Table 3.8. The mean fluoride content (mg F/kg dry wt) of the wings and bodies of dead bees (*Apis mellifera*) and cockroaches (*Periplaneta americana*) exposed to three concentrations of HF for 9 days under controlled conditions. Standard errors in parentheses. (Data from Davies, 1989.)

	Concentration of HF in air (µg/m³)					
	Controls < 0.1		1.85		14.6	
Species	Wings	Body	Wings	Body	Wings	Body
Honey-bee	ND	23.0 (1.4)	35.3 (4.9)	86.7 (6.1)	37.9 (5.8)	245.3 (36.5)
Cockroach	15.4 (5.6)	16.7 (4.3)	58.4 (2.2)	144.6 (22.6)	79.4 (34.9)	205.3 (42.4)

to data of Tong *et al.* (1975), who found that honey samples collected in New York State contained from 0.001 to 8.9 mg/kg. There were no significant differences between hive sites in terms of adult or brood populations (Table 3.10) or in brood survival (90–91% in all cases). The mean number of dead bees per day varied from 6 to 85 but there was no difference between sites. Finally, honey production was also not related to fluoride (Table 3.10) so overall, despite the high concentrations in the bees, there was no effect on the populations or productivity.

Perhaps the most intensively studied insect–fluoride interaction is the association of fluoride accumulated by mulberry leaves and feeding silkworm larvae (Colombini *et al.*, 1969; Kuribayashi, 1971, 1972, 1977;

Kuribayashi *et al.*, 1976; Alstad *et al.*, 1982; Wang and Bian, 1988). The toxicity of fluoride-contaminated mulberry leaves to silkworm larvae has been a challenge to commercial silk production in China, Japan, India and elsewhere. In addition to the large-scale sources of fluoride, the silkworm industry of China is plagued by fluoride emitted from brick and tile manufacturing, much of which is carried out by individuals or small groups with little or no control over emissions. A number of studies have indicated that silkworm larvae are extremely sensitive to fluoride. Metabolic studies in general indicate that fluoride affects not only larval metabolism (Chen, 1987), but also the nutritional quality of the larval food, the mulberry leaf. According to Das and Prasad (1973), nearly 70% of silk protein is derived directly from proteins present in mulberry leaves, so an effect of fluoride on plant metabolism might be detrimental to silk-protein production.

Fluoride salts depress feeding rates and cause softening of the insect cuticle, reduction in growth rate and/or death (e.g. Kuribayashi, 1971; Fujii and Hayashi, 1972; Fujii and Honda, 1972; Imai and Sato, 1974; Moshida and Yoshida, 1974; Kuribayashi *et al.*, 1976; Wang *et al.*, 1980; Qian *et al.*, 1984; Bian and Wang, 1987; Wang and Bian, 1988). By providing mulberry leaves containing known amounts of fluoride to larvae, Wang and Bian (1988) determined that the toxicity threshold concentration in mulberry leaves was about 30 mg/kg dry wt, and the lethal dose was between 120 and 200 mg/kg. No mortality occurred up to

Table 3.9. Mean fluoride concentration (mg/kg dry wt) in dead honey-bees collected at different locations in Washington State (after Mayer *et al.*, 1988). (Courtesy of *Fluoride* 21, 113–120, 1988.)

Site and degree of exposure	1984 Living bees	1984 Dead bees	1985 Living bees	1985 Dead bees
REC, low	86[a]	102[a]	68[a]	74[a]
VRF, medium	108[b]	144[b]	118[b]	130[b]
FH, high	170[c]	223[c]	219[c]	219[c]
CON	13[d]	15[d]	11[d]	–

[a,b,c,d]Means within a column followed by the same letter are not significantly different ($P \le 0.5$, Duncan's multiple range test). CON, control site; REC, 10 km south-south-east of smelter; VRF, 6.4 km east of smelter; FH, 0.8 km downwind of smelter.

Table 3.10. Performance of honey-bee populations at different locations in Washington State (after Mayer *et al.*, 1988). (Courtesy of *Fluoride* 21, 113–120, 1988.)

Site and degree of exposure	Mean number of frames per colony covered by adult bees, July 1984	1985	1986	Mean square cm of brood per colony, August 1984	1985	1986	Mean frames of honey in August 1984	1985	1986
REC, low	22[a]	16[a]	12[a]	5,321[a]	10,217[a]	3,483[a]	7.5[a]	2.8[a]	2.3[a]
VRF, medium	22[a]	17[a]	22[b]	8,727[ab]	10,210[a]	5,572[ab]	10.6[a]	1.5[a]	2.2[a]
FH, high	11[a]	18[a]	24[b]	7,359[ab]	10,681[a]	6,095[a]	9.2[a]	5.9[a]	7.3[a]

[a,b]Means within a column followed by the same letter are not significantly different ($P \le 0.5$, Duncan's multiple range test). REC, 10 km south-south-east of smelter; VRF, 6.4 km east of smelter; FH, 0.8 km downwind of smelter.

30 mg/kg, but between 30 and 50 mg/kg the mortality was more than 30%. Larvae fed leaves collected from the field and containing 20–30 mg F/kg weighed 15% less than controls after 6 days. At 50–60 mg F/kg, they weighed 60% less (Fig. 3.5). The weights of cocoons were also reduced in relation to fluoride exposure. On the basis of toxicity to feeding larvae and developing cocoons, the authors recommended that the atmospheric fluoride exposure should not exceed 1.2 μg/dm²/day. This refers to the rate of deposition on NaOH-treated filter papers. It is impossible to convert this into an accurate atmospheric concentration without on-site calibration but, assuming that their system

was comparable to that described in Chapter 6 (Davison and Blakemore, 1980), it implies a threshold of around 10 μg HF/m³.

Finally, the work of Aftab Ahamed and Chandrakala (1999) demonstrates the effects of sodium fluoride on food conversion efficiencies (Table 3.11). Clearly, silkworms were sensitive to fluoride in the diet but, unfortunately, there have been no comparative physiological investigations to determine why they are so much more sensitive than other larvae such as pine sawfly, large white butterfly (Davies et al., 1992; Port et al., 1998), Mexican bean beetle (Weinstein et al., 1973) and cabbage looper (Hughes et al., 1985).

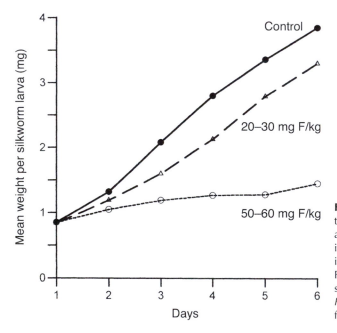

Fig. 3.5. The relationship between the weight gain of silkworm larvae and the concentration of fluoride in mulberry leaves. (Drawn from data in Wang, J.X. and Bian, Y.M. (1988) Fluoride effects on the mulberry-silkworm system. *Environmental Pollution* 51, 11–18, with permission from Elsevier.)

Table 3.11. Effect of sodium fluoride on wet food-conversion efficiencies (%) in silkworm (*Bombyx morae*) (after Aftab Ahamed and Chandrakala, 1999). (Reprinted from *Insect Science and Its Applications* 19, 193–198, 1999 with permission.)

Parameter	NaF concentration (mg/l)			
	Control	25	50	75
Wet food consumed/larva (5th instar)	11.0 ± 0.05^a	12.5 ± 0.16^b	12.1 ± 0.14^c	11.2 ± 0.05^d
Wet food to larval biomass conversion efficiency	26.3 ± 0.62^a	20.8 ± 0.94^b	16.5 ± 1.4^c	16.6 ± 1.6^c
Wet food to cocoon conversion efficiency	12.7 ± 0.8^a	6.9 ± 0.21^b	5.8 ± 0.06^c	4.9 ± 0.08^c
Wet food to pupa conversion efficiency	9.9 ± 0.78^a	5.3 ± 0.20^b	4.5 ± 0.03^b	4.1 ± 0.08^c
Wet food to silk conversion efficiency	3.2 ± 0.09^a	1.7 ± 0.08^b	1.3 ± 0.07^c	0.84 ± 0.04^d

[a,b,c,d]Means in the same row followed by different letters were significantly different.

Apart from bees and silkworms, there have been three or four other experimental studies of insects. Each demonstrates a feature of insect–fluoride relationships or raises unanswered questions. One, using Mexican bean beetle (*Epilachna varivestis*), is the only experiment in which the effects were studied over several (five) generations (Weinstein *et al.*, 1973). Beetles were cultured on untreated and fluoride-fumigated beans exposed to 7–10 µg F/m^3 (as HF) until the foliage contained about 1000 mg/kg dry weight. This is a very high concentration so the experiment was a severe 'worst possible case' test. Progeny of the second and fifth generations of beetles maintained on HF-fumigated plants were also cultured on control plants. After 10–14 days, larvae cultured on HF-fumigated beans had much lower wet or dry masses than those cultured on control beans in the five successive generations. The mean masses were 58, 49, 86, 47 and 50% of the control in each successive generation. Adult males were affected to a lesser degree and less frequently than females, but the reason for this is not known. When larvae that had been cultured on HF-fumigated plants for two generations were transferred to control plants, neither the larvae, the pupae nor the adults were significantly different in wet or dry mass from beetles cultured continuously on control plants, so there was no carry-over effect. Surprisingly, the beetles of larvae transferred to control plants after five generations of continuous culture on HF-fumigated plants were heavier than controls in both wet and dry mass at the end of the generation; again, there is no known mechanism for this effect.

The decreased mass of larvae from HF-treated plants was due to a delay in development as well as reduced growth. Egg masses laid and hatched at the same time, resulted in pupation and emergence 3–6 days later when they were fed on HF-fumigated plants, and the adults commenced reproductive activity with the same lag time. Beetles from HF-fumigated plants contained more fluoride than the controls, with females containing more fluoride than males in both treated and untreated individuals. This was an interesting finding because a comparative study may give useful information about the factors that control absorption from the digestive system. Between 8 and 20% of the fluoride was found in the exuvium. Eggs from beetles raised on HF-fumigated plants also contained more fluoride than control eggs. The fluoride content of beetles from the treated plants, however, was not commensurate with that found in the fumigated foliage upon which they fed. Generally, a 50- to 100-fold increase in fluoride in foliage was associated with a 7- to 10-fold increase in the concentration of fluoride in the adult beetle.

In addition to the delay in the onset of oviposition, there was also an obvious decrease in fecundity, displayed as fewer egg masses and fewer eggs per mass in the treated than in the control insects. When beetles that had been cultured for two generations were transferred to control plants for a life cycle, there was no significant difference in the number of eggs per mass between them and controls. However, when the transfer occurred after five generations, there was a significant reduction in eggs per mass compared with controls. The feeding activity of beetles cultured on HF-fumigated plants (measured as the amount of leaf tissue processed in a given time) was reduced in larval and adult stages. Adults cultured on treated plants had reduced flight activity and their elytra were a lighter orange than controls. The bean beetle appears to be a very useful model and the experiment raised many questions, but, unfortunately, there has not been any follow-up work.

The cabbage looper (*Trichoplusia ni*) was studied by Hughes *et al.* (1985) and it also produced some puzzling results that demonstrate the complexities of insect–fluoride relationships. Looper larvae were cultured by two methods: (i) on a defined medium containing either soluble fluoride salts or HF-fumigated cabbage leaves; or (ii) on intact plants fumigated with HF. When the larvae were raised on an artificial diet with sodium fluoride added, the results were as expected: a reduction in growth and survival. However, addition of cabbage leaves reduced toxicity (Table 3.12) and fluoride added as HF-fumigated cabbage

Table 3.12. The effects of fluoride (mg/kg) supplied in non-cabbage and cabbage diets on the mean weight of pupae (mg) of cabbage looper (*Trichoplusia ni*) (data from Hughes *et al.*, 1985). (Reprinted from *Environmental Pollution* 37, 1175–1192, 1985, with permission from Elsevier.)

Treatment	F content of diet	Non-cabbage diets		Cabbage diets	
		Male	Female	Male	Female
Control	10	232.7	215.5	238.9	217.5
NaF	60	219.9	208.5	236.8	216.5
NaF	210	186.8	178.5	218.8	212.5

had no effect, even when the concentration was as high as 438 mg F/kg. Both treatments reduced the concentration of fluoride retained by the insects so the difference was probably due to fluoride complexing with components of the leaf tissue and reducing availability. This is supported by Weismann and Svartarakova (1974, 1975, 1976), who found reduced toxicity of fluoride fed to insects in combination with calcium. There was a difference between the sexes, and fluoride added as HF-fumigated cabbage even enhanced growth and development in some combinations, but the reasons for these effects are not known.

Working with pine sawfly (*Diprion pini*) larvae taken from the field near an aluminium smelter, Davies *et al.* (1992) found that, although the fluoride content of the diet (pine needles) was as high as 170 mg/kg and whole larvae initially contained up to 219 mg/kg, pupae contained no detectable fluoride. There was no relationship between the fluoride content of the diet and pupal weight. A mass balance showed that 46% of the fluoride in the larvae was in transit in the digestive system and 23% was shed with the exuvium and was therefore assumed to be surface deposit. The lepidopteran larvae of *Pieris brassicae* showed a similar low rate of fluoride retention and there was no effect of up to 500 mg F/kg (Port *et al.*, 1998). These experiments and the work on cabbage looper highlight the difference in response compared with silkworms. Although it is clear that bioavailability plays an important part in these differences, a comparative study of the fluoride physiology of insects is sorely needed.

Aquatic invertebrates

Exposing aquatic invertebrates to controlled concentrations of fluoride is technically much easier than with terrestrial organisms, so there have been many studies reporting the effects of acute exposure. There are fewer long-term experiments and only one or two field studies. One of the greatest weaknesses is the lack of investigation of the bioavailability of different chemical species of fluoride in water and the effects of complexes involving aluminium and other cations. Most investigations used water with a single level of 'hardness' and few have systematically examined the factors that affect the toxicity of water that contains varying levels of calcium, magnesium, aluminium and organic matter. The literature was reviewed by Environment Canada (2001), WHO (2002a) and Camargo (2003).

The difficulty with field studies is getting truly comparable controls, but situations where there are gradients or seasonal changes have proved to be informative. Pankhurst *et al.* (1980) examined the diversity of marine organisms around the outfall from a fertilizer plant in Otago harbour, New Zealand. The rapid increase in diversity over a distance of about 400 m from the outfall appeared to be partly due to the elevated fluoride (about 20–200 mg/l) but some other factors were also involved because diversity remained relatively low even where the fluoride had been diluted to background levels. Samal *et al.* (1990) followed seasonal changes in a pond in which the fluoride was high in the dry season (mean 31 mg F/l) and relatively low for the rest of the year

(4.2 mg F/l). The abundance of aquatic insects was inversely related to the fluoride content, although it is very difficult to get adequate controls in this kind of situation and to rule out other factors related to season.

A few studies have used very low concentrations and they present some evidence that fluoride might be essential to some organisms and that it might stimulate growth under these circumstances. Dave (1984) worked with the water flea, *Daphnia magna*, an organism that is very convenient for toxicity testing. The 'no observable effect' level for growth and parthenogenetic reproduction was between 3.7 and 7.4 mg F/l, but, at the extremely low concentrations of < 0.007 mg/l, growth was stimulated. He suggested that the essential level was 0.004 mg F/l, which is lower than is found anywhere in the natural environment. Fieser *et al.* (1986) found a stimulation of egg production at much higher concentrations but there was a reduction in the production of live neonates.

Acute toxicity tests show that most aquatic invertebrates are only affected by concentrations that occur solely in industrial effluents. Typically they have LC_{50}s of tens to hundreds of mg F/l (WHO, 2002a; Camargo, 2003) but in tests some species respond to much lower concentrations, typically in the 1–5 mg/l range. Camargo (1996) has raised a question about the methodology for estimating the threshold for effects. In studies on caddis-flies, the LC_{50} decreased with exposure time, so an LC_{50} for a single, short time period may be misleading. For example, in the case of the most sensitive species, *Hydropsyche bronta*, the LC_{50}s for 48, 96 and 144 h were 52.6, 17.0 and 11.5 mg F/l, respectively (Camargo *et al.*, 1992). As a result, Camargo (1996) argued that the traditional short-term toxicity bioassay is inappropriate because it is based upon a single 96 h LC_{50}, which represents only one point in time and does not take into consideration the sequence of the concentration–response relationship through time that is usually a part of acute toxicity testing. Lee *et al.* (1995) recommended that multifactor probit analysis be used for fish because it solves the concentration–time–response equation simultaneously, using an iterative reweighed least-squares method. In this method the dependent variable is the probit of the proportion of organisms responding to each test concentration, and the independent variables are exposure time and toxicant concentration.

Finally, in a review, Camargo (2003) suggested that in soft waters with low ionic concentrations the safe level should be below 0.5 mg F/l. Apart from the fact that effects at such levels have not been confirmed by sufficient authors and that a very limited range of species and genotypes has been investigated, there has been insufficient research into the effects of cations like aluminium. Nor have there been any long-term field studies at the community level. It is dangerous to extrapolate from simple, acute laboratory tests to recommending a limit that is unattainable.

Note

[1] LC_{50} = concentration that kills approximately 50% of test organisms.

4

Effects of Inorganic Fluorides on Plants and Other Organisms

Introduction

In this chapter we first consider the physiological basis for fluoride toxicity and then discuss the effects on enzyme activities, photosynthesis and respiration, visible symptoms, growth, seed production and, finally, interactions with the biotic and abiotic environment. One of the problems that we have encountered, and it should be borne in mind throughout this chapter, is the age of much of the research. The peak of fluoride research was over 20 years ago, long before techniques like genomics and proteomics were available, and even before there were portable instruments that allow easy measurement of photosynthesis and stomatal conductance. The lack of information, particularly about stomatal conductance, makes it virtually impossible to interpret or compare many experiments in which light, temperature, humidity or mineral nutrition were altered. Many of the conflicting results could probably be explained in terms of differences in fluoride uptake if data on conductance were available. Another limitation is the fumigation systems that were used more than 30 years ago. Up to 1973, plants were fumigated in controlled-environment chambers and greenhouses. The former did not usually have adequate lighting compared with natural daylight and in both systems the rate of air exchange and movement was inadequate. The very low rates of air movement

led to such high canopy resistance that it controlled and limited pollutant uptake (Ashenden and Mansfield, 1977; Unsworth, 1982). In 1973, open-top chambers were introduced in an attempt to produce a fumigation environment that was as near to field conditions as possible (Heagle *et al.*, 1973; Mandl *et al.*, 1973). They were a vast improvement but, unfortunately, there has been only a handful of publications of research using them to investigate the effects of fluorides.

The Basis of Fluoride Toxicity

Concentrating mechanisms and interaction with cations

The toxicity of inorganic fluorides to plants, lichens, fungi and other organisms is due to mechanisms that concentrate the element either in leaves or within cells, and to its reactions with calcium, magnesium and other cations.

There are two mechanisms that result in the concentration and accumulation of fluoride in tissues and cells, transpiration and low pH (Chapter 2). The transpiration stream of vascular plants causes huge concentration gradients in leaves, so that the few mm near the tip or margins may have several hundred times more fluoride than the rest of the leaf. The result is areas with localized accumulation to toxic concentrations adjacent to areas that function normally. Mosses, lichens and fungi do not have this mechanism but, if they are in an acidic medium, the low pH results in the formation of HF, which will then promote uptake and accumulation in the cells by the mechanism suggested by Kronberger (1988a,b). Fungi living on leaf surfaces and mosses that are on poorly buffered substrates are exposed to low pH values during and immediately after rain, so they may be subjected to bursts of fluoride accumulation. There is surprisingly little published about the pH of the interhyphal space of lichen thalli but, if the pH is acidic, it would contribute to their sensitivity to fluoride.

Many of the visible symptoms of fluoride injury and the effects on metabolic processes can be explained in terms of interactions with calcium and magnesium and, to a lesser extent, with other cations (Weinstein and Alscher-Herman, 1982). Calcium is strongly compartmented in cells, with the highest concentrations in the walls, the outer surfaces of plasma membranes and the vacuoles. The concentration is also relatively high in the chloroplasts but, in contrast, it is maintained at a very low level in the cytosol. In some species, and especially when calcium is deficient or in storage tissues, as much as 50% of it may be in the cell walls (Marschner, 1995). Therefore, a major part of the plant calcium pool is exposed to fluoride as it is carried in the transpiration stream.

The very low concentration of calcium in the cytosol avoids precipitation of inorganic phosphate and competition with magnesium, but it is also vital to its role as a second messenger. Environmental signals (light, pathogens, abscisic acid, etc.) activate calcium channels in membranes and increase Ca^{2+} flux into the cytosol, which affects calcium-modulated proteins, such as calcium-dependent protein kinases, that phosphorylate other enzymes. In contrast, the high concentrations in the cell walls play an essential part in strengthening the tissue, providing rigidity by reversibly cross-linking the pectic chains of the middle lamella. Reversibility is important because

loosening is required during cell extension and it involves replacement of calcium links. One of the characteristic symptoms of calcium deficiency is disintegration of cell walls and collapse of the tissues. Conversely, spraying with calcium salts or postharvest dipping can increase the firmness of fruit, delay ripening and decrease fruit-storage disorders (Table 4.1). Clearly, the implication is that fluoride might complex with calcium in the middle lamellae, alter cross-linking and make the walls weak and prone to collapse. The calcium associated with the plasma membranes helps maintain stability by bridging phosphate and carboxylate groups, so fluoride-induced disruption of membrane calcium would be expected to produce a cascade of metabolic effects. Calcium also plays a vital role in pollen-tube growth because the latter depends on the presence of Ca^{2+} and the direction of growth is controlled by a calcium gradient in the stigma and style (Mascarenhas and Machlis, 1964). In fact, some of the best evidence of the importance of calcium–fluoride interactions comes from studies of the effects on fertilization and seed production (discussed later in this chapter).

Weinstein and Alscher-Herman (1982) reviewed fluoride–calcium relationships and cited a number of significant studies, including Pack (1966), Ramagopal et al. (1969), Bligny et al. (1973b), Garrec et al.

Table 4.1. The effects of a calcium spray on the wastage of Cox apple during storage. Adding calcium greatly reduced a number of storage disorders, illustrating the importance of the element in determining the structural integrity of cell walls. (From Sharpless and Johnson, 1977.).

	Calcium content (mg/100 g fresh wt)	
	Unsprayed 3.35	Sprayed 3.9
Storage disorders	Wastage (%)	
Lenticel blotch pit	10.4	0
Senescence breakdown	10.9	0
Internal bitter pit	30	3.4
Gleosporium rots	9.1	1.7

(1974) and Garrec and Vavasseur (1978). The first two studies showed that low calcium increased HF injury. Conversely, the application of lime to protect plants from fluoride injury has been a practical control method for many years (e.g. Benson, 1959; Brewer *et al.*, 1969). It was generally assumed that the application of lime promoted a reaction with fluoride on the plant surfaces, forming relatively insoluble CaF_2 (Allmendinger *et al.*, 1950), but, because $CaCl_2$ sprays were effective, Benson (1959) stated that calcium was absorbed, 'thus increasing the tolerance of the tissue for fluoride', and it is now known that the protective effect of calcium is due to a Ca–F interaction on and within plant tissues (Lévy and Strauss, 1973; Garrec *et al.*, 1974, 1978). Bligny *et al.* (1973b) used ^{18}F as a tracer to examine the effect of calcium on the transport of fluorides in cut maize leaves. They showed that the amount of fluoride accumulated at the tip was inversely related to the amount of calcium present in the leaves, which suggests that as fluoride anions are carried in the transpiration stream they react with calcium in the cell walls and outer plasma membranes. Thus, calcium may increase the retention of fluoride and reduce its rate of transport in the transpiration stream. This idea was supported by the work of Garrec *et al.* (1974), who found that fluoride disrupted the normal calcium gradient that is found in silver-fir needles.

Magnesium is also present in the cell walls bound to pectin, but from 5 to 50% is in the chloroplasts (Marschner, 1995). It is also present in the vacuoles and cytosol. The functions of magnesium are related to its capacity to react with ligands, such as phosphoryl groups, to act as a bridging element and to form complexes. It helps regulate pH and cation–anion balance and it is required by many enzymes. Most importantly, it is the central atom in the porphyrin 'head' of chlorophyll molecules and it establishes the precise geometry of certain enzymes, such as the key enzyme of photosynthesis, ribulose bisphosphate carboxylase. The cellular location of magnesium means that interactions with fluoride are more likely under circumstances where the pollutant accumulates in the cytosol or chloroplasts.

Effects of fluoride on enzyme activities

Over the last 50 years there have been many investigations of the effects of fluoride on enzyme activities, the sizes of metabolic pools and rates of photosynthesis and respiration. Tables 4.2–4.4 summarize the results of representative studies to give an indication of the range of species, conditions and results. Some studies used fluoride salts and detached tissue slices or discs. While they have been valuable in determining the relative sensitivities of specific metabolic reactions in different species, they do little to advance our understanding of fluoride as an atmospheric pollutant. Other studies exposed whole plants to HF for varying periods, but often the fluoride concentrations were so high that tissues may have been in a state of advanced morbidity, where every aspect of metabolism was affected. Many different species have been investigated under a range of conditions but, on the other hand, there have been few investigations of effects on cell walls or plasma membranes – the two organelles that are probably the first sites of attack by airborne fluoride. The published data give isolated snapshots of effects that might be caused directly or indirectly, and they are often contradictory. The result is a very confusing collection of data that cannot be used to build a step-by-step picture of the progression of events when a plant is exposed to an excess of fluoride at environmentally realistic concentrations. At present, the best that can be done is to draw out a few general principles.

Table 4.2 shows some of the effects of fluoride on enzyme activities and other metabolic functions. If fluoride reaches toxic concentrations in a plant tissue or organelle, it may be expected that enzymes that are activated by divalent cations would be inhibited, so there have been many studies of enzymes such as enolase and phosphoglucomutase. Fluoride inhibition of enolase

Table 4.2.   Some effects of fluoride (as HF or F⁻) on enzyme activities and other metabolic functions. (Based on Horsman, D.C. and Wellburn, A.R. (1976) Guide to the metabolic and biochemical effects of air pollutants on higher plants. In: Mansfield, T.A. (ed.) *Effects of Air Pollutants on Plants*, Cambridge University Press, Cambridge, pp. 185–199.)

Genus	Experimental condition	Enzymic/metabolic function	Effect	Authors cited by Horsman and Wellburn (1976)
Glycine, Phaseolus	13 mM NaF	Chlorophyll synthesis	Inhibited	McNulty and Newman, 1961
Zea	0.5–5 mM NaF	Ribosomal activity	Reduced in number	Chang, 1970
Lycopersicon, Phaseolus	1.3 μg/m^3 for 8 days	Free sugars	Increased	Weinstein, 1961
Phaseolus	1.7–7.6 μg/m^3 for 10 days	Keto-acid levels	Decreased	McCune *et al.*, 1964
Phaseolus	35 mM KF for 0.5–2.5 min	Hill-reaction activity	Decreased	Ballantyne, 1972
Avena	5 mM NaF for 2 h	Cellulose synthesis	Inhibited	Gordon and Ordin, 1972
Chenopodium, Polygonum	5 μg/m^3 for 5–6 days	Pentose phosphate pathway	Stimulated	Ross *et al.*, 1962
Glycine	25 μg/m^3 for 3–5 days	UDP-glucose-fructose transglycosylase	Inhibited	Yang and Miller, 1963a
Glycine	25 μg/m^3 for 3–5 days	Phosphoenolpyruvate carboxylase	Stimulated	Yang and Miller, 1963b
Glycine	25 μg/m^3 for 3–5 days	Phosphoglucomutase	Inhibited	Yang and Miller, 1963a
Avena	10 mM NaF for 1 h	Phosphoglucomutase	Inhibited	Ordin and Altmann, 1965
Pisum	0.1–10 mM NaF	Enolase	Inhibited	Miller, 1958
Phaseolus	1.7–2.6 μg/m^3 for 10 days	Enolase	Stimulated	McCune *et al.*, 1964
Glycine	8.2 μg/m^3 for 24–144 h	Enolase	Stimulated	Lee *et al.*, 1966
Sorghum	5 μg/m^3 for 11 days	Enolase	Stimulated	McCune *et al.*, 1964
Sorghum	5 μg/m^3 for 11 days	Catalase	Initially stimulated, then inhibited	McCune *et al.*, 1964
Phaseolus	1.7–2.6 μg/m^3 for 10 days	Catalase	Stimulated	McCune *et al.*, 1964
Glycine	51–96 μg/m^3 for 24–144 h	Catalase	Stimulated	Lee *et al.*, 1966
Sorghum	5 μg/m^3 for 10 days	Pyruvate kinase	Stimulated	McCune *et al.*, 1964
Pisum	0.5–5 mM NaF	Glucose-6-phosphate dehydrogenase	Stimulated	Arrigoni and Marré, 1955
Glycine	51–96 μg/m^3 for 24–144 h	Glucose-6-phosphate dehydrogenase, cytochrome oxidase, peroxidase	Stimulated	Lee *et al.*, 1966
Glycine	51–96 μg/m^3 for 24–144 h	Polyphenol oxidase	Inhibited	Lee *et al.*, 1966
Gycine	51–96 μg/m^3 for 24–144 h	Ascorbate oxidase	Initially stimulated, then inhibited	Lee *et al.*, 1966
Solanum pseudo-capsicum	2 μg/m^3 for 24 h	Peroxidase, glucose-6-phosphate dehydrogenase, acid phosphatase, phosphoglucomutase	Stimulated	Weinstein, 1977
Pinus strobus	1.6 μg/m^3 for 28 days	Plasma-membrane ATPase	Drastically reduced	Rakowski and Zwiazek, 1992

UDP, uridine diphosphate; ATPase, adenosine triphosphatase.

Table 4.3. Some reported effects of fluoride on net photosynthesis (NP) of higher plants (measured as changes in CO$_2$ assimilation). (After Amundson, R.G. and Weinstein, L.H. 1980 Effects of airborne F on forest ecosystems. In *Proceedings. Symposium on Effects of Air Pollutants on Mediterranean and Temperate Forest Ecosystems*. Pacific Southwest Range and Forest Experiment Station, Berkeley, California).

Genus or species	Concentration and duration	Response	Authors
Exposure to fluoride salts			
Cornus florida, Liquidambar styraciflua, Platanus occidentalis, Liriodendron tulipifera, Acer rubrum, Oxydendrum arboreum, Pinus strobus, Pinus taeda, Pinus echinata	0.1 mM NaF, 24 h	NP ↓ in older needles of *P. taeda* and *P. echinata*	McLaughlin and Barnes, 1975
	1 mM NaF, 24 h	NP ↓ in all species	
	10 mM NaF, 24 h	NP ↓ in all species	
Glycine max	10 mM NaF, 10 min	Photophosphorylation ↓	Giannini *et al.*, 1985
Azalea cvs	20 mM KF, 22 h	NP ↓	Ballantyne, 1972
Exposure to HF (μg/m³)			
Gladiolus	0.8–8.0, 7 days	% ↓ in NP = % ↑ in injury	Thomas and Hendricks, 1956
Hordeum	32, 4–8 h		
Medicago	200, 4–8 h	Total inhibition of NP with recovery in hours or days	
Fruit-trees	16–40, 4–8 h		
Lycopersicon	1.4–5.2, 4 weeks	No effect	Hill *et al.*, 1958, 1959
	0.9–11.2, 3 weeks	No effect	
Fruit-trees	2.1, 183 h	14% ↓ in NP, 10% injury	Thomas, 1958
Gladiolus	3.1–5.2, 30–205 days	% reduction in NP proportional to % injury	
Gossypium	13.6, 138 h	No effect	
Citrus	0.32–0.77, growing season	No effect	Thompson *et al.*, 1967
Gladiolus	0.8, 39 days	No effect	Hill, 1969
	1.2, 27 days	3% ↓ in NP greater than % injury	
Fragaria	2.3, 63 days	No effect	
	38, 1 day	50% ↓ in NP	
Lycopersicon	5.1, 12 days; 17, 21 days	No effect	
Prunus	1.6, 42 days	No effect	
Zea	2.7, 38 days	No effect	
Hordeum, Medicago	32, 2 h	NP ↓ during exposure, recovered afterwards	Bennett and Hill, 1973
Sorghum	0.7, 14 days	No effect	McCune *et al.*, 1976
	2.2 then 1.7, 12/2 days	NP ↓ with recovery later	
	3.5 then 5+, 7/7 days	NP ↓ during 3.5 exposure, then severe injury, little recovery	
Pinus sylvestris, Pinus nigra, Pinus strobus, Larix leptolepis, Larix decidua, Pseudotsuga menziesii, Picea excelsa, Quercus borealis, Alnus incana, Sorbus aria, Acer pseudoplatanus	Ambient near source, November–April	↓ NP of whole plant due to loss of foliage with injury on remaining foliage	Keller, 1977

Table 4.3. *Continued.*

Genus or species	Concentration and duration	Response	Authors
Vicia faba	41, 7 days	NP ↓ in 24 h, followed by partial recovery	Horvath *et al.*, 1978
Pinus sylvestris (susceptible strain)	20, 2 days	NP ↓	Lorenc-Plucinska and Oleksyn, 1982
Pinus elliottii var. *elliottii* *Pinus caribaea* var. *caribaea*	1.2–4, 125 days	NP not greatly affected, NP ↑ in older needles at 1.2 μg/m³	Doley, 1988
Pinus strobus	0.4–1.6, 1–28 days	Little effect on NP	Rakowski and Zwiazek, 1992
Mangifera indica	12–24, time unknown	↓	Zhang and Huang, 1998

Table 4.4. Effects of fluoride on dark respiration (R) in plants. (After Horsman, D.C. and Wellburn, A.R. (1976) Guide to the metabolic and biochemical effects of air pollutants on higher plants. In: Mansfield, T.A. (ed.) *Effects of Air Pollutants on Plants*, Cambridge University Press, Cambridge, pp. 185–199.)

Species	Concentration (μg/m³) and time	Response	Authors
Phaseolus vulgaris	24.6, 18 days	R ↑ by *c.* 70%	McNulty and Newman, 1957
Gladiolus sp., *Lycopersicon esculentum*, *Fragaria* sp., *Apium graveolens*, *Prunus* sp., *Hordeum vulgare*	0.31–45.9, 21–73 days	R ↑ in *Gladiolus*, no effect on other species	Hill *et al.*, 1959
Phaseolus vulgaris cv. Burpee	2.2, 24 h and growing season	R ↑ after 24 h, ↓ after 8 days, then ↑	Applegate and Adams, 1960a
Phaseolus vulgaris cv. 'Tendergreen', *Lycopersicon esculentum* cv. 'Bonny Best'	1.3, 8 days	R ↑	Weinstein, 1961
Chenopodium murale	4.9, 5–6 days	R ↑ + visible injury	Ross *et al.*, 1962
Pinus strobus, *Pinus taeda*, *Pinus echinata*, *Cornus florida*, *Liquidambar styraciflua*, *Platanus occidentalis*, *Acer rubrum*, *Liriodendron tulipifera*, *Oxydendrum arboreum*	0.1, 1, 10 mM NaF, 24 h	R ↑	McLaughlin and Barnes, 1975
Vicia faba cv. 'Inovec'	41, 24 h–7 days	R ↓ after 1 h, then recovered. R ↑ 24 h after exposure	Horvath *et al.*, 1978

(2-phosphoglyceric acid > phosphoenol-pyruvic acid) is perhaps the best known of the effects of fluoride *in vitro* and was first studied in yeast (Warburg and Christian, 1942). The fluoride concentration needed for 50% reduction in enzyme activity was found to increase with increasing Mg^{2+} concentration. Mg^{2+} was required to activate the enzyme and, in the presence of fluoride and phosphate, an inactive dissociable magnesium fluorophosphate complex was postulated to be formed and to be inhibitory to the enzyme, although fluorophosphate itself was shown not to be inhibitory (Peters *et al.*, 1964), and it may be necessary for the complex to be formed on the enzyme to have any effect. In plants, the sensitivity of enolase to fluoride was described by Miller (1958); the pea-seed enzyme also had an Mg^{2+} requirement, while Mn^{2+}, Co^{2+}, and Zn^{2+} were less effective. Later studies using intact, fumigated plants generally showed stimulation (McCune *et al.*, 1964; Lee *et al.*, 1966). This apparent contradiction may be explained in a number of ways. Apparent stimulation may be an artefact caused by the extraction or assay procedures, but it may simply be that the fluoride did not reach toxic concentrations in the mitochondria or that there was increased respiration (Table 4.4) due to an extra demand for maintenance. There may also be several isozymes that differ in sensitivity. Multiple forms of enolase have been found in muscle (Tsuyuki and Wold, 1964), but this has not been confirmed for plants, nor has the relative sensitivity of the isozymes to fluoride been tested. An important observation made by McCune *et al.* (1964), after exposure of bean plants to relatively low concentrations of HF (1.7 and 7.6 $\mu g/m^3$), was that, although fluoride increased the activities of enolase and catalase, they tended to approach the control levels after a recovery period. Unfortunately, this is one of the few studies that examined recovery.

The mechanism of fluoride inhibition of phosphoglucomutase (glucose-1-phosphate $\rightleftarrows$ glucose-6-phosphate) appears to be similar to that for enolase (Chung and Nickerson, 1954). A magnesium–fluoride complex is formed with glucose-1-phosphate and the enzyme. Like enolase, inhibition depends upon the Mg^{2+} concentration, but, in the one study in which it was measured in plants that had been exposed to HF, phosphoglucomutase was inhibited. The HF exposure, however, was extremely high, at 25 $\mu g/m^3$, for 3–5 days.

Fluoride also combines readily with many haem enzymes and there has been some research on catalase (McCune *et al.*, 1964; Lee *et al.*, 1966). In the case of cytochrome oxidase, catalase and peroxidase, the reaction is with the enzyme in the ferric iron state, which is freely reversible and involves one atom of iron and one molecule of fluoride (Hewitt and Nicholas, 1963). With many other metalloenzymes the reaction with fluoride is weak. Thus, the zinc-containing enzyme, carbonic anhydrase, and the molybdenum-containing enzymes, nitrate reductase and other molybdo-flavoproteins (which also contain iron), are not very sensitive to fluoride. From *in vitro* studies, it is apparent that multiple sites of fluoride action exist, so the distribution and concentrations of fluoride within

the cells determine the degree to which metabolism is affected.

Photosynthesis

The experiments summarized in Table 4.3 give an indication of threshold concentrations above which photosynthesis is inhibited, the persistence of inhibitory effects and, in one case, the effects on carbohydrate translocation. A few indicate the relationship between effects on photosynthesis and visible symptoms. The combinations of concentration and duration of exposure that the authors used are plotted in Fig. 4.1. Some concentrations were so high that it is not surprising that net photosynthesis was affected, but the interesting feature of Fig. 4.1 is that, below concentrations of about 2–3 $\mu g/m^3$, given for as much as 5–100 days, most studies reported no effect. This may be just due to chance but it may also suggest that photosynthesis is not particularly sensitive to HF. This could be due to the sharp gradients in fluoride concentrations within leaves or possibly chloroplasts are not such a strong sink for fluoride as models suggest.

There is some evidence that effects on photosynthesis are causally linked to visible symptoms (Thomas and Hendricks, 1956; Hill and Pack, 1983; Doley, 1988). Thomas and Hendricks (1956) found that the reduction in photosynthesis in *Gladiolus* was proportional to the injury and Doley (1988) found that differences in photosynthetic rates of two pine species (*Pinus elliottii*, *Pinus caribaea*) were correlated with concentrations of chlorophylls a, b and a + b. In Hill and Pack's (1983) experiments, HF treatments either had no significant effect on the rate of photosynthesis or the effect was proportional to the amount of leaf necrosis or chlorosis. Continuous fumigation of strawberry plants for 18 days at average daily fluoride concentrations ranging from 1 to 9 $\mu g/m^3$ had no effect on photosynthesis, although a small amount of necrosis was induced. But, when the HF concentration was increased to 36 $\mu g/m^3$ for 24 h, there was an abrupt drop in the photosynthetic

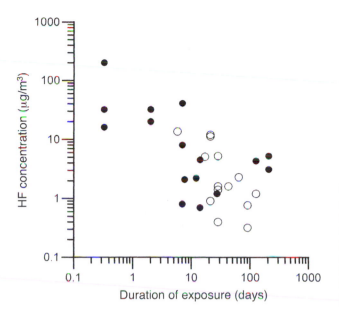

Fig. 4.1. Combinations of HF concentration and exposure time used in experiments on the effects of fluoride on net photosynthesis (sources cited in Table 4.3). ● = experiments in which net photosynthesis was reduced, ○ = experiments in which there was no effect on net photosynthesis.

rate of about 50%. As the HF concentration was decreased, the rate returned to about 75% of the previous rate within 24 h and then recovered more slowly for about 3 weeks until the rate was about 95% of the control. Considering that 4 to 5% of the leaf area was necrotic, this was considered to be complete recovery. In maize, the photosynthetic rate was reduced by 12 to 14% after exposure to 5.1 μg/m³ for 15 days. This treatment induced chlorotic mottling typical for the species and was considered to be the reason for the photosynthetic decline. Several other authors have reported that the effects of fluoride were reversed after exposure (Thomas and Hendricks, 1956; Bennett and Hill, 1973; McCune *et al.*, 1976; Horvath *et al.*, 1978), but this is difficult to understand if the inhibition was caused by accumulation of fluoride in the chloroplasts. Perhaps the inhibition was caused by greatly reduced stomatal conductance, which limited CO_2 uptake, or, more indirectly, through some feedback system.

Respiration

Mitochondrial respiration provides the energy for synthesis of new biomass,

translocation of photosynthates, ion uptake, assimilation of elements, such as nitrogen, protein turnover and maintenance of ion gradients (Amthor, 2000). It is relatively easy to measure the O_2 uptake or CO_2 emission by plant tissues in the dark but it is difficult to measure respiration in the light because of photosynthesis and photorespiration. It is even more of a problem interpreting differences in respiration between treatments, because so many factors regulate the process. However, the majority of published studies have reported increased respiration on treatment with fluoride. This was the case for intact plants (Applegate and Adams, 1960a; Applegate *et al.*, 1960) and in tissues removed from plants after fumigation with HF in the absence (Applegate and Adams, 1960a; Weinstein 1961; Miller and Miller, 1974; McLaughlin and Barnes, 1975) or presence (Hill *et al.*, 1959; Yu and Miller, 1967) of foliar lesions. Several investigators have shown that low concentrations or short durations of fluoride exposure result in stimulated O_2 uptake, while the exposure becomes inhibitory as the concentration or duration is increased (McLaughlin and Barnes, 1975). In contrast, two studies showed that respiratory activity was relatively insensitive to fluoride (Hill *et al.*,

1959; Givan and Torrey, 1968) and in others it was inhibited, the effect depending upon the age of the plant or tissue (Béjaoui and Pilet, 1975), the duration of exposure (Applegate and Adams, 1960a), nutrient status (Applegate and Adams, 1960a) and concentration of fluoride in the tissue (Applegate *et al.*, 1960).

The relationship between inhibition and the age of the tissue can perhaps be explained by the shift from glycolysis to the pentose phosphate pathway, which occurs with ageing of tissues (Gibbs and Beevers, 1955), and by the greater sensitivity of some glycolytic enzymes to fluoride. Determination of $^{14}CO_2$ production from specifically labelled glucose has been used to assess the relative importance of glycolysis and its sensitivity to fluoride. The production of $^{14}CO_2$ from carbons 1 and 6 of glucose is equal if metabolized via the glycolytic pathway and Krebs cycle, but the production of $^{14}CO_2$ from the carbon 1 of glucose is greater than from carbon 6 if oxidized via the pentose phosphate pathway. Thus, the C_6/C_1 ratio has been used as an index of the relative activities of the two pathways. In wheat leaves this shift was shown by a decrease in the C_6/C_1 ratio and was accompanied by a decreased sensitivity of O_2 uptake or $^{14}CO_2$ production to fluoride (Luštinec and Pokorná, 1962). At low concentrations of fluoride Luštinec *et al.* (1962) found that there was a stimulation of wheat-leaf respiration, which was accompanied by a higher C_6/C_1 ratio, suggesting that the activity of the glycolytic pathway was greater.

A better estimate of the pathways of glucose catabolism is obtained if glucose is used that is specifically labelled on each carbon. Employing glucose labelled in the 1, 2, 3–4 or 6 positions, it was demonstrated with unifoliolate leaves of bean plants (McCune *et al.*, 1967; McCune and Weinstein, 1971) that there is a normal change in the pattern of glucose catabolism with increasing age of the leaf. In normal ageing, there is a progressive rise and fall in catabolism of glucose during the course of leaf maturation. The effects of HF were modified by the normal, progressive sequence in metabolism (McCune *et al.*,

1967), the greatest relative changes being found in $^{14}CO_2$ liberated from carbons 3 and 4 of ^{14}C-labelled glucose, with carbon 2 being the next greatest, suggesting altered glycolytic and Krebs cycle activities. As the leaf matured and aged, Krebs cycle activity decreased and was accompanied by an increase in use of the pentose phosphate pathway. After 6 days of HF exposure, there was a decrease in $^{14}CO_2$ produced from all carbons, especially carbon 2. When HF fumigation was superimposed on the normal metabolic changes, it appeared that the pre-existing metabolic pattern was affected by a limitation in the plasticity of the metabolic system and constraint on the adaptive mechanisms of the plant. It is possible that the relative contributions of each carbon of glucose and the changes produced by both age and HF reflected the pool sizes of various intermediates produced by respiratory or photosynthetic activity and the rates of exchange between active and inactive pools of metabolites, as well as turnover rates and the relative activity of various metabolic pathways. It is apparent that patterns of glucose catabolism change with age and that the effect of HF is greatest initially and decreases with increasing HF concentration, a recovery period and age (McCune *et al.*, 1967).

Effects on other metabolic functions

Despite the obvious potential for fluoride to affect the functioning of plasma membranes, there have been very few studies of this important aspect of fluoride toxicity. Rakowski and Zwiazek (1991, 1992) investigated effects on membranes in white pine (*Pinus strobus*). Seedlings were pretreated with a 12 h photoperiod to induce dormancy and then they were exposed to HF at 0.4–1.6 µg/m³ for 1–28 days. Needles that did not show visible symptoms were selected and plasma membranes were isolated. In plants treated for 2 days, ratios of plasma membrane free sterols : phospholipids, sterols : proteins and plasma membrane adenosine triphosphatase (ATPase) activity were higher than in control plants. Exposure to HF at 1.6 µg/m³ for 28 days produced a drastically reduced activity of plasma membrane ATPase activity. Changes at 1.6 µg/m³ in sterol and phospholipid levels after only 2 days of fumigation suggested that plasma membrane composition was affected even during dormancy and that these membranes may be among the early sites of HF injury (Rakowski *et al.*, 1995; Rakowski, 1997).

Effects of HF on partitioning of ¹⁴C between transport and non-transport photo-assimilates from source (treated) leaves to sink regions were studied in soybean plants (*Glycine max* cv. 'Hodgson 78') fumigated with HF at 0, 1 and 5 µg/m³ for 8–10 days at three stages of development (vegetative, flowering and early fruit set, and pod filling) (Madkour and Weinstein, 1987). In plants exposed to HF, there was a greater retention of ¹⁴C-labelled assimilates in the treated leaf and reduced export of these compounds to other plant parts (sink tissues). There was also a greater incorporation of ¹⁴C into ethanol-insoluble assimilates at the expense of ethanol-soluble assimilates in the treated leaf. Compared with control plants, HF also increased the proportion of ¹⁴C incorporated into sugars in treated (source) tissues and a decrease in ¹⁴C sugars in sink tissues. Results suggest that fluoride inhibits the processes involved in assimilation and transport of sugars, i.e. phloem loading (Madkour and Weinstein, 1987).

Effects on Different Taxa

Lichens

Lichens obtain their mineral nutrients by wet and dry deposition on the thalli. They have large numbers of ion-exchange sites, which results in their capacity to accumulate some elements in high concentrations. Although absorption of deposited elements is essential for lichen survival, it is not selective, so there is no discrimination between essential elements and those that are potentially harmful. There have been several field studies around fluoride sources that demonstrate the visible symptoms and changes in species composition of lichen communities in response to fluoride (Gilbert, 1971, 1973; LeBlanc *et al.*, 1971, 1972; Nash, 1971, 1973, 1976; Skye, 1979; Holopainen, 1984; Palomäki *et al.*, 1992; AMS, 1994). Gilbert (1973) summarized the information about the relative sensitivity of lichens.

LeBlanc *et al.* (1971) described the effects of pollution in the vicinity of an aluminium smelter on transplanted discs of *Parmelia sulcata*. After 4 or 12 months of exposure at distances up to 15 km from the source, the thalli ranged from brown to pink and yellow to grey shades (the normal colour), respectively, at the nearest (100 m) to the farthest (15 km) distances. Microscopic examination of thalli near the smelter 'had their margins upturned, showed partial detachment from the underlying bark substratum, had developed cracks on their upper surfaces, lacked soredial structures, and their exposed rhizines were often curved and entangled'. The algal cells displayed varying degrees of plasmolysis and loss of chlorophyll. At a distance of 5 km, soredia were present, the incidence of plasmolysis was low, the chlorophyll content was about normal and most of the thalli had retained their normal grey colour. Although the effects close to the smelter were undoubtedly caused by fluoride, there were other pollutants in the area, notably SO₂, so they may have had some effect. The fluoride content of the thalli decreased with distance but the concentrations appear to be unusually high (e.g. *c.* 700 mg F/kg) compared with data reported by others (Gilbert, 1971; Nash, 1971). Even the control site at 40 km apparently contained as much as 70 mg F/kg, which suggests that their analytical method was unreliable.

Nash (1971) compared three lichen species, the terricolous (living on soil, etc.) species, *Cladonia cristatella* and *Cladonia polycarpoides*, and the saxicolous (living on rock) species (*Parmelia plittii*) to airborne fluoride near an industry and to HF in controlled environment laboratory fumigations.

Lichens growing near the industry or in the fumigation chambers exhibited chlorosis, followed by elimination of all pigments, including chlorophyll *a* and *b* and carotenoids. Chlorosis in *Cladonia* species first appeared in the upper part of the podetia and gradually spread downward to the squamules on the ground. Small necrotic patches also appeared occasionally on the squamules. In *P. plittii*, chlorosis appeared first as a series of intermittent green and chlorotic areas. As the injury symptoms progressed, chlorosis extended over the entire thallus and this was followed by death. The critical level of fluoride in the lichen to cause injury was estimated to be between 30 and 80 mg/kg. Roberts and Thompson (1980) reported that the minimum damage to *Cladina* sp. occurred at a fluoride concentration of 25 mg/kg, with structure loss occurring between 50 and 70 mg/kg. Based upon washed thalli of *Usnea subfloridana* and *Parmelia saxatilis*, Gilbert (1971) found the range to be 20–47 mg/kg. Davies (1982) examined *Xanthoria parietina* growing on calcareous substrates in the Bedfordshire brickworks area of England. He reported that, when the fluoride concentration reached about 70 mg/kg, the thalli exhibited small, scattered, brown patches, but, at about 160 mg/kg, larger grey-brown patches were present and the margins and tips of the thalli were curled. At 90 and above, algal plasmolysis within the thalli was noted. No visible effects were present at concentrations around 7–8 mg/kg. However, the brickworks also emitted large quantities of SO_2 and there was a degree of correlation between the fluoride and sulphur contents of the thalli, so some of the injury may have been caused by SO_2. Furthermore, *Xanthoria* grows in nitrogen-rich habitats, often as a result of ammonia emission from livestock. Ammonia increases local SO_2 deposition, so there was an unusual combination of circumstances that makes interpretation far from straightforward. Typical ultrastructural changes were found to occur when the fluoride content of thalli of *Hypogymnia physoides* and *Bryoria capillaris* reached 30 mg/kg. Visible symptoms appeared at about 70 mg/kg.

Gilbert (1973) studied the pattern of injury to *Ramalina farinacea*, *Ramalina fraxinea*, *H. physoides* and *Parmelia subaurifera* on the trunks of European beech (*Fagus sylvatica*) as a result of emissions from an aluminium smelter at Invergordon (Scotland) in 1969. After the first year, injury was observed up to 3 km from the source and it was evident up to a height of 6 m on the bole, being especially concentrated on the side facing the source (Fig. 4.2). Fruticose and foliose lichens were the first to be affected, turning a brown or whitish colour, beginning at the distal end. This was followed by the appearance of a whitish halo around the crustose species. The author also noted the great variability in response among lichen species, which he attributed to genetic differences in susceptibility, though it is impossible to determine the relative contributions of genetic and environmental variability to lichen response. Palomäki *et al.* (1992) described symptoms on *H. physoides* and *B. capillaris* growing on branches of Norway spruce (*Picea abies*) as a bleaching of the lobe tips of the thalli, followed by complete bleaching and, finally, browning and death.

Ultrastructural changes in lichens have been studied by a few investigators. They were described in the algal cells of both of the epiphytic lichens *B. capillaris* and *H. physoides* in the area near a fluoride-emitting ceramic factory (Holopainen, 1984). At the early stages of injury, there was a dilatation of thylakoid interspaces and swelling of thylakoids. Plastoglobuli appeared between the thylakoids. The injury continued with granulation and the breakdown of thylakoid membranes. Aggregates of pseudocrystalline materials were present in the chloroplast or cytoplasm. The mitochondria and cytoplasm were highly degenerated. Palomäki *et al.* (1992) stated that fluoride injury to the photobiont of *B. capillaris* was characterized by loss of structure of thylakoids and other cellular membranes, forming a granulated, white-stippled mass; lipid-like globules formed between the thylakoids in the chloroplast. In general, it appears that the algal or cyanobacterial partner is more sensitive than the fungus.

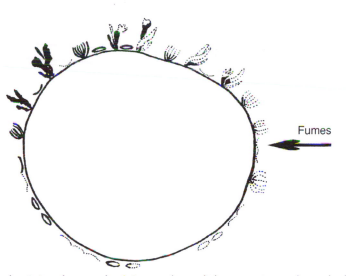

Fig. 4.2. Effects of emissions from an aluminium smelter on lichens growing on the trunk of a beech (*Fagus sylvatica*) tree that was 400 m downwind. Black = living thallus; dotted lines = visibly damaged thallus. Tufts = *Ramalina farinacea*; ribbons = *Ramalina fraxinea*; flattened ellipses = *Hypogymnea physodes*; arcs = *Parmelia subaurifera*. Crustaceous lichens not shown. (Redrawn from Gilbert (1973). Reprinted by permission of the Continuum International Publishing Group.)

Bryophytes

Bryophytes include mosses and liverworts, but the effects of fluoride have been reported only for mosses (Nash, 1973; Paterson and Kenworthy, 1982). According to Nash (1973), the sensitivity of mosses to air pollution is about the same as in lichens, but it has been the experience of the authors that in areas where lichens were damaged or killed by fluorides many species of mosses were present, mostly without visible symptoms. In transplant experiments, LeBlanc *et al.* (1971) reported that colonies of *Pylaisiella polyantha* exhibited colour changes from green or golden green to brown or dark brown after exposure to fluoride downwind of an aluminium smelter for 4 and 12 months, but the atmospheric concentrations were not measured. Comeau and LeBlanc (1972) exposed the moss *Funaria hygrometrica* to acute concentrations of 13, 65 or 130 p.p.b. (10.7, 53.3 and 106.6 μg/m³) for 4, 8 or 12 h, or 13 p.p.b. (10.7 μg/m³) for 36, 72 or 108 h. At 65 p.p.b. for 12 h, the moss exhibited chlorotic lesions. Interestingly, after 3 weeks of recovery, *Funaria* lost between 26 and 36% of the fluoride accumulated, but no mechanism was postulated.

Pteridophytes

Pteridophytes include ferns, clubmosses, quillworts, spike mosses and horsetails. Field observations by the authors suggest that the clubmosses, quillworts, spike mosses and horsetails are tolerant to HF, but there is no experimental evidence to support this. Fluoride injury, consisting of marginal necrosis of fern fronds, has been observed many times in the field, but there has been no published research on this group of plants.

Gymnosperms

The following descriptions of symptoms shown by vascular plants are based on Treshow and Pack (1970), Weinstein and McCune (1971), Weinstein *et al.* (1998) and the authors' field experience in many countries.

In conifers, such as pine (*Pinus*), fir (*Abies*), spruce (*Picea*) and larch (*Larix*), the initial symptom is often chlorosis of the needle apex and, with time or higher concentrations, the apical tissues may become necrotic. The necrotic areas are usually light brown to reddish brown and the injury progresses towards the base with continued exposure. The dead tissue is often, though not always, delineated from healthy tissue by a dark band, usually subtended by a narrow chlorotic band. Banding can be seen developing in lodgepole pine (*Pinus contorta*) in Plate 6. Multiple or intermittent exposures may produce a series of darker bands that proceed basipetally. The dark band is thought to be due to a general stress response, which results in the leaf producing phenolic compounds to isolate the cause of the stress (e.g. a disease or insect). In the case of pine, needles within the same fascicle tend to be marked to the same extent (Plate 6). When exposed to very high concentrations, injury may occur in medial portions of the needle, rather than the tip. Needles of conifers are most sensitive when they are emerging from the fascicle and elongating, and they become less and less sensitive as elongation proceeds. Needles from previous years are highly resistant and are rarely injured by a current fumigation, which is illustrated in Plate 7, where the current year's needles of mugo pine (*Pinus mugho*) are completely necrotic but the previous year's needles are mostly unaffected. Note that the adjacent blue spruce (*Picea* sp.) is also unaffected.

Often, the pattern of injury on an individual leaf or leaves or the pattern of injury during the development of a branch can have forensic value in establishing when a fumigation occurred. Because the youngest leaves are most susceptible, intermittent fumigations injure young leaves as they develop, providing markers along the branch. This may also be useful in conifers, where needles from a single shoot emerge from the fascicle at different times, so their condition can be a qualitative measure of changes in fluoride air quality.

Angiosperms

Monocotyledons

The initial symptom shown by monocotyledons (lilies, iris, grasses, etc.) is often chlorosis at the tips and margins of elongating leaves. This may develop into necrosis if the concentration of duration of the exposure is sufficiently high. Necrosis is caused by the total collapse of the plasma and/or vacuolar membranes, so the first sign of impending necrosis is the appearance of dark green areas that look 'water-soaked'. This often resembles the early stages of frost injury and it can be seen at the base of the injured area in an *Iris* leaf on the left-hand leaf in Plate 8 and at the tip of a bluebell (*Hyacinthoides non-scriptus*) in Plate 9. As the dead tissues dry out, the colour changes so the necrotic areas may eventually be white (e.g. *Allium* spp., including garlic), tan, dark brown (*Iris*, Plate 8; common in many species), reddish brown (as in redbud) and black (more common in some dicotyledons, such as poplar). The speed with which the necrosis develops depends not only on the dose but also on the weather.

The initial symptom in *Gladiolus* and several other members of the *Iridaceae* is tip necrosis, which progresses more or less evenly down the leaf at low concentrations but which is distributed more irregularly over the leaf at higher concentrations (Plate 10). The colour of the necrotic area ranges from ivory or light tan through various shades of brown, but it is characteristically light in colour with a narrow, dark brown margin. When necrosis is produced by several exposures, each increment may be separated from the previous one by these narrow bands (Plate 8). The necrotic area may also be sharply delineated from the healthy portion of the leaf blade by a narrow band of chlorotic tissue, which is sometimes streaked with red. The bracts of *Gladiolus* that subtend the flowers are also very sensitive and exhibit necrosis at the tips and along the margins.

In maize (*Zea*), *Sorghum*, sugarcane (*Saccharum*) and some other grasses, the symptoms begin as scattered chlorotic flecks at the tips and upper margins of maturing, but still elongating, leaves. This can be seen in the right-hand photograph of Plate 11, which was taken using transmitted light. As the symptoms progress, the flecking becomes more intense and extends downward, especially along the margins. The amount of chlorosis diminishes from the tip downward and from the margins towards the midrib (left-hand photograph, Plate 11). A transverse chlorotic band often appears at the arch of leaves that are bent downward. With continued exposure, tip and marginal necrosis may appear. At higher concentrations of fluoride, there is more of a tendency to produce tip, marginal and interveinal necrosis, with a transverse necrotic band appearing at the arch of the leaf. In some species of monocotyledonous plants (such as some varieties of *Sorghum*), necrotic lesions produced by high concentrations are often accompanied by reddening caused by anthocyanin pigments.

In banana (*Musa* spp. and hybrids) symptom development is similar to that in other monocotyledons, with fumigation leading to a diffuse chlorotic band at the apex, followed by the formation of tan to pale brown necrotic areas, which are often delineated by a dark brown line. Successive fumigations lead to a series of transverse bands separated by dark lines, but the necrotic areas often fall off, leaving a tattered, obtuse apex to the leaf (Plate 12). The initial symptom on the pinnae of most palm species resembles the chlorotic lesions described for grasses. In tucumá palm (*Astrocarpyum tucuma*), a very fluoride-sensitive species, injury appears as necrosis, which may be at the tip, margin or intercostal areas of the pinnae (Weinstein and Hansen, 1988).

Dicotyledons

The initial symptom of injury in leaves of dicotyledons exposed to chronic doses of fluoride is commonly chlorosis of the leaf tip, which later extends downward along the margins and inward towards the midrib. The chlorosis becomes more intense and

extensive with prolonged exposure, until the midrib and some veins stand out as a green silhouette against a chlorotic background. This gives the green part the shape of a spruce or fir tree, so it is sometimes referred to as a 'Christmas tree' symptom (Plate 13). In some species, such as *Eucalyptus*, anthocyanin pigments are produced, giving a diffuse red colour to the affected areas. With continued exposure, the leaf tip and other chlorotic areas may become necrotic. In many species, the tip falls off, leaving the leaf notched or ragged and misshapen (Plate 14). As the symptoms become more advanced, necrosis may appear on the leaf margins, which may result in a tattered, ragged leaf. The eventual colour of necrotic areas depends on the species; most frequently it is tan to dark brown but in some the tissue becomes black. Poplar (*Populus* spp.) is a good example of this and the Australian *Persoonia laevis* (broad-leaved geebung) is another (Plate 15). Note that the necrosis in this case is in the centre of the lamina and that some of the leaf tips are also chlorotic.

There are some exceptions to the sequence of symptom development outlined above. For example, in apricot (*Prunus armeniaca*), Italian prune/European plum (*Prunus domestica*) and redbud (*Cercis canadensis*), the initial symptom is wilting and then desiccation of tissue at the tip or along the margins of the leaf, which then develops into light to dark brown or reddish brown necrotic lesions. The lesions become brittle and may fall out, as in apricot and Italian prune, or may remain on the leaf, as in redbud (Plate 16).

When young, developing leaves are exposed to fluoride, concentration of the pollutant in the tip and margins may inhibit longitudinal and lateral expansion of the lamina, resulting in leaves that are crinkled and/or cupped, upwards or downwards. Often, it results in leaves that are rolled. A range of distortions can be seen in alder, *Alnus glutinosa* (Plate 17) and willow (Plate 18). Note the variation in symptom expression within each photograph. Some of the older alder leaves in Plate 17 are crinkled in such a way as to resemble a savoy cabbage ('savoying'). The younger, expanding leaves vary, some being undistorted, but one of the smallest (upper-left quadrant of the photograph) has almost half of the lamina necrotic, so it would go on to develop an abnormal shape. Several of the willow leaves in Plate 18 are cupped, making them slightly concave, and most are chlorotic and have dead tips. If the dead tissue falls off the tip, the leaf is left with a notched shape. The dependence of the effect upon the stage of development is often demonstrated graphically when a branch is exposed to a series of fumigations during development. Plate 19 shows this in mock orange (*Philadelphus* sp.). The older leaves (marked 1) have normal pointed tips, but the slightly younger ones (2) were distorted by a fumigation when they were expanding. It did not last because the next two or three leaves have normal tips (3). A second fumigation distorted the next youngest leaves, resulting once again in rounded tips (4).

Petals are rarely injured except where there is exposure to high concentrations. Thomas (1958) listed *Petunia* petals as intermediate in sensitivity, although personal experience suggests that the concentrations of fluoride generally found today are not injurious to petunia petals. Spierings (1968) reported that petals of *Cyclamen* were more sensitive than the leaves, but the reason for this is unknown. Despite the great sensitivity of *Gladiolus* leaves and bracts, Guderian *et al.* (1969) reported that sepals and petals were uninjured even when 50% or more of the leaf was necrotic. Petals of apple, cherry, apricot and peach are reportedly tolerant. Symptoms shown by peach and strawberry are described later, but some other fruits are also injured by fluoride, but at high concentrations. In maturing fruits of apricot, a brownish black injury appears at the stylar end of the fruit, often encircled with a red halo (Bolay *et al.*, 1971b). Plum and cherry fruits can be deformed and 'pinched' at the calyx end, along with tissue collapse around the seed. Pears produce deformed fruit with occasional dark collapsed tissue at the calyx end (Dässler and Grumbach, 1967; Bonte, 1982). In apples, reddening occurs at the calyx end, resulting in reduced growth and a ring-shaped deformation (Bolay *et al.*, 1971b; Bonte, 1982).

Dose–Response Relationships

An important advance in defining the relationships between visible symptoms and fluoride exposure was made in 1969 when McCune collated all the available information on the dose–response relationships for different groups of species. He constructed a series of curves that define the thresholds for foliar lesions in tomato, maize (*Zea*), *Sorghum*, conifers, *Gladiolus* and certain tree fruits (Fig. 4.3). The variability in the original data and the differences between varieties and effects of the weather mean that the curves cannot be regarded as sharp boundaries but more as indications of the doses above which there is an increasing probability of an effect. For example, the data suggest that, if a sensitive variety of *Gladiolus* is exposed to 10 µg F/m^3 for 1 day, the dose is above the threshold and there is therefore a high risk that it will develop lesions. On the other hand, if the exposure is 1 µg/m^3 for 5 days, the probability is that it will not develop lesions. The relationships have proved to be of value for establishing air quality criteria and they have formed the basis for standards in some jurisdictions. However, as will become clear later, there is a very poor relationship between sensitivity in terms of visible lesions and effects on growth or yield, so, in order to evaluate the full effects of fluoride, it is desirable to provide a parallel set of curves for other parameters – growth reduction, seed set and so on (Weinstein, 1977).

Stresses that Mimic Visible Fluoride Injury

Unfortunately, the symptoms induced in vascular plants by fluorides are not specific, and similar or identical symptoms may be caused by other abiotic and biotic stresses (Table 4.5; Brandt and Heck, 1968; Treshow and Pack, 1970; NAS, 1971; Weinstein,

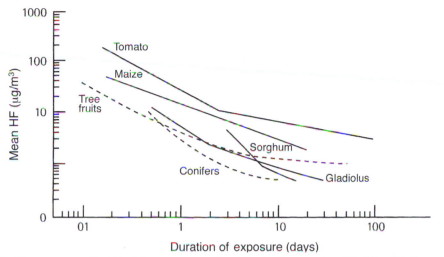

Fig. 4.3. Dose–response for foliar injury to a range of species. (Redrawn and modified from McCune, D.C. (1969) *On the Establishment of Air Quality Criteria, with Reference to the Effects of Atmospheric Fluoride on Vegetation.* Air Quality Monographs, Monograph #69-3, Courtesy of the American Petroleum Institute, New York.)

Table 4.5. Biotic and abiotic stresses that may resemble fluoride-induced injury.

Abiotic stresses	Biotic stresses
Freezing	Fungal diseases
Nutrient deficiency	Virus and mycoplasma
Nutrient excess	diseases
Salt (sodium chloride)	Insects (borers, miners,
Boron	etc.)
Heavy metals	Mites
Drought	
Anoxia	
Heat	
Pesticides	

1977). Mimicking symptoms have been described in detail in a number of reviews (Thomas, 1961; Thomas and Alther, 1966; Brandt and Heck, 1968; Guderian *et al.*, 1969; Hindawi, 1970; Treshow and Pack, 1970; NAS, 1971; Weinstein and McCune, 1971; Chang, 1975; Weinstein, 1977). The result is that diagnosis of the cause of injury, even when there is a known source of fluoride, is not always easy or even possible. It is particularly difficult in situations where there are only one or two specimens of a species growing in only one location or where there are other stresses, such as dry soil or exposure to wind.

Foremost among the agents or factors that mimic fluoride-induced injury are other air pollutants, such as SO_2, O_3, chlorine and chloride, but the degree to which they resemble fluoride injury depends on the species and the pollutant exposure. Photographs of symptoms produced by other pollutants are available in Flagler (1998). Symptoms are most similar after exposure to high concentrations; after relatively low exposures, the symptoms caused by each pollutant tend to be more specific. For example, O_3 affects the upper surface of older leaves or the oldest part of the leaf and is characterized by small lesions and patches of yellow, red, purple, brown or black. Chlorine causes symptoms that are similar. Sulphur dioxide generally affects middle-aged leaves first and chlorotic or necrotic injury is usually confined to the interveinal area of the leaf. Chloride is a special case because it affects the same portions of the leaf as fluoride and is commonly seen in colder countries along roads that are treated with salt during winter (Hall *et al.*, 1979). It also occurs where there is marine salt spray and accidental release of salt aerosols from industry. The symptoms can easily be confused with fluoride injury, as Plates 20–22 show. Plate 20 shows the younger needles of a pine (*P. contorta* × *Pinus banksiana*) with fluoride-like necrosis caused by an accidental release of salt aerosol (note that the previous year's needles are chlorotic because of chronic exposure to SO_2).

Deficiency symptoms of manganese, iron, zinc, magnesium, calcium, potassium, copper and boron may resemble fluoride injury in some species. This is especially true in *Eucalyptus*, where it can be extremely difficult to distinguish fluoride injury from deficiencies of manganese, zinc, magnesium, calcium and potassium in some species (Dell *et al.*, 1995). Individual elements that are in excess and the overzealous applications of fertilizer can also induce fluoride-like symptoms.

Water stress in broad-leaved plants often leads to marginal necrosis, but it generally lacks a sharp line of demarcation between the necrotic and healthy tissue. Instead, there is often a diffuse yellow-green or yellow band (Plate 23). There is also a tendency for the necrosis to extend along the interveinal areas and along the mid-vein of the leaf. However, in some species, such as maple, aspen, willow and elm, and in conifers, such as pine, spruce and fir, sharply delineated tip necrosis may occur. Because the response of conifer needles to many abiotic and parasitic stresses, including frost, wind desiccation and excessive water, is realized as tip necrosis, diagnosis becomes even more complex (Treshow and Transtrum, 1964). In apricot, the symptoms may be virtually identical to those of fluoride injury.

Massey (1952) discussed the similarities of foliar symptoms produced by HF, SO_2, Cl_2, H_2S, NH_3, winter injury, high-temperature scorch and biotic diseases, such 'as scalds, blights, scorches, leaf spots, shotholes, blotches, silvering, tip burn, and chlorosis'. Many of these do not resemble chronic or acute fluoride injury but the similarities between such a disparate group of stresses reinforce the difficulty of field identification of symptoms and the great care that has to be taken with diagnosis.

Massey did not discuss salt injury to vegetation along roadsides, which occurs particularly in eastern white pine and sugar maple, or airborne sea-salt injury to many species during windstorms or hurricanes (Plates 20–22).

Abnormal suture diseases of peach, which have symptoms closely resembling those of suture red spot (p. 106), have been related to herbicides of the phenoxyacetic acid type (Benson, 1959), physiological stress (Dorsey and McMunn, 1944) and viruses, such as western X decline and peach mosaic (Richards and Cochran, 1957). Hildebrand (1943) reported a condition on peaches grown in Wayne County, New York, that resembled fluoride-induced suture red spot but he ascribed it to an unknown aetiology.

Effects on the Ultrastructure, Growth and Reproduction of Vascular Plants

Histology and ultrastructural effects

Understanding the mechanisms underlying the production of visible symptoms and effects on growth would be greatly promoted by a knowledge of the sequence of changes in histology and ultrastructure in plants that are exposed to realistic concentrations of HF. Huttunen and Soikkeli (1984) were in agreement with this when they said that 'the greatest potential for understanding the resistance or sensitivity of plants to air pollution and for elucidating the mechanisms of stress metabolism comes from the combination of electron microscopy with biochemical and histological investigations'. Although there have been several detailed studies using light and electron microscopy (Table 4.6), this ideal is far from being realized. Many studies used HF concentrations that were very high, far in excess of those that are typically found in the environment, and many have described the pathological events in the later stages, when membrane and cell integrity are breaking down. Nevertheless, some general conclusions can be made.

Table 4.6. Summary of reports of histological and ultrastructural effects of HF.

Species	Comments	Authors
Angiosperms		
Glycine (soybean)	Changes in endoplasmic reticulum; disruption of vacuolar membrane; detachment of ribosomes; changes in chloroplast shape; cellular collapse	Wei and Miller, 1972
Gossypium (cotton)	Within 0.5 mm of necrotic lesions: various stages of plasmolysis; less chlorophyll; chloroplasts and mitochondria swollen; eventually degradation of membranes and cellular collapse	Timmerman, 1967
Zea (maize)	In subnecrotic zones shrinkage of chloroplasts; fewer chloroplast lamellae	Lhoste and Garrec, 1975
Saccharum (sugarcane)	In chlorotic areas: decrease in size of chloroplasts, number of grana; increase in osmiophilic globules; swelling of middle lamellae. In red-brown areas destruction of chlorophyll and increase in other pigments	Englebrecht and Louw, 1973
Olea (olive)	Progressive changes over months including: large starch grains; increased plastoglobuli in chloroplasts; changes in shape of chloroplasts and mitochondria; dilatation of thylakoids; electron-dense granules in vacuoles	Eleftheriou and Tsekos, 1991
Gymnosperms		
Several species including *Phaseolus* (bean), *Malus* (apple), *Prunus* (apricot)	Before complete collapse of cells, disintegration of chloroplasts occurred. This could be seen macroscopically as pale areas in the normally dark green leaves	Solberg and Adams, 1956
Prunus (prune, apricot)	Effects similar to above in area within 2 mm of necrotic tissue	Treshow, 1957
Pinus ponderosa	Transitional zone adjacent to necrosis: enlargement of xylem and phloem parenchyma; transfusion parenchyma cells swollen/collapsed. Resin canal epithelial cells hypertrophied, plasmolysed; epithelial cells enlarged within the resin canal until they completely occluded the openings	Solberg *et al.*, 1955
Pinus	Hypertrophy of phloem and transfusion parenchyma cells; granulosis of chloroplasts; occlusion of resin canals	Carlson and Dewey, 1971
Pinus (various species)	Detailed comparison of effects of several pollutants and abiotic stresses. All produced similar effects, including changes in the phloem and occlusion of resin canals	Stewart *et al.*, 1973
Pinus	Abundant starch granules, dilatation of thylakoids, increase in endoplasmic reticulum and cell vacuolization	Soikkeli and Touvinen, 1979; Soikkeli, 1981
Abies (fir)	Young needles emerging from bud: chloroplasts smaller, thylakoids dilated; delayed formation of epicuticular waxes	Bligny *et al.*, 1973a
Pinus strobus	Membrane integrity affected; plasma membranes among first sites affected	Rakowski and Zwiazek, 1992

Although there are limited data, it appears that histological changes and ultrastructural effects are restricted to the tissues in which visible symptoms are developing. For example, Treshow (1957) found that effects were observed only after the appearance of visible necrosis and they were limited to a 2 mm band of cells adjacent to the necrotic tissue. This is what would be expected, given the steep concentration gradients in leaves, and it is an important observation when interpreting the effects on processes such as photosynthesis.

Several authors have compared the effects of a range of pollutants and abiotic stresses on leaf and tissue structure. One of

Table 4.7. Histological responses of pine needles to abiotic stresses. (After Stewart *et al.*, 1973, *Canadian Journal of Botany* 51, 983–988, 1973. Reprinted with permission.)

| | Initial symptoms before necrosis | | | |
Causal agent	Mesophyll collapse	Resin duct occlusion	Transfusion parenchyma hypertrophy	Phloem abnormality
Natural senescence	–	–	+	+
Sodium chloride (salt)	+	+	–	+
Boron (Na₂B₄O₇)	+	+	–	–
Water stress	–	+	+	+
Winter injury	–	+	+	–
Anoxia (waterlogged soil)	–	+	–	+
Fluoride (HF)	–	+	+	+
Ozone (O₃)	+	+	+	–
Sulphur dioxide (SO₂)	+	+	–	+
HF + SO₂	+	+	+	+
HF + O₃	+	+	+	+

the first studies was by Solberg and Adams (1956), who concluded that microscopic injury was indistinguishable between the various stresses (HF and SO_2) in all species examined. Similarly, Huttunen and Soikkeli (1984) pointed out that pollutants other than HF may cause many of the same ultra-structural changes. For example: swelling and curling of thylakoids is also identified with NO_2 exposure; curling of thylakoids might be caused by the combined action of frost and air pollutants; swelling of thylakoids is a general symptom of acute injury; and lipid bodies are also associated with natural senescence. Stewart *et al.* (1973) examined needles with tip necrosis from Scots (*Pinus sylvestris*), lodgepole (*P. contorta*), ponderosa (*Pinus ponderosa*) and white (*P. strobus*) pines and from Douglas fir (*Pseudotsuga menziesii*), collected from sapling trees exposed under greenhouse conditions to several pollutants and other environmental stresses (Table 4.7). These included HF, SO_2, O_3, HF + SO_2, HF + O_3, winter injury, drought, normal senescence, waterlogging, excess salt and excess boron. Needles were also collected from trees exposed to HF or SO_2 under field conditions. Macroscopically, the external appearance of the needles from the different stresses was similar in most cases. Stewart *et al.* (1973) suggested that the 'senescence and death of pine needles is probably a result of a precise

and programed [*sic*] series of events which can be induced by a number of pathogens, each of which initiates or triggers much the same mechanism of morbidity'. Fink (1988) reviewed the sequence of cellular injury from pollutants and listed the responses of the chloroplasts to various stresses:

1. Granulation of the stroma and crystal-line inclusions have been reported for HF, SO_2, O_3, peroxyacetylnitrate, water stress and nutrient deprivation.
2. Curling of internal membranes, resulting in an undulated appearance of thylakoids, has been described following treatments with HF, SO_2, O_3 and cold treatment.
3. Stretching of the chloroplast envelope, as a consequence of shrinkage of the chloroplast, has been reported for HF, O_3 and other pollutants.
4. Swelling of thylakoids by dilatation of their membranes has been associated with HF, iron deficiency and premature senescence induced by mineral deficiency.
5. Dilatation of the parallel alignment of the thylakoid systems and formation of electron-transparent areas of the swollen stroma have been related to HF, SO_2, O_3, frost and potassium deficiency.
6. Incomplete differentiation of the plastids, with reduced membrane systems, caused by HF, O_3, etiolation and phosphorus or iron deficiency.

7.   Accumulation of electron-dense plasto-globuli induced by HF, senescence and mineral deficiency.

Thus, it appears that, for conifer needles at least, microscopical examination is of very little use as a forensic tool for diagnosis.

Fluoride and the growth of vascular plants

Although the effects of fluorides on growth have been studied for over 50 years, because of the shortcomings of the older fumigation systems the only reliable information comes from post-1973 studies using open-top chambers and from studies of seed production and tree rings. Nevertheless, there is sufficient reliable information to draw several important conclusions and to indicate some deficiencies in our knowledge.

We shall use three open-top chamber studies to illustrate several important points, the first involving research on wheat. MacLean and Schneider (1981) grew wheat in the field in an area where there was only background fluoride. At the stage where the spikes were first emerging (= the 'boot' stage), the plots were enclosed in open-top chambers. Four days of continuous fumigation provided three treatments: 0, 0.9 and 2.9 μg F/m^3. The same treatments were applied to another group of plots set up at anthesis. Although these were short fumigations, there were significant effects (Table 4.8). At the boot stage, 0.9 μg/m^3 reduced yield because the spikes were smaller, but 2.9 μg/m^3 did not affect yield because the smaller spikes were offset by an increase in the number of spikes per plant at harvest. The later fumigation at anthesis reduced yield because of smaller, fewer spikes. These are quite striking results for such a short fumigation and they reveal the importance of the stage of development and effects on seed production. The grain fluoride concentration was not increased by the fumigation, which the authors interpreted as indicating that the fluoride was not translocated from the leaves to the developing seeds. Murray (1983) found that the same was true for grape

Table 4.8.   The yield of field-grown wheat after exposure to HF for 4 days at two stages of development. Means within a column followed by the same letter were not significantly different ($P = 0.05$). (After MacLean and Schneider, 1981.) (Reprinted from *Environmental Pollution* 24, 39–74, 1985, with permission from Elsevier.)

HF (μg/m^3 for 4 days)	Yield per plant (g)	Spikes per plant	Weight per spike (g)
Exposed at boot stage			
0	1.29 a	3.46 b	0.364 a
0.9	1.03 b	3.45 b	0.306 b
2.9	1.21 a	4.65 a	0.254 c
Exposed at anthesis			
0	1.57 a	3.55 a	0.488 a
0.9	1.07 b	3.26 ab	0.333 b
2.9	0.93 b	2.99 b	0.326 b

berries that had leaf fluoride concentrations as high as 85 mg F/kg. These are important observations because they mean that the grain or berries can be used with no risk to the consumer. An equally important finding was that there were no visible symptoms of injury. The authors (MacLean and Schneider, 1981) commented on the significance of this in relation to assessments of relative sensitivity. They pointed out that, on the basis of foliar injury, young wheat plants are usually classified as having intermediate sensitivity and mature plants as being resistant, but their data show that this classification is wrong when the criterion is yield. They recommended that classification should be based on measures that are related to the use of the plant. This is an important consideration when plants are used as bioindicators (Chapter 6) or when devising air-quality standards.

MacLean et al. (1977) reported results similar to those of MacLean and Schneider (1981) but using an even lower HF concentration. They exposed bean plants in open-top chambers to 0.6 μg F/m^3 continuously for 43 and 99 days. The vegetative growth of the bean was unaffected and there were no visible symptoms of injury but there was a significant reduction in the number and mass of pods (Table 4.9). The authors

Table 4.9. The yield of bean plants (*Phaseolus vulgaris*, cv. 'Tendergreen') after fumigation with 0.6 µg F/m³ for 43 days from 2 days after emergence. (After MacLean *et al.*, 1977.) (Reprinted from the *Journal of the American Society for Horticultural Science* 102, 297–299, 1977, with permission.)

	Control	HF-fumigated
Dry wt tops (g)	80.9	81.8
Dry wt leaves (g)	41.5	44.3
Dry wt stems (g)	39.4	37.5
No. of pods harvested	88.4 *	77.3
No. of marketable pods	70.4 *	56.5
No. of unmarketable pods	18	20.8
Fresh wt pods (g)	424 **	337
Fresh wt marketable pods (g)	391 **	296
Fresh wt unmarketable pods (g)	34.3	40.5

*Significantly different ($P = 0.05$); **Significantly different ($P = 0.01$).

concluded that HF affects leaf injury and fruiting independently. They also commented on the fact that the fumigation was continuous at 0.6 µg F/m³ and that such a condition does not always exist in the field. They went on to study the effects of variable fumigation on wheat and sorghum (MacLean *et al.*, 1984). In these experiments they gave the plants three successive 3-day exposures, in which the mean concentration in each period was about 1.6 or 3.3 µg/m³. There were eight permutations of the levels, plus a control. The yield of wheat was not related to the mean concentration over all the exposure periods but it was related to a weighted contrast between the first and the later two periods. In the case of *Sorghum*, the yield was related to a weighted mean of the second and third exposures. Analysis of the data showed that exposure during anthesis was most effective in reducing yield, indicating an effect on the fertilization process. This experiment demonstrates the complexity of dose–response relationships and that we are still far from being able to predict the effects of fluoride on yield under field conditions, where the concentrations vary in much more complex ways and where there are so many other variables in operation.

Effects on fertilization, seed set and fruit yield

One of the best sources of evidence of the importance of calcium–fluoride interactions comes from studies of pollen germination, fertilization and fruit production (Pack, 1966, 1971, 1972; Sulzbach and Pack, 1972; Facteau *et al.*, 1973; Pack and Sulzbach, 1976; Facteau and Rowe, 1977; Bonte, 1982; Weinstein and Alscher-Herman, 1982). Effects of fluoride on fruiting were reviewed by Pack and Sulzbach (1976) and Bonte (1982). Pack (1966) grew tomato plants at two levels of calcium (40 and 200 mg/l) and exposed them to 6.4 µg HF/m³ for 22 weeks. Fruit size was related to both calcium and fluoride levels, the smallest fruit being in the low calcium + HF treatment (Table 4.10). The same treatment resulted in a high seedlessness rating. Pack (1966) concluded that calcium played an essential role in fertilization and that fluoride interfered with it. There was also more injury on the leaves of low-calcium plants. An interactive effect of fluoride and calcium on tomato pollen germination and growth were confirmed by Sulzbach and Pack (1972), but others have reported mixed effects of fluoride on pollen germination.

Although most fruits are relatively tolerant to HF concentrations up to 2–3 µg/m³, some, such as peach fruits, are exceptionally sensitive. A condition known as 'black tip' has been described in peaches in an area in which fluoride was present (McCornack *et al.*, 1952), but more commonly HF induces a condition known as 'suture red spot' or 'soft suture'. It is characterized by premature ripening of the flesh on one or both sides of the suture towards the stylar (blossom) end of the fruit (Plate 24). The ripening of this tissue precedes that of the normal fruit and is characterized by external and internal reddening of the suture area. The affected area may enlarge faster for a short time before ripening. The suture symptoms are often accompanied by splitting of the flesh along the suture line. At harvest, the affected areas are soft and often decomposing (Griffin

Table 4.10. The effects of 22 weeks of exposure to 6.4 µg HF/m³ on tomato production (data from Pack, 1966). (Data used courtesy of the Air & Waste Management Association from Pack, M.R., *Journal of the Air Pollution Control Association* 16, 541–544, 1966.)

	Controls		HF		
	40 p.p.m. calcium	200 p.p.m. calcium	40 p.p.m. calcium	200 p.p.m. calcium	LSD
Fruits/plant	35	30	25	41	11
% of flowers with fruit	92	88	78	89	NS
Wt of fruit per plant (g)	1680	2650	406	2570	557
Mean wt per fruit (g)	48	88	20	63	19
Seedlessness rating[a]	0.9	0.62	2.32	1.01	0.86

[a]Seedlessness was rated on a 0–1 scale: 0 = normal seed development, 3 = completely seedless.
LSD, least significant difference, $P = 0.05$; NS, not significant.

and Bayles, 1952; Benson, 1959; Drowley *et al.*, 1963; Bolay *et al.*, 1971b; Mezzetti and Sansavini, 1977; MacLean *et al.*, 1984). MacLean *et al.* (1984) studied the effects of very low concentrations of HF on the development of symptoms on Elberta peach fruits from trees grown in the field and enclosed in open-top chambers. Continuous exposure to concentrations above about 0.3 µg F/m³ for periods from 70 to 108 days resulted in a high proportion of abnormal fruit (MacLean *et al.*, 1984). An association between the fluoride treatment and calcium was shown by the use of a ⁴⁵Ca tracer and by analysis of the suture and the rest of the fruit. Although the tracer showed that little calcium was transported from leaves adjacent to the fruit, HF treatment led to a significant increase in tracer found in the fruit: 1.04% versus 0.034% in controls. Analysis of the fruit showed that HF altered the distribution in the fruit, significantly decreasing the percentage found in the suture.

Strawberry (*Fragaria* sp.) fruit are also affected by fluoride. Pack (1972) showed that exposure to HF as low as 0.55 µg/m³ caused a small increase in fruit deformation, predominantly at the apical end of the fruit. This was brought about by lack of development of seeds and associated receptacle tissue. The degree of deformation of the receptacle was proportional to the number of undifferentiated achenes and was greater at higher HF concentrations. Because receptacle growth takes place only if fertilization and normal embryo development have

occurred, this was considered to be evidence that fluoride is inhibitory to fertilization and/or embryo development. Although fruit weights were reduced at 2.0 µg/m³ and higher, no foliar symptoms were produced. These results corroborated the evidence, in earlier studies with tomato and bean (Pack, 1966, 1971), that HF affects fruiting by interfering with fertilization and seed development. A link with calcium was made by Bonte *et al.* (1980), who examined the effects of HF on strawberry fruit development. They confirmed that the flowering stage was most sensitive to HF. Strawberry plants were exposed to HF at 5.4 µg/m³ at three stages of development: (i) preflowering, from the appearance of the first leaf to the appearance of petals (corolla not yet opened); (ii) flowering and fertilization, from the appearance of the closed corolla to the dropping of the first petal; and (iii) postflowering, from the dropping of the first petal to fruit harvest. The least amount of fruit deformity was found where plants were not exposed during the flowering and fertilization stage (Table 4.11). Further experiments demonstrated that deformity occurred much more frequently when carpels were exposed to HF than with the pollen-producing anthers (74% deformed fruit when carpels were fumigated vs. 11% if the anthers were fumigated). The authors concluded that the stigmatic surface was altered by exposure to HF, affecting pollen-tube growth and subsequent fertilization (see also Sulzbach and Pack, 1972). Analysis

Table 4.11. The effects of periods of fumigation with 5.4 µg HF/m³ on the % of malformed fruit and wt per fruit of strawberry. Plants were fumigated at different times during development to give six combinations of treatment: HF = stages when plants were fumigated and CA indicates when they were in clean air. Means within a column followed by the same letter were not significantly different ($P = 0.05$). (Reprinted from Bonte, J. (1982) Effects of air pollutants on flowering and fruiting. In: Unsworth, M.H. and Ormrod, D.P. (eds) *Effects of Gaseous Pollutants on Agriculture and Horticulture*. Butterworths, London, pp. 207–223, with permission from Elsevier.)

Combination of treatments	Period of treatment			Results	
	Before anthesis	Flowering/ fertilization	Maturation	% malformation	Mean wt per fruit (g)
1	HF	HF	HF	57	3.47 a
2	HF	HF	CA	58	4.49 b
3	HF	CA	CA	1.3	5.58 c
4	CA	CA	CA	2.7	5.71 c
5	CA	CA	HF	5.4	5.56 c
6	CA	HF	HF	42	5.45 c

with an electron microprobe of the style and stigma of fumigated plants showed that there was a significant accumulation of fluoride on and just inside the stigmatic surface and that the fluoride disrupted the calcium gradient in the stigma and style (Bonte and Garrec, 1980).

Analysis of tree growth rings

There have been several productive studies of fluoride effects using tree rings. Although such studies are time-consuming and difficult to analyse statistically they do offer something that cannot be done using open-top chambers and that is an estimate of the effects of prolonged exposure over many years. One of the first was by Lynch (1951), who showed that growth of ponderosa pine near Spokane, USA, was depressed close to an aluminium smelter. Later, Høgdahl *et al.* (1977) studied tree growth in the vicinity of an aluminium smelter at Mosjøen, Norway, in order to solve claims for damages between 1964 and 1973. A forest area of 10,000 acres (4000 ha) was involved, consisting mainly of Norway spruce (*Picea abies*). Samples of five to 15 trees were cored in each of 147 plots, giving 1332 samples in total. The cores covered 13 years before and 13 years after the

smelter started operation. The situation was complicated by the advanced age of many of the trees and by the fact that there were also SO₂ emissions. It was considered that fluoride was the dominant pollutant, but the role of SO₂ in growth reduction was not determined, which is unfortunate in view of the known sensitivity of Norway spruce to SO₂. Two scientists analysed the data independently, using different methods, but their results were essentially the same. One comment made was that there was great variability between individual trees within plots and between adjacent plots, which is the reason why such studies must sample large numbers and have robust statistical analysis. There were few records of atmospheric fluoride concentrations for the forest area, but needle fluoride was measured. Although the areas with the greatest growth reduction usually had relatively high needle fluoride contents, one of the investigators noted that there was no clear functional relationship between the two. Growth was reduced by about 20% up to about 4 km downwind, but there was no effect beyond 8 km. The case was settled and, since then, some of the smelter plant has been converted to prebake technology and dry scrubbing has been installed (AMS, 1994).

Taylor and Basabe (1984) made a similar study in Douglas fir (*P. menziesii*) forests in Washington, USA. As in the Norwegian case, the source was an aluminium smelter, but there were also two sources of SO_2. Their data show the immense variation and low correlations between the pollutant and growth, but there were statistically significant effects. Unlike the findings in Mosjøen, there was a relationship between the fluoride content of needles and growth reduction, but the authors did not give the regression equation or the probability. Figure 4.4 shows the mean ring widths at two distances from the smelter for a period before and during operation. In the area lying from 0 to 8 km from the smelter the growth reduction was 33% compared with the preoperational years. In discussing the results, the authors drew attention to the complicating factor of the presence of SO_2. They assumed that the pollutants interacted and implied that it was responsible for large growth reductions in areas where the fluoride in the needles was relatively low. However, SO_2 concentrations were not measured, so it was impossible to try to separate the effects of the pollutants and determine the specific effects of fluoride. This is, of course, the weakness of field studies – isolating the effects of one component in a complex system.

Several aspects of the health of forest trees growing in the vicinity of the Alcan smelter at Kitimat (British Columbia, Canada) are described in Chapter 5, but the research included an illuminating study of ring growth (Bunce, 1979). Reductions in the growth of western hemlock (*Tsuga heterophylla*) were found for the period up to 1973, but the most intriguing result came when the growth was analysed over a later period, after emissions were greatly reduced (Bunce, 1985). In the first study (Bunce, 1979), growth reductions (basal area) were estimated in three zones, the inner, outer and surround, as being 28.1, 19.0 and 2.2%, respectively (Table 4.12). These were associated with airborne fluoride concentrations estimated to be 3.42, 2.05 and 1.3 µg/m³. After a reduction in emissions of 64%, the

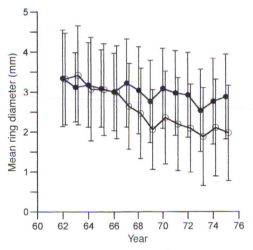

Fig. 4.4. Mean ring diameter in Douglas fir (*Pseudotsuga menziesii*) growing at two distances from an aluminium smelter over a period before (1962–1966) and during operation (1967–1975). ○ = mean for trees growing within 8 km of the smelter and ● = trees growing at 8–16 km. Vertical bars are 1 standard deviation. (Redrawn from Taylor, R.J. and Basabe, F.A. (1984) with permission from Elsevier.)

Table 4.12. Fluoride concentrations and estimated effects on western hemlock (*Tsuga heterophylla*). (Reprinted from Bunce, H.W.F., *Journal of Air Pollution Control Association* 35, 46–48, 1985 with permission of the Air & Waste Management Association.)

Period	Zone	Estimated HF (µg/m³)	[F] in foliage (mg F/kg)	Change in basal area (%)	Change in wood volume (m³)
1954–1973	Inner	3.42	271	−28.1	−12,703
	Outer	2.05	163	−19.0	−41,098
	Surround	1.3	104	−2.2	−4,212
1974–1979	Inner	1.09	87	−45.3	−3,087
	Outer	0.34	29	+2.8	+1,192
	Surround	0.26	22	+13.6	+4,616

fluoride concentrations fell but growth in the inner zone over the period 1974–1979 was still reduced. However, in the two other zones, where the fluoride was 0.34 and 0.26 µg/m³, respectively, growth was apparently increased by 2.8 and 13.6%, respectively. Bunce (1985) pointed out that there have been a number of reports of low levels of fluoride stimulating growth (reviewed by Treshow *et al.*, 1967, and Weinstein, 1977), but the cause is not known. If there is a genuine stimulation of a component of plant performance, it may be due not to an increase in the total mass of the plant but to a reallocation of resources. For example, it might involve an increase in stem wood at the expense of root growth, or perhaps a reduction in seed production allows more carbon to be allocated to the wood. Whatever the cause, the tree-growth studies all illustrate how much remains to be learned about fluoride effects in the field.

Abiotic and Biotic Interactions

Any environmental factor, abiotic or biotic, that alters the uptake or movement of fluoride in leaves, the concentrations of key elements, such as calcium, or the rate of development through the sensitive stage of growth or resource allocation can potentially affect plant response to fluoride. The list includes: light, temperature, humidity, other pollutants, mineral nutrition, pests and diseases.

Light

Light would be expected to have an effect because it plays a major part in determining stomatal conductance, but little is known about the effects of light intensity, light quality or the length of the photoperiod. Early experiments, which were done before instruments were readily available for measuring stomatal conductance, investigated the effects by fumigating plants either in the light or in the dark. Some, such as Daines

et al. (1952), found what might be expected, that there was less injury when plants were exposed in the dark, but in several cases the onset of foliar injury in dark-exposed plants occurred with an HF dose that was only slightly higher than that required in the light. Also, in general, injury occurred in dark-exposed plants that had accumulated about half as much fluoride as that required in the light (Adams *et al.*, 1957). Lucerne plants exposed to HF in the dark accumulated about 40% as much fluoride as in the light, despite the fact that the stomata appeared to be closed (Benedict *et al.*, 1965). These observations are difficult to explain without information about stomatal conductance but, bearing in mind the fact that controlled-environment systems were poor in the 1950s, it is possible that stomatal conductance was not very high during the light period and that stomata may have remained partly open at night (Musselman and Minnick, 2000). The sequence of dark and light exposures may also be important. When a fluoride-susceptible variety of Jerusalem cherry (*Solanum pseudo-capsicum*) was fumigated in the light, the plants exhibited mild injury, which was not affected by a postexposure light or dark period (MacLean *et al.*, 1982). Plants fumigated in the dark also showed mild injury, but it became very severe shortly after exposure to light even though the plants accumulated less foliar fluoride. These results suggested that metabolic changes associated with the dark treatment rendered the plant more susceptible to a small dose of HF after exposure to light, but the mechanism is unknown.

Paradoxically, shaded plants have been reported to be more (Hitchcock *et al.*, 1964) and less (Wiebe and Poovaiah, 1973) susceptible to injury than plants exposed in direct sunlight. Wiebe and Poovaiah (1973) exposed soybean plants to a high concentration of HF for a relatively short duration, after which the plants were shaded to give either 58% or 17% of full sunlight. After a 3-day postexposure period, they reported a remarkable decrease in foliar injury with increasing amount of shade. The cause of this difference has not been identified.

Temperature

There has not been an extensive amount of research on the relationship between temperature and response to fluoride, and yet air and soil temperature are generally recognized to have an influence on this response. For example, MacLean and Schneider (1971) exposed the susceptible *Gladiolus* cultivar 'Snow Princess' and the tolerant sunflower *Helianthus annuus* cultivar 'Mammoth Russian' to HF at a concentration of 4.7 µg/m^3 for 104 h (4.3 days). This was followed by a 7-day period without HF to allow plants to fully develop symptoms. In *Gladiolus*, both the severity of the foliar necrosis and the accumulation of fluoride were affected by temperature. By the end of the third day of exposure, there was distal leaf-tissue collapse at all temperatures, but it appeared to be more intense at the two higher temperatures than at 16°C. At the end of the experiment, there was a profound increase in foliar necrosis in *Gladiolus* plants exposed at 21°C, but at 26°C the degree of foliar injury was about the same or less than at 21°C. This effect was sustained in young, middle-aged and the oldest leaves (Table 4.13). It appeared, however, that fluoride accumulation was controlled by different factors from those controlling necrotic development. There was no significant difference in

Table 4.13. Effect of temperature on HF-induced necrosis in *Gladiolus* ('Snow Princess') leaves of different ages. Data presented as % of total leaf area necrotic. Means followed by the same letter are not significantly different ($P = 0.05$). (From MacLean, D.C. and Schneider, R.E. (1971) Fluoride phytotoxicity: its alteration by temperature. In: Englund, H.M. and Beery, W.T. (eds) *Proceedings. 2nd International Clean Air Congress*, Academic Press, New York, pp. 292–295.)

Leaf age	Temperature (°C)		
	16	21	26
Youngest	13.9 b	28.1 f	20.8 d
Intermediate	10.9 a	21.8 d,e	24.0 e
Oldest	8.2 a	19.9 d	16.8 c
Total per plant	32.7	69.8	61.7

accumulation at temperatures between 16 and 21°C, but at 26°C it was significantly depressed. There was also a significant interaction between temperature and leaf age with respect of foliar necrosis, with the oldest leaves being the most tolerant to HF-induced necrosis at all temperatures and the youngest leaves being the most susceptible at 16 and 21°C. At 26°C, however, there was a significant decrease in necrosis.

Wiebe and Poovaiah (1973) also found a positive correlation between temperature and foliar injury in soybean (*Glycine max*). Plants were exposed to high HF concentrations, ranging from 100 to 300 nl/l (82 to 246 µg/m^3) for exposures of 8–18 h. After subjecting the plants to postexposure day/night temperature regimes of 23/16°C, 32/23°C or 43/30°C, it was found that increasing temperatures at all concentrations resulted in greater foliar injury. Soybean plants maintained at day/night temperatures of 32°/23°C (or higher) after exposure to HF developed more severe foliar injury than plants maintained at 23°/16° C.

Relative humidity/vapour-pressure deficit

The stomata of many species respond to the 'dryness' of the air and several studies have examined the effects of differences in relative humidity (RH) on fluoride accumulation and effects. Daines *et al.* (1952) was one of the first groups to investigate RH and they showed that there was more foliar injury and greater fluoride accumulation in *Gladiolus* (cv. 'Margaret Fulton') and tomato (cv. 'Rutgers') at high than at low RH. Unfortunately, they do not specify the RH values used. A somewhat different result was reported by McCune *et al.* (1977), who found that an increase in RH from 75% to 85% had no effect on fluoride uptake or foliar injury in mandarin orange, spinach or rice foliage, but this was probably because the range of RH used was very small. MacLean *et al.* (1973) used potted plants of the *Gladiolus* cultivar 'Oscar' which were acclimatized to RH values of 50, 65 or 80% for 24 h, and then fumigated with HF at

4.5 µg/m^3 for 144 h (6 days). After exposure, the plants were maintained at their respective humidities for an additional 24 h for symptoms to be expressed. Tissue collapse occurred in leaf tips at all RHs by 36 h after fumigation began, but it occurred earlier and was significantly greater at 80% than at 65 and 50% RH (Table 4.14). MacLean *et al.* (1973) proposed four mechanisms for the influence of RH on exposure to fluoride: altered stomatal conductance and absorption of HF; altered translocation of fluoride in the leaves; changes in the threshold for cellular injury by fluoride; and the influence of tissues injured by fluoride, or feedback, on the other mechanisms.

Pollutants

It is rare for there to be only a single pollutant in an air parcel and commonly HF is accompanied by variable concentrations of SO_2, O_3 and NO_x. There may also be cryolite or several other fluorine-containing particles, and considerable amounts of

Table 4.14. Severity of HF-induced necrosis on *Gladiolus* (cv. 'Oscar') leaves as affected by relative humidity and leaf age. (From MacLean, D.C., Schneider, R.E. and McCune, D.C. (1973) Fluoride phototoxicity as affected by relative humidity. In: *Proceedings of the Third International Clean Air Congress*: VDI-Verlag GmbH, Dusseldorf.)

Relative leaf age	% of leaf length necrotic		
	Relative humidity (%)		
	50	65	80
5 = youngest[†]	32.5 a	29.4 a	41.7 b*
4	40.9 a	39.9 a	48.2 a*
3	29.2 b	33.9 a	37.2 c
2	6.4 c	8.3 b	12.8 d
1 = oldest	3.9 c	6.2 b	14.4 d*
Average per plant	20.1	22.2	28.2*

*Significantly greater than corresponding values at other RHs ($P = 0.01$). Means within a column followed by different letters were significantly different ($P = 0.05$).
[†]All plants did not have leaves of this age.

chlorine, chlorides and hydrocarbons may be present in the emissions from aluminium smelting. Emissions from phosphate-fertilizer processing include not only two forms of gaseous fluoride (HF and SiF_4) and SO_2, but also rock dust and a number of calcium-containing particles. Concentrations of all of the pollutants vary continuously, so fumigation may consist of concurrent or sequential exposures and the number of combinations is infinite. For example, near aluminium smelters emissions may co-occur with high concentrations of atmospheric O_3 during the summer months but not when the weather is cool, cloudy or wet, so the sequence may be high fluoride–low ozone in late winter and spring, followed by high fluoride–high ozone in summer, with high fluoride–low ozone during rainy periods. Each plant species tends to have its own spectrum of responses to air pollutants and it is very uncommon for a species to be equally sensitive to two different pollutants. So, for example, ragweed (*Ambrosia artemisiifolia*) and lucerne (*Medicago sativa*) are sensitive to SO_2 but relatively tolerant to HF, whereas *Gladiolus* and apricot (*P. armeniaca*) are tolerant to SO_2 but sensitive to HF. One exception to this general rule is some of the pines, such as Scots pine, *P. sylvestris*, which is relatively sensitive to both SO_2 and HF.

The two obvious mechanisms for fluoride–pollutant interactions are via effects on uptake and enzyme activities and metabolite pools. The effects of HF on stomatal conductance are unclear but both SO_2 and O_3 are well known to alter conductance. There are many reports of both pollutants increasing and decreasing conductance but the effect depends on the species and concentration, so they might increase or decrease fluoride uptake (Darrall, 1989). An interactive effect of fluoride on plant response to SO_2 or ozone might be expected if fluoride affects enzymes or metabolites involved in scavenging free radicals, especially in the apoplast. Conversely, if SO_2 or ozone altered the disposition of calcium or magnesium in cells, it would alter fluoride toxicity.

Although there has been research on the effects of fluoride in combination with other air pollutants (reviewed by Ormrod, 1982; McCune, 1983; McCune *et al.*, 1984; Mansfield and McCune, 1988), it has been relatively limited and most of what is available is on the joint action of HF and SO_2. The results are often inconsistent and effects appear to be related to concentrations of the individual components of the mixture, differences between species or cultivars of the plant, environmental conditions, exposure dynamics and probably a host of other factors. What has been lacking are quantitative measures of effects. Qualitative comparisons have been made on symptom development, growth and reproduction and the relative susceptibility of forest species exposed to HF, SO_2 or their combinations (Pollanschütz, 1969; Bohne, 1970, 1972; Roques *et al.*, 1980; Buckenham *et al.*, 1982).

In experiments in which plants were exposed to fluoride combined with one or more pollutants, fluoride accumulation was variable, exhibiting no change, an increase or a decrease, but in most cases accumulation was decreased by the presence of SO_2 (Table 4.15). Although few authors measured

stomatal conductance, the effect was probably caused by SO_2 reducing stomatal apertures (Darrall, 1989). Despite the fact that the reduction of fluoride accumulation would be expected to decrease the effects of HF, there are very few reports of interactive effects on the production of foliar lesions. Murray and Wilson (1988b) reported that, in *Eucalyptus* species exposed to 0.39 or 1.05 μg F/m^3, the presence of 271 μg SO_2/m^3 increased foliar lesions, but there were no effects at 122 μg SO_2/m^3. The increased injury, however, was due to an additive effect of the two pollutants and was found in only two of the three *Eucalyptus* species studied (*E. calophylla* and *E. gomphocephala* but not *E. marginata*).

There are mixed reports of the effects of pollutant interactions on growth and yield. Bonte (1982) reported that HF had no effect on SO_2-induced changes at or above 5 μg/m³ but, in a field study using maize, Mandl *et al.* (1980) found that the weight of stalks, number of mature ears and fluoride accumulation were reduced by the combination of HF and SO_2. However, the combination had no

Table 4.15.   Summary of the effects of SO_2 on fluoride accumulation.

Species	Effect	Authors
Citrus	↓ F accumulation	Matsushima and Brewer, 1972
Sweet corn (*Zea mays*)	↓ F accumulation by *c.* 30%	Mandl *et al.*, 1975
Sweet corn (*Zea mays*)	↓ F accumulation by *c.* 40%	Mandl *et al.*, 1980
Lucerne (*Medicago sativa*)	↓ F accumulation by *c.* 30% ↓ S accumulation by *c.* 18%	Brandt, 1981
Millet (*Setaria italica*)	↓ F accumulation by *c.* 20%	
Italian ryegrass (*Lolium multiflorum*)	↓ F accumulation	
Gladiolus	↓ F accumulation by *c.* 34%	McCune, 1983
Lolium	↓ F accumulation by *c.* 24%	
Maize (*Zea mays*)	No effect	
Lucerne (*Medicago sativa*)	↓ F accumulation by *c.* 24% ↓ S accumulation by *c.* 18%	Mandl, cited by McCune, 1983
Wheat (*Triticum aestivum*)	No striking effect on F	Davieson *et al.*, 1990;
Barley (*Hordeum vulgare*)	accumulation, but SO_2 offset	Murray and Wilson 1988a,b,c,
Soyabean (*Glycine max*)	some of the effects of HF.	1990
Maize (*Zea mays*)	Concentrations of F low	
Groundnut (*Arachis hypogea*)		
Bean (*Phaseolus vulgaris*)		
Eucalyptus spp. (4)	↓ F accumulation in two species ↑ F accumulation in one species	

effect on fresh- or dry-weight yields of leaves, husks or tassels, height of plants or number of kernels per ear, which shows that detecting effects depends very much on which parameters are measured. In contrast, Davieson *et al.* (1990) and Murray and Wilson (1990) have produced some evidence that negative effects of HF may be offset by SO_2. The former authors showed that the inhibitory effect of HF on grain protein content was offset by SO_2 while Murray and Wilson (1990), working with soybean, maize, groundnut and navy bean, concluded that effects were less than additive – this was despite the fact that there was no effect on foliar fluoride concentrations (all of which were surprisingly low).

Mineral nutrition

In view of the evidence presented at the start of this chapter it is not surprising that mineral nutrition affects the toxicity of

fluoride. Research on this subject was reviewed by Cowling and Koziol (1982), Weinstein and Alscher-Herman (1982) and MacLean (1983), and is summarized in Table 4.16. The main elements with which interactions might be expected are those with which it forms complexes (Ca, Mg, Fe, Mn, Cu and Zn). There is evidence that chronic fluoride injury symptoms are associated with a lower foliar content of Mn and Mg in citrus (Brewer *et al.*, 1967) and symptoms of chronic fluoride injury in many broad-leaved species resemble deficiency of Mn, Zn or Mg; for this reason, Brewer *et al.* (1960) suggested that chronic fluoride injury was synonymous with a deficiency of certain nutrient elements. Interactions with other elements (e.g. N, P, K, Mo) are likely to be indirect and due to changes in the uptake, accumulation or movement of the fluoride.

Several of the reports cited in Table 4.16 found that sensitivity was increased by low calcium, but among the earliest studies were those of Brennan *et al.* (1950) and Daines

Table 4.16. Relationship between mineral nutrition and sensitivity to HF. (From Cowling, D.W. and Koziol, M.J. (1982) Mineral nutrition and plant responses to air pollutants. In: Unsworth, M.H. and Ormrod, D.P. (eds) *Effects of Gaseous Pollutants on Agriculture and Horticulture*. Butterworths, London, pp. 349–376, courtesy of Elsevier.)

Species	HF ($\mu g/m^3$)	Exposure time (h)	Nutrient regime	Effect on sensitivity	Authors cited by Cowling and Koziol (1982)
Lycopersicon (tomato)	420	3.5	[N] 14–448 mg/l [Ca] 10–240 mg/l [P] 0–62 mg/l	Greatest at medium N Greatest at 40 mg/l ↓ by P deficiency	Brennan *et al.*, 1950
Phaseolus (bean)	2	> 240	N, P, K or Ca deficiency	No injury. Uptake ↑ with P or K deficiency	Applegate and Adams, 1960b
Phaseolus	43	15	N deficiency	↑ by N deficiency	Adams and Sulzbach, 1961
Lycopersicon	3, 6	3696	Ca at 40 or 200 mg/l	↑ by Ca deficiency	Pack, 1966
Gladiolus	2	24	P, K or Mg deficiency N or Ca deficiency	↑ by deficiency ↓ by deficiency	McCune *et al.*, 1966
Lycopersicon	5–12 1–15 3	120 120 1536	Mg deficiency Ca deficiency P deficiency	↑ by deficiency ↑ by deficiency ↑ by deficiency	MacLean *et al.*, 1969
Lycopersicon	5–10	168	[Mg] from 2 to 384 mg/l	↑ by deficiency, ↓ by excess	MacLean *et al.*, 1976
Avena (oat)	12	192	Various P, K, Ca	↑ by deficiency of P, K and especially Ca	Guderian, 1977
Spinacea (spinach)	Not stated		N-deficient soil	↑ with deficiency	Guderian, 1977

↑ = an increase in sensitivity or F uptake, ↓ = a decrease.

et al. (1952). In tomato, for example, they found that low or deficient supplies of N, Ca and P reduced the absorption of toxic amounts of fluoride, while very high concentrations of N and Ca prevented fluoride injury. Conversely, when Adams and Sulzbach (1961) exposed bean plants to acute HF concentrations, symptoms were exhibited only in those plants grown in an N-deficient medium, although the tissue fluoride levels were equal to or less than those of plants grown in other nutrient media and exposed to the same HF concentrations. Guderian (1977) reported that moderate levels of N fertilizer resulted in reduced fluoride injury but not reduced uptake in potted spruce trees. Excess fertilizer decreased fluoride uptake and enhanced foliar injury. McCune *et al.* (1966) grew two *Gladiolus* cultivars ('Snow Princess' and 'Elizabeth the Queen', the latter more tolerant than the former) in hydroponic culture in nutrient solutions deficient in N, P, K, Ca and Mg, and exposed them to 2.1 μg F/m^3 for 24 h once weekly for 6 or 7 weeks. Increased tip necrosis was associated with deficiencies of K and P in 'Snow Princess' and Mg deficiency in 'Elizabeth the Queen'. Surprisingly, reduced tip necrosis was found with calcium and nitrogen deficiencies. None of the nutrient levels affected the leaf fluoride contents.

Water stress

Water stress usually results in reduction in stomatal conductance and lower gas exchange so it is not surprising that there is some evidence of an interaction with fluoride. However, there have been few controlled experiments and none in which stomatal conductance, uptake and plant water status were recorded. In laboratory experiments, plants grown under conditions of water stress are generally more tolerant to fluoride exposure than those receiving an adequate supply of water (Daines *et al.*, 1952; Benedict and Breen, 1955). Wiebe and Poovaiah (1973) indicated that plants under water stress were more tolerant to fluoride because of stomatal closure. On the other hand, Treshow (1971) stated that under field conditions plants that are poorly managed and without an adequate water-supply show more injury than those in adjacent irrigated fields (Treshow, 1971). MacLean (1983) suggested that the contradiction might be explained by the fact that vegetation adapted to semi-arid conditions can accumulate high concentrations of fluoride with little injury (Ares, 1978) and that the appearance of symptoms requires a period of wet weather. Because of its potential importance, the whole area of fluoride–water-stress interactions needs further investigation, using modern instruments and methods.

Biotic interactions

Several aspects of plant–invertebrate interactions were discussed in the previous chapter, so this section will focus on plant–pathogen interactions.

It is well known that pathogens and the severity of the diseases they produce can be altered when the host is exposed to certain air pollutants (Heagle, 1973; Hughes and Laurence, 1984; Flückiger *et al.*, 2002). There are three possible explanations for this. The pollutant may alter the success of the pathogen by: (i) a direct effect on the growth, development and reproduction of the organism; (ii) an indirect effect on the pathogen through the altered physiology and biochemical processes of the host; or (iii) alteration in the microbiota or microenvironment of the plant surface (McCune *et al.*, 1973). There is a considerable body of evidence that a change in the incidence and severity of plant disease is associated with air pollution in the field, but little attention has been devoted to fluoride. There are no striking examples of effects of fluoride that are comparable to SO$_2$–disease interactions. All of the publications available on fluoride are laboratory studies conducted under some degree of control, so, although they establish that fluoride may affect diseases, it is not known if such effects occur in the field. The subject was reviewed some years ago by Heagle (1973), McCune *et al.* (1973), Treshow (1975) and Laurence (1981, 1983). The majority of studies are summarized in Table 4.17. One criticism of many of these experiments is that the atmospheric concentrations were higher than normally occur near well-controlled industries, but not all industries are as well controlled as might be desired and foliar concentrations similar to those used in experiments do still occur. Also, in one study, Laurence and Reynolds (1984, 1986) found effects at levels of HF

that are commonly found under field conditions.

In general, experiments in which fluoride was incorporated into growth media showed that it may have direct effects on the pathogen, usually decreasing growth. For example, the fungus *Verticillium lecanii* grew on a medium containing up to about 0.2 M NaF (Leslie and Parberry, 1972), but, as the concentration of fluoride was increased, growth decreased. Similarly, Treshow (1965) showed suppression of growth of several fungi at concentrations in defined media as low as 5×10^{-4} M NaF

Table 4.17. Effects of fluoride on plant–pathogen interactions (After Laurence, J.A. (1981) Effects of air pollutants on plant–pathogen interactions *Zeitschrift für Pflanzenkrankheiten und Pflanzenschutz* 88, 156–173.)

Plant/pathogen	Typical common name of disease	Exposure type	Effects on disease	Authors cited by Laurence (1981)
Fungal diseases				
Alternaria solani	Blight	*In vitro*	Suppressed growth	Treshow, 1965
Botrytis cinerea	Mould, rot	*In vitro*	Suppressed growth	Treshow, 1965
Colletotrichum lindemuthianum	Anthracnose	*In vitro*	Suppressed growth	Treshow, 1965
Cytospora rubescens	Canker	*In vitro*	Suppressed growth	Treshow, 1965
Helminthosporium sativum	Leaf blotch	*In vitro*	No effect	Treshow, 1965
Pythium debaryanum	Damping off, rot	*In vitro*	Suppressed growth	Treshow, 1965
Verticillium albo-atrum	Wilt	*In vitro*	Suppressed growth (high conc.)	Treshow, 1965
Verticillium lecanii	Wilt	*In vitro*	Suppressed growth	Leslie and Parberry, 1972
Phaseolus bean/*Erysiphe polygoni*	Powdery mildew	HF fumigation	Reduced disease	McCune *et al.*, 1973
Phaseolus bean/*Uromyces phaseoli*	Rust	HF fumigation	Reduced disease	McCune *et al.*, 1973
Tomato/*Alternaria solani*	Early blight	HF fumigation	Reduced disease	McCune *et al.*, 1973
Tomato/*Phytophthora infestans*	Late blight	HF fumigation	No effect	McCune *et al.*, 1973
Viral diseases				
Phaseolus bean/tobacco mosaic virus		*In vitro*	Generally stimulated disease	Dean and Treshow, 1966
Phaseolus bean/tobacco mosaic virus		HF fumigation	Stimulated disease	Treshow *et al.*, 1966
Bacterial diseases				
Xanthomonas campestris pv. *phaseoli*	Bacterial blight	*In vitro*	Suppressed growth (high conc.)	Laurence and Reynolds, 1984
Phaseolus bean/ *Pseudomonas phaseolicola*	Halo blight	HF fumigation	Increased stem collapse	McCune *et al.*, 1973
Kidney bean/*Xanthomonas campestris* pv. *phaseoli*	Bacterial blight	HF fumigation	Longer latent period, smaller lesions	McCune *et al.*, 1973

(for *Pythium debaryanum*). One species was stimulated by 1×10^{-3} M NaF at 24°C (*Colletotrichum lindemuthianum*). The one bacterial pathogen tested *in vitro*, *Xanthomonas campestris*, was also suppressed by fluoride.

McCune *et al.* (1973) observed that, at certain times of the year, bean plants grown in environmental chambers and exposed to charcoal-filtered air developed infections of the fungus powdery mildew (*Erysiphe polygoni*). For example, the average number of lesions on the unifoliolate leaf of these plants was 48.7, while those on HF-fumigated plants (containing 399 mg/kg) was only 4.4. When the total exposure was partitioned into two periods, both the first and second exposure to HF significantly reduced the number of lesions on the first trifoliolate leaf from about 24.5 lesions per leaflet, compared with 38.2 on controls, and the combined effect of both exposures was additive (11.7 lesions per leaflet). The results suggested that HF altered the infective capacity of the pathogen itself because the reduction in the disease symptoms was proportional to the length of exposure, infection was continuous during the exposure period and the pathogen was an epiphyte. Working with bean plants and bean rust (*Uromyces phaseoli*), McCune *et al.* (1973) obtained mixed results when they exposed plants to HF before or after inoculation. One experiment is shown in Table 4.18. An analysis of variance showed that the only significant effect of HF was with post-inoculation exposure, but in other experiments the results were different, showing a significant reduction in the severity of the disease in plants exposed to HF before inoculation. Overall, because both pre- and postinoculation exposures were effective and additive, the effect of HF appeared to be related to the accumulated fluoride having direct or indirect effects on the pathogen.

In the case of early blight of tomato (*Alternaria solani*) (McCune *et al.*, 1973), there was a significant effect of exposure to HF before inoculation on the youngest leaves, whereby HF decreased injury, and the effect was greatest as leaf age decreased. Median ranks for HF and control were 24.5 and 16.5 for the seventh leaf and 25.2 and 15.8 for the eighth leaf. HF had no effect on postinoculation exposures on the incidence of disease on any leaf. It appeared that there was a direct effect on early blight that was related to foliar fluoride accumulation, since preinoculation exposures were effective. An indirect effect was indicated by the fact that, when a change in the incidence of the disease occurred, it was on the youngest leaf, which is also the most susceptible to fluoride.

There are very few reports of research on viruses, but the number of local lesions produced by tobacco mosaic virus on pinto bean was increased significantly when leaves contained up to 500 mg F/kg and decreased to below the control level when leaves contained more than 500 mg/kg (Dean and Treshow, 1965; Treshow *et al.*, 1966). In HF-fumigated bean plants (0.2–2.5 µg/m^3), the number of local lesions increased with foliar fluoride concentrations up to about 500 mg/kg, above which the number decreased below that of controls (Treshow *et al.*, 1967). The authors also

Table 4.18. The effect of HF and rust inoculation on the number of rust uredia in pinto bean and the concentration of fluoride in the leaves. (From McCune, D.C., Weinstein, L.H., Mancini, J.F. and van Leuken, P. (1973) Effects of hydrogen fluoride on plant–pathogen interactions. *Proceedings of the Third International Clean Air Congress*. VDI-Verlag GmbH, Düsseldorf, pp. A146–149).

Treatment		Number of uredia per leaflet	Fluoride concentration after 2nd exposure F (mg/kg dry wt)
Preinoculation exposure	Postinoculation exposure		
Filtered air	Filtered air	249.7	70
Filtered air	HF	162.9	864
HF	Filtered air	239.5	578
HF	HF	191.9	1148

noted that plants exposed to the high concentration of HF had fresh and dry masses about 12% higher than controls, an apparent case of hormesis (Calabrese and Baldwin, 2002). The authors stated, however, that 'The tremendous variation encountered in lesion development, and in the striking influence of environmental factors, such as light, temperature, and moisture, make it highly improbable that any measurable modifications of TMV activity could be detected in the field.' Wolting (1975) found that a virus-like leaf necrosis of *Freesia* was aggravated by continuous exposure to 0.5 µg HF/m^3 for 6 weeks or intermittent exposures (three to four 6 h exposures to 0.3 µg HF/m^3/week) for 4 months. Not only did HF cause more severe symptoms, but it reduced the time required for symptoms to develop and eliminated the chlorotic lesion stage.

The bacterial disease halo blight (*Pseudomonas phaseolicolus*) was studied by McCune *et al.* (1973). Fumigation with HF had no consistent effect on the severity of symptoms, despite the high accumulation of fluoride in the foliar tissue. There was no effect of HF before or after inoculation on the incidence of foliar curling or stem collapse, but there was a significant difference in the frequency of stem collapse between plants that received only filtered air and all other treatments (3.9 and 19.2%, respectively). It appeared that the effect of fluoride in this case was indirect, because the site of an effect was spatially removed from the sites of fluoride accumulation.

The only other experiments on bacterial diseases are those of Laurence and Reynolds (1984). Exposure of red kidney beans ('California Light') to 1 or 3 µg HF/m^3 for 5 days continuously after inoculation with *X. campestris* pv. *phaseoli* resulted in longer latent periods and smaller initial lesion size. When the fluoride exposure was applied before bacterial inoculation or in intermittent exposures of 12 h/day for up to 4 days after inoculation, there were no measurable effects recorded. Growth of the bacterium in its resident phase on the leaf surface was slowed by continuous but not by intermittent exposures. Later, pre- and post-inoculation exposures of HF and SO$_2$, alone and combined, were studied (Laurence and Reynolds, 1986). Postinoculation exposure of HF alone resulted in significantly longer latent periods. When the two gases were applied concurrently, the joint action resulted in smaller lesions and longer latent periods. Neither pollutant alone or in combination altered the growth of the pathogen in its resident phase.

Plate 1.

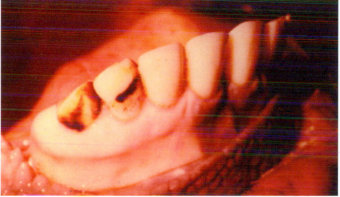

Plate 2.

Plate 3.

Plate 4.

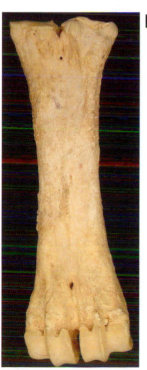

Plate 1. Trees and shrubs showing brown, dead foliage shortly after an accidental release of HF in 1987 (Marathon Oil Co.).
Plate 2. Fluoride-induced dental lesions in the most recent two teeth (1st and 2nd from the left) of a cow (AWD).
Plate 3. Bovine metatarsal showing exostoses – rough chalky white deposits (AWD).
Plate 4. Fractured bovine pedal bone (AWD).

Plate 5.

Plate 6.

Plate 8.

Plate 7.

Plate 5. An animal with characteristic cross-legged stance due to bilateral fracture of the inside foot (J. Alcock).
Plate 6. Necrosis in developing needles of lodgepole pine (*P. contorta*) (LHW).
Plate 7. Current year's needles of *P. mugho* killed by HF but previous year's needles unaffected (AWD).
Plate 8. HF injury on *Iris* leaves. Recent injury (left), successive fumigations (right) (AWD).

Plate 9.

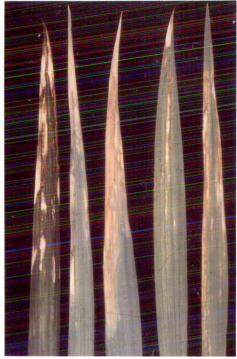

Plate 11.

Plate 10.

Plate 12.

Plate 9. 'Water soaked' area at the tip of a *Hyacinthoides non-scriptus* leaf due to recent HF injury.
Plate 10. Degrees of HF injury in varieties of *Gladiolus* (LHW).
Plate 11. HF-induced chlorosis in *Zea mays* (LHW/AWD).
Plate 12. HF injury to banana (*Musa* sp.) growing near a brick factory (C. Umponstira).

Plate 13.

Plate 15.

Plate 14.

Plate 16.

Plate 13. 'Christmas tree' chlorosis in *Eucalyptus grandis* (LHW).
Plate 14. *Eucalyptus globulus* leaves showing ragged appearance due to loss of necrotic tissue (AWD).
Plate 15. Black necrotic areas in the middle of the laminae of *Persoonia laevis* (AWD).
Plate 16. Reddish-brown necrotic lesion shown by redbud (*Cercis canadensis*) (LHW)

Plate 17.

Plate 18.

Plate 19.

Plate 20.

Plate 17. *Alnus glutinosa* with crinkled, distorted leaves (AWD).
Plate 18. A willow (*Salix* sp.) with cupped, chlorotic leaves and some necrosis (AWD).
Plate 19. *Philadelphus* leaves showing the effects of successive fumigations (LHW).
Plate 20. Pine (*P. contorta* x *banksiana*) needles with necrosis caused by salt aerosol (A.H. Legge).

Plate 21.

Plate 22.

Plate 23.

Plate 24.

Plate 21. Aspen (*Populus tremuloides*) leaves with chlorosis and necrosis caused by salt aerosol (A.H. Legge).
Plate 22. Necrosis of the leaf margins of dogwood (*Cornus florida*) caused by salt aerosol (A.H. Legge).
Plate 23. Necrosis and chlorosis in *Pleione* leaves caused by water stress (AWD).
Plate 24. 'Suture red spot' or 'soft suture' of peach caused by exposure to HF (LHW).

Plate 25.

Plate 26.

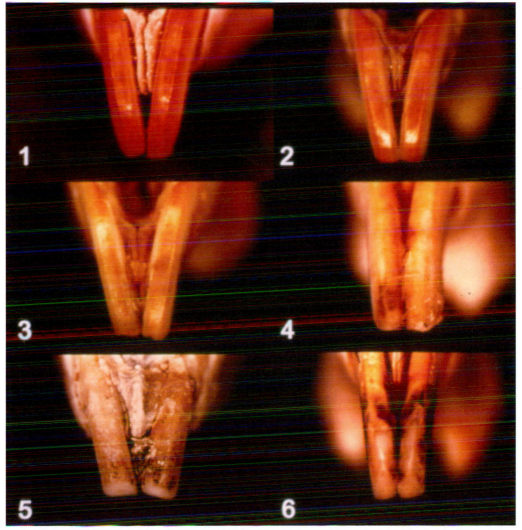

Plate 25. Mature and over-mature trees in the vicinity of the Kitimat smelter killed by insect infestation (AWD).
Plate 26. Effects of fluoride on the incisors of *Mastomys natalensis* and the scale used for scoring teeth (J.A. Cooke).

Plate 27.

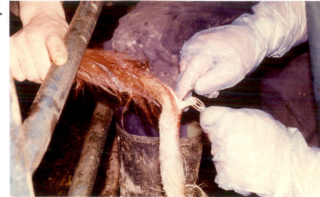

Plate 28.

Plate 29.

Plate 27. A tail bone being removed for fluoride analysis as part of a monitoring programme (AWD).
Plate 28. Cattle killed by eating toxic plants of *Acacia georginae* (T. McCosker).
Plate 29. Distortion of soybean (*Glycine max* cv. 'Ranson') caused by uptake of trifluoroacetate (AWD).

Some Case Histories Involving Fluoride Contamination

Introduction

Examples of the environmental effects of fluoride emissions can be found in most countries, but some of the more high-profile cases are worth considering in detail because they demonstrate the magnitude of potential effects and the improvements in investigative techniques that have taken place in the last 50 years. Some have played a vital part in changing public and political attitudes and some have led to better environmental protection. Others have raised interesting scientific questions. Selecting case histories that demonstrate the diversity of problems encountered in the field, scientifically, forensically and even politically, was difficult, not because there was a shortage of possibilities, but mostly because it was difficult to acquire sufficiently detailed information to weave a comprehensive story. In some cases, one or more key elements were inaccessible. There were a number of excellent prospects, including: aluminium smelting in the Rhône Valley of Switzerland and injury to apricots, grapes, other fruits and forests over many years; the massive HF release from a petroleum refinery in Texas in 1987; cherry and apricot damage in The Dalles, Oregon, decided (in part) by a federally appointed scientific arbitration panel; the widespread fluorosis of livestock in the Bedfordshire, Yorkshire

and Staffordshire areas of England; injury to citrus from phosphate-fertilizer industry emissions in Florida during the 1950–1970 period; steel smelting and plant and animal damage in Utah; aluminium smelting and allegations of damage within Glacier National Park; and several others. The five cases we have chosen represent fluoride problems in all their 'glory' – damage to farmlands, fruits, forests and indigenous plant life, and to farm animals and humans. Some cases were clear-cut, some equivocal and some unproved. The problems were resolved by lawsuit, public outcry, international cooperation, a Queen's Inquiry and time.

Aluminium Smelting at Fort William and Kinlochleven, Scotland

Background

This case occurred in the 1940s, after the publication of Roholm's (1937) seminal work, and therefore at a time when knowledge of fluoride effects on humans and livestock was just developing. Industrial workers in the aluminium smelters at Fort William and Kinlochleven were exposed to high concentrations of several pollutants and concern grew about their health and the potential effects of emissions on the local area. The inquiry was one of the most comprehensive of its time, but, reading the report of the case now (Agate *et al.*, 1949), the investigative techniques seem haphazard at times, probably because the scientists were finding their way in uncharted territory, and the report itself falls far short of current standards in that there is missing detail about some of the procedures. At the time, analytical techniques for fluoride determination were subject to significant errors, so great care has to be taken in interpreting some of the data. Nevertheless, it was an important study that brought fluoride to the attention of British scientists and the public. It also had an influence on the second case history in the USA.

The environmental concerns about the aluminium smelters near Fort William and Kinlochleven, operated by the British Aluminium Company, were principally related to fluoride toxicity to humans and other animals. Effects on plants appear to have been considered as a peripheral problem. In 1945, the Department of Health for Scotland selected a group of scientists to investigate concerns and to report to the Fluorosis Committee. The purpose was to examine the complaints from areas near the factories and 'to determine effects of exposure of fluorine compounds on workers employed in factories manufacturing aluminium in Inverness-shire by an electrolytic process, and also to determine any similar effects on those living in the neighbourhood of such factories' (Agate *et al.*, 1949). Members or designees of the Fluorosis Committee visited the sites of the smelters twice in 1945 and once in 1946, conducting medical examinations and collecting blood, urine, atmospheric, grass and soil samples for fluoride analyses. The main clinical and environmental aspects of the study were carried out by the staff of the Medical Research Council's Department for Research in Industrial Medicine at the London Hospital, the Chief Dental Officer of the Department of Health and Professor G.F. Boddie of the Royal Veterinary College, who studied the effect of fluorine-containing compounds on livestock in the area.

Location of the smelters and the aluminium reduction processes

The smelters were located in the Western Highlands of Scotland, the Kinlochleven smelter being about 15 km south-south-east of Fort William. Both smelters were situated in valley systems, with lochs to the west of both Fort William and Kinlochleven (Fig. 5.1). The Fort William smelter utilized both prebake and Söderberg 'green paste' anodes. The oldest pot room, referred to as 'Furnace room A' and dating from 1929, used prebake anodes but had no mechanical ventilation, depending on natural draughts to carry the fumes out through the side and roof vents.

> Radiant heat from the furnaces makes it very hot in this furnace room and the air is heavily charged with fine dust which settles everywhere as a white deposit; gases produced by the furnaces are also discharged directly into the air of the room.
> (Agate *et al.*, 1949)

Furnace room B dated from 1938 and the furnaces (or pots) were referred to as old-type Söderberg units, a technology which was notoriously dirty, although there was a ventilation unit that carried air at 855,000–956,000 ft³/min (24,000–27,000 m³/min), resulting in an air exchange time of 3.5–4 min. This resulted in a lower concentration of dust and fumes but 'a considerable fog of pitch fumes from the Söderberg electrodes is usually present'. Furnace room C was the most modern, dating from 1942 and fitted with the latest Söderberg technology. Air was withdrawn from each pair of pots at the rate of 7500–9000 ft³/min (212–255 m³/ min), theoretically resulting in a complete exchange of air in the pot room once every 2.2–2.6 min. In relation to the external environment, the important fact is that the fumes extracted from all three pot rooms escaped to the outside atmosphere unscrubbed.

The smelter at Kinlochleven was older than the one near Fort William, and the two furnace rooms were similar to Furnace room A and Furnace room C at Fort William. Unfortunately, the situation at Kinlochleven

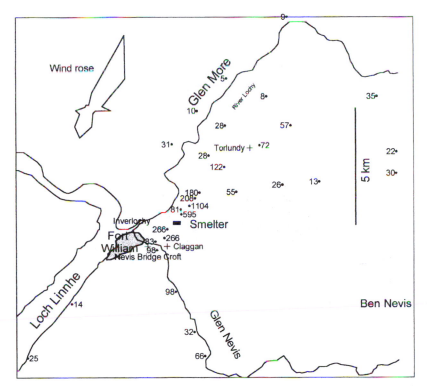

Fig. 5.1. Map of the Fort William area in 1949 showing the location of the aluminium smelter, three farms (Nevis Bridge Croft, Claggan and Torlundy) and grass fluoride concentrations (mg F/kg dry wt). (Redrawn from Agate *et al.* (1949) Agate, J.M., Bell, G.H., Boddie, G.F., Bowler, R.G., Buckele, M., Cheeseman, E.A., Douglas, T.H.J., Druett, H.A., Garrad, J., Hunter, D., Perry, K.M.A., Richardson, J.D. and Weir, J.B. de V. (1949) *Industrial Fluorosis: a Study of the Hazard to Man and Animals near Fort William, Scotland.* Medical Research Council Memorandum No. 22, H.M. Stationery Office, London, according to licence number C02W0002815.)

was not documented in detail by the Medical Research Council Inquiry (Agate *et al.*, 1949), so the subsequent discussion will refer primarily to Fort William.

Meteorology and the fume path

The average wind speed during the period from 1932 to 1944 was from about 0.4–5.4 m/s about 90% of the time, and there was little difference between months of the year. Monthly ambient temperatures never exceeded 60°F (15.6°C), even in June, July and August, and averaged 47.7°F (8.7°C). During this period, rainfall averaged about 77 inches (nearly 200 cm) per year, making it one of the wetter districts in

the British Isles. The relative humidity was fairly constant throughout the year, with a mean value for the period of 80.8%. An obvious plume of dense white smoke drifted from the pot rooms at Fort William and was carried in a north-north-easterly direction, with secondary flow to the south-south-west. Flow in other directions was minor (Fig. 5.1).

Pot-room atmospheres

Workers in Furnace room A, which was the worst of the furnace rooms, with no forced ventilation, complained that, after a period in this room, their eyes and skin became inflamed. The atmosphere in Furnace room

B, although fitted with an exhaust system, was not as clear as in Furnace room A, 'because there is a considerable fog of pitch fumes produced by the baking of the "green paste" in the Söderberg electrodes'. Furnace room C was described as clean, relatively cool and free of fumes. The fluoride content of the furnace-room atmospheres is shown in Table 5.1. Although the concentrations are probably of the right order of magnitude, some of the authors' comments about the technique indicate that they were not very accurate. They had difficulty measuring the volume of air, and the use of a single absorbent bubbler led to loss of some of the fluoride. Furthermore, the proportions of gaseous to particulate fluorides should be interpreted with caution because the technique used to separate the two fractions (an untreated filter-paper) would have overestimated the particulate and underestimated the gaseous components. When the sampling began, the Second World War blackout restrictions were in effect, so both forced and natural ventilation were reduced, resulting in very high fluoride concentrations. By the time the second set of air samples was collected, the war was over, blackout devices had been removed, ventilation was much improved and the fluoride concentrations were reduced several-fold in most cases. In general, the data showed that Furnace room A was somewhat worse than B, and C was the cleanest of the three. The atmosphere in pot rooms at Kinlochleven was about on a par with that of Furnace room B at Fort William. Unfortunately, each air sample was taken at a different location in the pot rooms, resulting in wide variations from sample to sample.

The ambient atmosphere

Immediately outside the Fort William pot rooms, the total fluoride in the ambient atmosphere was reported as being from

Table 5.1. Concentration of fluoride (mg/m^3) in the pot-room atmospheres at Fort William. Each sample was collected at a different location in the pot rooms. (From Agate, J.M., Bell, G.H., Boddie, G.F., Bowler, R.G., Buckele, M., Cheeseman, E.A., Douglas, T.H.J., Druett, H.A., Garrad, J., Hunter, D., Perry, K.M.A., Richardson, J.D. and Weir, J.B. de V. (1949) *Industrial Fluorosis: a Study of the Hazard to Man and Animals near Fort William, Scotland.* Medical Research Council Memorandum No. 22, Her Majesty's Stationery Office, London, according to licence number C02W0002815.)

Date	Pot room	Total F	Gaseous F	Particulate F
21–5–45	A	3.43	2.54	0.89
28–5–45	A	2.15	1.05	1.10
28–5–45	A	2.12	1.04	1.08
17–5–45	B	2.72	1.77	0.95
29–5–45	B	1.49	0.80	0.69
18–5–45	C	–	(Lost)	1.50
Postblackout				
22–7–46	A	1.08	0.58	0.50
23–7–46	A	0.77	0.39	0.38
23–7–46	A (night shift)	1.00	0.44	0.56
25–7–46	A	0.37	0.18	0.19
22–7–46	B	0.64	0.32	0.32
22–7–46	B	0.41	0.19	0.22
23–7–46	B (night shift)	0.70	0.27	0.43
25–7–46	B	0.44	0.25	0.19
23–7–46	C	0.60	0.36	0.24
23–7–46	C	0.14	0.04	0.10
23–7–46	C (night shift)	0.34	0.21	0.13
25–7–46	C	0.33	0.05	0.28

16.4 µg/m³ 100 m east-north-east to 164 µg/
m³ 200 m south-west of the smelter.
Concentrations fell with distance from
the works but, even in the centre of Fort
William, the air was reported as containing
39.4 µg total fluoride/m³. It was assumed
that the average person in the town
was exposed to 7.2 µg/m³ and in nearby
Inverlochy to 25.2 µg/m³. In environmental
terms, these are extremely high concentrations and, if they were anywhere near accurate, there would have been severe damage
to vegetation, which should have been obvious even to lay people. There would have
been very few plants left unmarked. The
authors' experience suggests that concentrations were probably lower, not least
because of scrubbing by the heavy rainfall. In addition, there was a discrepancy
between the atmospheric and grass concentrations (see below), so it is difficult
to avoid the conclusion that the analyses
overestimated the atmospheric concentrations outside the factory. This is under-
standable because quite low volumes of
air were sampled and the analytical techniques that were available were prone to
interference.

Soil contamination

The soils consisted mainly of deep,
acid peat overlying a rock base. About 40
samples were collected at distances up to
about 10 km from the smelter. They were
collected by cutting plugs of turf and
discarding the top inch (2.5 cm), and about
six subsamples were combined from
an area of about 10 m². The analytical
techniques were not described, so it is
not known if the soils were subjected to
alkali fusion to ensure release of all of
the fluoride. However, the data do show
evidence of considerable surface contamination (Table 5.2), which extended
for as far as about 10 km from the smelter.
The contamination at such distances was
probably detectable because of the low
background concentration in the peat (see
Chapter 2).

Table 5.2. Concentration of fluoride in soil samples (mg/kg, dry wt) at various depths in the north-west corner of the Fort William smelter property. (From Agate, J.M., Bell, G.H., Boddie, G.F., Bowler, R.G., Buckele, M., Cheeseman, E.A., Douglas, T.H.J., Druett, H.A,, Garrad, J., Hunter, D., Perry, K.M.A., Richardson, J.D. and Weir, J.B. de V. (1949) *Industrial Fluorosis: a Study of the Hazard to Man and Animals near Fort William, Scotland*. Medical Research Council Memorandum No. 22, Her Majesty's Stationery Office, London, according to licence number C02W0002815.)

Depth of soil sample (cm)	Fluoride content
0–2.5	1010
2.5–7.6	372
7.6–17.8	400
17.8–27.9	268
27.9–38.1	161

Fluoride in forage

The fluoride content of forage is the key to
determining the risk of fluorosis in livestock and other herbivorous animals. Grass
sampling was not described except to state
that 'Handfuls of grass were taken.' However, the individuals who processed the
samples were aware that much of the
fluoride found in grass samples could
be deposited on the leaf surfaces and they
were interested in determining if particulate fluorides were taken into the plant
cells. Therefore they attempted to wash the
grass but abandoned it because of unidenti-
fied 'practical difficulties'. Although in
the original plan the fluoride contents of
samples of fruits and vegetables were also
to be determined, only a few were found
and this phase was abandoned.

Regarding the composition of the indigenous vegetation, the report states:

Coarse grasses, sphagnum moss, heather
and bracken, and clumps of spruce, fir and
pine trees are to be seen on the low ground
about a mile north-east of the factory. The
streams are lined mainly with common
alder and dwarf birch. The hillsides are
bare of trees except for the western side of
Glen Nevis, where the Forestry Commission
has planted large numbers of fir trees.

Apparently none of this vegetation was sampled or analysed, nor were there any descriptions of foliar symptoms or of plant mortality.

Figure 5.1 shows the fluoride concentrations in grass. As mentioned earlier, although they are high, they do not appear to be in accordance with the reported atmospheric concentrations. They were much lower than would be expected if the atmospheric concentrations were correct. The discrepancy might have been due to analytical errors or to the fact that the grass was collected on only one occasion, possibly at a time just after heavy rain. Nevertheless, the values are high enough to indicate that there would have been problems with fluorosis over a considerable area, probably up to distances of 4–5 km in the downwind direction.

Effects on livestock

In 1943, a farmer at Nevis Bridge Croft (south-west of the smelter) complained that his cattle and sheep were suffering from an affliction that he had never previously observed. His animals had been grazing in an enclosed pasture immediately behind the town of Fort William and less than 1 km south of the smelter. Clinical evaluation revealed that tooth formation was abnormal: the enamel of incisors was mottled and the teeth were malformed. This condition was found in all adult sheep but not in any of the 6-month-old lambs. The two cows present had mottled incisors that were worn to stubs at the gums. Cheek teeth were badly worn and abnormal in form. By the spring of 1944, there were complaints from other farmers for damage to sheep at the Achentee Farm, a large holding. In a group of 116 sheep grazing about 1.2 km from the smelter, abnormal teeth were found in 72 animals; in another group of 85, 39 were found to be abnormal. The incisor teeth were not only mottled but also brittle, and many of them had broken. In other animals the abrasive resistance of teeth was so low that many had been worn to the gums.

Cheek teeth were often so abnormal in form that the animal could not close its mouth completely. The worn teeth became a pathway for infection, resulting in purulent inflammation of gums and sinuses. In addition to the drastic effects on teeth, an unusual number of animals had rib, pelvis and mandible fractures. Because milk production was also reduced, there was a secondary effect on the normal development of lambs. Forage contained 44 mg/kg fluoride (on a dry-weight basis). (This is an unusually low value considering the extreme effects on sheep, so it should not be considered as reliable.) No clinical fluorosis was found in cattle at Achentee, probably due to the farmer's practice of purchasing pregnant cows that were not kept on the pastures for more than 1 or 2 years.

In 1946, the herd of 45 cattle at Torlundy Farm, about 2.8 km north of the smelter, was examined. Most of the dairy cows:

> were not in the bodily condition that might be expected at a time of the year when grazing was good . . . the coats of the cows were lacking in lustre and their skins were hard. In contrast to the cows, the young stock were in reasonably good condition.

Incisor teeth in adult animals were mottled, deformed and badly worn. The cheek teeth were affected, as described earlier. In adult animals there was also a relationship between the age of the animal and the degree of dental lesions, with the oldest animals being the least affected. This was due in part to the increase in exposure to fluoride after the smelter began operations. Peak production and fluoride emissions occurred in 1942. Thus, the older animals had been exposed to lower amounts of fluoride during their most vulnerable periods.

In 1945, several cows and sheep were purchased for autopsy. Teeth were in poor condition, as described earlier, and rib bones were often fractured or weak, but other bone deformities were not evident. In both sheep and cattle, there were massive increases in bone fluoride compared with controls (Table 5.3). There was evidence of elevated fluoride in milk of cows and sheep

Table 5.3. Fluoride content (mg/kg) of sheep tissues collected near the Fort William smelter. (From Agate, J.M., Bell, G.H., Boddie, G.F., Bowler, R.G., Buckele, M., Cheeseman, E.A., Douglas, T.H.J., Druett, H.A., Garrad, J., Hunter, D., Perry, K.M.A., Richardson, J.D. and Weir, J.B. de V. (1949) *Industrial Fluorosis: a Study of the Hazard to Man and Animals near Fort William, Scotland.* Medical Research Council Memorandum No. 22, Her Majesty's Stationery Office, London, according to licence number C02W0002815.)

	Rib	Mandible	Radius and Ulna
Achintree, 1945			
Black-face ewe	12,000	14,100	10,300
Cheviot wether 2 years	10,400	17,200	8,980
Black-face gimmer	7,000	9,420	8,170
Cheviot gimmer	660	2,140	1,010
Control, Ettrick Valley, 1945			
Black-face ewe	478	415	279
Black-face ewe	83	111	99
Black-face ewe	382	238	222
Grey-face ewe	343	269	211

(average of 0.44 mg/l in cow's and 0.62 mg/l in ewe's milk), but these values were not considered to be a public health hazard. The inquiry concluded that normal dairy farming and successful sheep farming would not be possible near Fort William.

Effects on humans

For this study, employees at Fort William (and a few from Kinlochleven) were separated into two groups, based upon their degree of fluoride exposure. Group I, the most exposed group, contained 220 men and 22 women. Group II was less exposed; there were 44 men and 25 women in the group. In the residential areas near the smelter, two additional groups were formed – Group III had 27 men and 52 women and Group IV consisted of schoolchildren from the Inverlochy and Fort William secondary schools. Each person received a complete

physical examination, with particular emphasis on movements of the thoracic cage, vertebral column and joints. Each child was given a complete dental examination. All participants received a detailed radiographic examination (with some defaulters for various reasons). Blood samples for fluoride analysis and cell counts and 24 h urine samples were taken from a smaller number of people. The workers in Group I were further subdivided into groups based upon the degree of exposure, with furnace men, stub squad and tappers being in the most exposed group. When questioned about the incidence of back pain, digestive disturbances or dyspnoea (laboured breathing), there did not appear to be an excess incidence of back pain, but there was evidence of more complaints of digestive disorders and coughs. There was also no evidence of an increased incidence of bone fractures among the most exposed group. Fluoride in urine was highest in the most exposed workers, commensurate with their degree of exposure. Older workers exposed for a number of years exhibited bone abnormalities typical of workers elsewhere who had been exposed to fluoride. These radiographic abnormalities were mainly seen as increased bone density, but none of the workers was found to suffer clinical disability. However, the panel stated:

> It does not follow that the risk to health of workers in the older type of furnace room is negligible, for it is well known that the progressive course of fluorosis is slow; rather are the changes to be regarded as a warning that the conditions under which these men work are such as to call for constant vigilance and for determined efforts to reduce the amount of fluorine to which they are exposed.

It is worth noting that the study of teeth suggested that the local children were less prone to caries than children from other areas.

Regarding effects on human health in areas outside the smelter, the total fluoride intake of the population was not recorded, apart from an abandoned attempt to determine the fluoride content of vegetables. However, the report stated that although a clinical evaluation of a small number of residents 'has shown no sign of injury to health, it is only prudent to site new developments in such a way that, so far as possible, residents are kept out of the zone known to be most liable to contamination.' The final sentence in the panel's conclusions was: 'It is important that everything practicable should be done to reduce the amount of fluorine discharged from the factory.' Both smelters continued operation but Kinlochlevin closed in 2000 for economic reasons. Over the years the technology at Fort William has been improved, the pots were changed to use centre-worked, prebake technology and scrubbing was improved. The result is that the current emissions meet 'new plant' standards. Emissions to the outside are dramatically lower and today the Fort William smelter is surrounded by mature forest, with conifers and deciduous trees growing within 200 m or so – something that would have been impossible 50 years ago.

Martin v. Reynolds Metals

This case history has a connection with the Fort William and Kinlochleven inquiry in that much of the background information associated with the Martins' allegations of damage to their health as well as their livestock by fumes from the Reynolds smelter in Troutdale, Oregon, was supported by one of the authors of the Fort William report (Dr Donald Hunter, a Fellow of the Royal College of Physicians and a director of research in industrial medicine of the Privy Council's Medical Research Council). The 'Martin case' appeared in major litigations several times, twice in Federal proceedings (US Court of Appeals, 9th Circuit, 1958, 1964) and once in Oregon State proceedings (Supreme Court of Oregon, 1959).

Background

Reynolds Metals acquired an aluminium smelter in Troutdale, Oregon, from the US government in 1946. That same year, the first pot line was put into operation. The technology employed was Söderberg. Each pot room was equipped with roof ventilators that drew air through louvred doors at the base of the room, carrying it upward to the ventilators. At the ventilators, a system of water sprays ran the full length of each building. A baffle system condensed mist from the exiting air and returned it to the scrubbing water. Allegedly, 60% of the effluent fluoride was captured by this system. In 1950, hoods were placed over each pot and the air was forced through spray towers with a stated efficiency of 90% (at least of the fluoride released in the process).

In December 1946, Paul Martin, his wife, Verla Martin, and his daughter, Paula Yturbide, moved into a cattle farm (Troutdale Ranch) about 1.6–2.4 km from the smelter. They resided there until November 1950, when they filed lawsuits claiming that during their years in residence on the ranch they were exposed to toxic levels of fluorides from Reynolds. Their claims included not only human-health effects, but also fluorosis in cattle and damage to vegetables growing in their garden. These lawsuits emphasize the woeful lack of knowledge at the time regarding the routes of uptake of fluoride by humans, the inadequate design of experiments to examine vegetation effects and the lack of comprehensive ambient-air monitoring at the Martin ranch.

The legal actions

The legal actions began in 1952. In the course of the proceedings, the Martins moved for a directed verdict against Reynolds for personal injuries as a result of negligence, stating that:

> the poisonous fluoride compounds emanating from the plant of the defendant and carried by the prevailing winds upon and over, and settling upon, the lands upon which the plaintiff resided and was present, were inhaled and ingested by plaintiff, and said poisonous fluoride compounds rendered plaintiff permanently ill and caused

plaintiff to suffer and to undergo serious and permanent personal injury.

(*Martin* v. *Reynolds*, 1958)

The motion was denied, but the court rendered judgements for the plaintiffs of $47,135 (P. Atkins, personal communication). Reynolds then filed an appeal against these judgements, which was denied (*Martin* v. *Reynolds*, 1958).

The crux of this case from a legal standpoint was whether or not there was sufficient proof that the damages claimed were caused by emitted fluorides and if there was evidence of negligence on the part of Reynolds. For its part, Reynolds argued that there was insufficient proof that there was damage to the Martins' health or that there was any negligence or 'breach of duty' on the part of Reynolds.

A plant scientist from Oregon State College (now University) established plots of *Gladiolus* and buckwheat, both HF-sensitive species, in 1948, 1949 and 1950, at various distances from the smelter on and near the Troutdale Ranch, the closest being 1.8 km. Plants grown in plots closest to the smelter had the greatest amount of fluoride, with lower concentrations being found with increasing distance, as one would expect. The court found, however, that these studies did not establish the amounts of fluoride that would be found in garden vegetables grown and used for domestic purposes by the Martins. In fact, the Martin family would have had to consume huge quantities of vegetables to have significantly increased their fluoride intake because the fluoride content of vegetables remains low even in areas with very high atmospheric concentrations. Grain, fruit and root crops contain essentially background levels and careful washing removes much of the surface deposits from leafy vegetables.

Reynolds did not contest the allegations that large quantities of fluoride were released into the air each day. In 1947, this amount averaged 2845 pounds (1290 kg) per day with comparable amounts in 1948, 1949 and the first half of 1950. This average rate of emission has been known to cause vegetation injury and fluorosis of livestock

at other smelters. Examination of the Martin cattle indicated damage due to fluorosis. At around the same time, Reynolds began the installation of a more efficient control system.

During the examination of witnesses, there was frequent reference to the inquiry conducted at Fort William (Agate *et al.*, 1949), and the court was provided with information that cattle grazing in the vicinity of Fort William exhibited 'osteodystrophia [deformed bones], dental lesions, and "some cases of osteomalacia [softening of bones] due to fluorine intoxication"; also severe loss of condition and loss of milk in affected animals' (*Martin* v. *Reynolds*, 1958). The court noted that cattle would receive excessive fluoride from the forage, which represented a major part of their diet, while the human diet would receive only a small portion of its total intake from plants. Nevertheless, the British physician Dr Donald Hunter (from the Fort William inquiry), testified as follows:

> Q. Why, Doctor, are both cattle and animals as well as men, humans, affected by effluents from an aluminium factory?
> A. Because fluorides – fluorine compounds are deadly poisonous to mammalian tissues, and man is a mammal just as much as a cow or sheep.

Dr Hunter also testified to the existence of etched glass from the windows in the Martin house, contending that this was evidence of excessive quantities of fluoride contamination in the atmosphere. Dr Capps (see below) testified, 'I think that if there is enough fluorine to etch a window, it should be able to etch a lung' (*Martin* v. *Reynolds*, 1958). This is the kind of sound bite that plays well with the media but it displays an ignorance of chemistry and routes of uptake that should have damaged Dr Capps' credibility. (Although the occurrence of etched glass in the vicinity of aluminium smelters was once common, it is rarely seen today because of greatly reduced fluoride emissions.)

The Martins' case regarding effects on their health was impeded by the lack of medical literature and of expert witnesses who had intimate knowledge of persons living outside fluoride-emitting industries with the same symptoms as those who worked inside. Nevertheless, there were appearances by Dr Hunter and Dr Capps, a Chicago specialist on diseases of the liver, who had been referred to by Dr Hunter as 'probably the world's greatest expert on this subject'. Dr Capps examined the Martins in November 1951, and Dr Hunter examined them shortly before the trial commenced and also reviewed the results of tests by Dr Capps and other physicians. Dr Capps's findings included the onset of diarrhoea, indigestion, heartburn and bloating, as well as skeletal symptoms manifested as back pain and sciatica, the latter so severe that Mr Martin was unable to bend to put on or tie his shoes. Other symptoms included a dry cough, shortness of breath and an abnormal frequency of urination. At the trial, Dr Capps's diagnosis was presented as follows:

> In view, then, of this potential history of the exposure and in view of the fact that we were advised – that we were faced with a rather bizarre group of symptoms – here is an involvement of a number of symptoms in the body which was unusual and definitely bizarre, and it corresponds exactly to the description of published cases of fluorosis; furthermore, there is no other explanation. One cannot make another single diagnosis that would cover all these symptoms. One could make four or five diagnoses, but, of course, that is always obviously a very poor thing to do. This is the picture of fluorosis. There is a history of the potential exposure. We are unable to find any explanation after a careful search, and under the circumstances one is justified, and not only justified, but you are forced to make the diagnosis of poisoning with fluorine.
>
> (*Martin* v. *Reynolds*, 1958)

Pursuing a similar theme, Dr Hunter testified that 'fluorine compounds are deadly poisons to mammalian tissues', that, when they enter the body, they inhibit metabolic enzymes and that fluorides fed to experimental animals result in degeneration of the liver and kidney, symptoms seen in the Martin family. His conclusion was that the Martins were suffering from subacute fluorosis. He also recounted having examined a family in England with the same symptoms as the Martins who lived in a house with etched windows near an ironworks. He also referred to a study conducted in 1946 that concluded with the warning to industry 'against throwing into the atmosphere an effluent which would etch glass'. These were quite remarkable statements in a situation where there was virtually no information available about the fluoride intake of the family.

Another major issue in the case was the allegation that Reynolds Metals had not installed the latest equipment to capture fluoride emissions and had operated the smelter from 1946 to 1950 with an inefficient system. Reynolds denied this and said that it:

> was utilizing and had installed the best known means of protecting against the escape of fluorides; that the company, continually studying the problem, installed an experimental pilot plant to develop a new process and then changed over to the improved process as soon as the feasibility developed.

The previously mentioned Dr Hunter testified as follows:

> Q. Doctor, are you familiar with an induced air system or dust control or fume control? A. Naturally, I haven't been trained as a ventilating engineer. But since I first studied industrial hygiene in the Harvard School of Public Health with the famous Phillip Drinker – he is a ventilating engineer – and because of the work which I have to do and so teach, which is industrial medicine, I am, naturally, interested in systems of ventilation. And in all of the chapters of that book there is an exact description of systems of ventilation which are used in various parts of the world. And in 1946 it was well-known to all industrialists the world over that exhaust ventilation could be effectively applied to remove these effluents.

Was Reynolds aware that an 'induced air system' was available in 1946 but was not installed until 1950, or could the original system be considered as an 'induced air system'? The judge allowed this opinion regarding negligence to be decided by the jury on the theory that this was a case correctly permitting the application of the doctrine of *res ipsa loquitur* (the thing speaks for itself). The court also instructed the jury as follows:

> Under the law and the facts of these actions, the defendant was sole operator and in exclusive possession and direct control of the aluminium plant involved during the time involved, namely, between on or about September 23, 1946 and November 30, 1950. Further, that under the ordinary course of events, it is unexpected that persons being in the vicinity of such a plant would be injured or harmed by fluorine compounds emanating therefrom, and that such a mishap would not occur.

The first instruction was obvious, but the second instruction conformed to the defendant's contention as well as to much of the evidence produced at the trial that the smelter operating as it was would not cause fluorosis to persons who had not been employed at the smelter itself, nor had any evidence been presented regarding the amount of fluoride necessary to cause fluorosis in such persons (*Martin* v. *Reynolds*, 1958). The court viewed the plaintiff's case 'in the most favorable light, as we are required to do' and stated that the jury was warranted in accepting the testimony of the witnesses that, in fact, they were exposed to excessive amounts of fluoride and that they subsequently suffered fluoride poisoning. But, and here is a crucial point in the case, the plaintiffs were unable to establish the percentage or concentration of fluorides in the emissions that actually reached them. The defendants had already conceded that excessive amounts of fluorides were poisonous. Thus, the question was whether or not Reynolds was guilty of negligence or other breach of duty.

The Martins claimed that their land was unfit for raising livestock and sought damages of $450,000 for the loss of the use of their land for grazing and for deterioration of plant life growing on their land, and $30,000 punitive damages. The jury returned a judgement of $91,500, of which $71,500 was for the loss of use of their land and $20,000 for the deterioration of their land. Human health effects and punitive damages were rejected (*Martin* v. *Reynolds*, 1958, 1959).

During the course of the pleadings, Reynolds asserted that the incursion of fumes from the smelter was at most a nuisance and not trespass. In the former case, liability would be only for the 2-year period preceding the date at which the Martins instituted their lawsuit, whereas in the case of trespass there would be a 6-year statute of limitations. If there were an actionable invasion of the Martins' interest in the exclusive possession of the land, it would be a trespass; if it were an actionable invasion of the Martins' land that interfered with their use and enjoyment of the land, it would be a nuisance. The court held that the intrusion of fluorides fulfilled all the requirements under the law of trespass and that the Reynolds fumes 'rendered plaintiffs' land and the drinking water on the land unfit for consumption by livestock grazing thereon' (*Martin* v. *Reynolds*, 1959). This was the first time that a court held that an invasion of private property by an air pollutant constituted a trespass rather than a nuisance.

In response to the continued emissions of fluorides from the smelter, albeit at a lower rate, the Martins erected a large billboard on US Highway 30 as traffic passed the Troutdale Ranch and approached the Reynolds Metals smelter. The sign read as follows:

> THIS RANCH IS CONTAMINATED
> 831 Cattle Killed in past six years
> FLUORIDE POISON from REYNOLDS METALS CO.
> kills our cattle***endangers human health
> CONTROLS MUST BE ENFORCED
> THIS STATEMENT PAID FOR BY PAUL R. MARTIN

Although not mentioned in the litigation, beneath the billboard was a stuffed black cow, lying on its back with legs reaching skyward. Reynolds demanded an injunction requiring removal of the billboard, based first upon libel and secondly upon tort, referred to as injurious falsehood and malicious interference with Reynolds's business relations. They argued further that there was no competent evidence on record to indicate that fluoride endangered human health and they objected to a portion of the finding that:

> fluoride is toxic and the fluoride compounds emanated from defendant's plant are poisonous; insofar as such compounds settle upon plaintiffs' land, they contaminate the same; to that extent the portions of the sign which refer to poisons and contamination are true.

Upon considering the opinion of a Special Master, the court ruled that the sign contained false and untrue material and that the language of the sign constituted an actionable libel of Reynolds and 'is tortious conduct not in the public interest or for the public interest'. The sign was removed (*Martin* v. *Reynolds*, 1964).

The Martins commenced an action to recover actual damages in the amount of $300,000 and punitive damages in the amount of $100,000 in the State of Oregon Circuit Court in 1961, but, because of 'diversity of citizenship', it was removed to the US District Court. A second amended complaint was filed that was seeking actual damages of $1,428,342 and punitive damages of $1,000,000, but this grew to $4,250,000 (P. Atkins, personal communication). Martin also asked for injunctive relief against continued smelter operations until adequate controls were installed. In 1962, Reynolds purchased 456.9 acres (185 ha) and, in 1968, 1407.94 acres (570 ha) (P. Atkins, personal communication) for an undisclosed amount.

Kitimat, British Columbia, Canada

Background

The environmental calamity at Kitimat, British Columbia (BC), is among the best-documented cases. Much of the information for this case history came from publications by Silver (1961), Reid Collins and Associates (1976, 1981), Alcan Surveillance Committee (1979), Bunce (1979, 1984, 1985, 1989), Weinstein and Bunce (1981), Gurnon

and Smart (1990), and from comparison with the forests at Long Harbour, Newfoundland (Sidhu, 1979, 1980; Thompson *et al.*, 1979). The aluminium smelter at Kitimat, BC, was inaugurated by the Aluminium Company of Canada in 1954. At the time, it was the second largest smelter in Canada, the USA or Europe (exclusive of the Soviet Union or other Eastern bloc countries), with a production capacity of 800 tons a day, or 300,000 tons per year. The smelter was located at the head of Kitimat Arm, a fjord-like encroachment with direct access to the Pacific Ocean, and it lies on the west side of a wide valley, which runs in a north–south direction. The residential area is located about 5 km north-east of the smelter and out of the main path of the prevailing winds.

The forests

Kitimat is in the Pacific coastal rain forest, which, prior to the construction of the smelter, was described (in forestry terms) as an uneven-aged, overmature, decadent and stable climax forest. It passed from full maturity to overmaturity when individual trees died of old age and were replaced in the natural course of events by young, naturally regenerated trees. About 40% of the hemlocks and 20% of the firs were classed as culls, and there was little loss from disease in trees less than 250 years old. The average age of firs and hemlocks exceeded 300 years, with more than one-third of the total wood volume of these species contained in trees more than 425 years old. The main soil groups found in the Kitimat Valley and the lands adjacent to the estuary are podzols and brown podzolic soils associated with the forested areas. Chernozems are found in patches of semi-open grasslands, regosols on recently deposited river alluvium and organic soils with decaying vegetation. The industrial area of Kitimat is located on the flood-plain of the Kitimat River.

During the first three decades of smelter operation, logging was an important commercial activity. In 1945, an intensive forest survey was made in the Kitimat River valley. The overstorey was found to comprise old-growth trees of large diameter and with a substantial number of defective (i.e. diseased, etc.) trees. The understorey of younger, smaller trees was in better condition, although suppressed. The tree species native to the area, arranged in order of importance, were western hemlock (*Tsuga heterophylla*), balsam fir (*Abies amabilis*), western red cedar (*Thuja plicata*), Sitka spruce (*Picea sitchensis*), yellow cedar (*Chamaecyparis nootkatensis*), mountain hemlock (*Tsuga mertensiana*), red alder (*Alnus rubra*), black cottonwood (*Populus trichocarpa*) and lodgepole pine (*Pinus contorta*). Lichens and mosses were abundant but not extremely varied, a condition usually found in the coastal rain forests. As quoted from a BC Forest Service report (Reid Collins, 1976):

> South slopes are often predominantly occupied by hemlock, while creeks, shady sites and deep soils favour balsam [true fir]. Red cedar will form a major part of the stand wherever soil moisture is abundant. Sitka spruce is at present mostly confined to the rich swamps and alluvial soils of the valley bottom, where it grows often with cedar in a dense ground cover of Devils club . . . On young soils along the river and its arms grow stands of alder and cottonwood. Finally, lodgepole pine is able to grow on the rounded rock knolls of the lower elevations.

Fume dispersion

The smelter utilized the vertical-stud Söderberg process to reduce alumina to aluminium. From 1955 to the period of 1965–1969, emissions of gaseous fluoride increased, from 1971 to 1974 there was a slight decrease and from 1975 to 1980 emissions were less than half of those of the previous period (Table 5.4).

The path of fluoride emissions from the smelter is predominantly in a northerly direction in spring, summer and early autumn, and in a southerly direction in the

Table 5.4. Gaseous fluoride emissions from the Kitimat smelter and annual mean fluoride contents of mixed-species foliage (mixed species collected from 12 locations). Data for 1970 omitted because of strike. (Modified from Reid Collins and Associates Ltd (1976) *Fluoride Emissions and Forest Growth*. Aluminium Company of Canada, Kitimat, B.C.)

Year	Gaseous F emissions (t/day)	Mean F content of foliage (mg/kg)
1955–1959	4.5	142
1960–1964	5.2	185
1965–1969	5.9	238
1971–1974	5.5	178
1975	2.3	50
1976	2.3	60
1977	2.9	107
1978	2.5	59
1979	2.0	43
1980	1.7	40

winter. Winds in the valley carry the emissions into several smaller valleys that branch to the west of the general north–south wind flow. Insect injury is a common feature of all forests but, by 1960, excessive insect injury was noted in the forests north and south of the smelter. Silver (1961), referring to an unusual infestation by the saddleback looper (*Ectropus crepuscularia*), stated:

> There was no gradual decrease in defoliation or number of larvae towards the edge of the infestation but rather an abrupt line, in places only 1/4 mile wide, separating infested from non-infested stands. The area of heavy population coincided remarkably well with the extent of the 'fume' cloud from the smelter. The reason for this is not understood yet.

This remarkable correspondence was also noted 10 years later by Weinstein (Alcan Surveillance Committee, 1979), who witnessed the overlapping of the areas of primary insect attack with the pattern of fume dispersion by aerial survey and concluded that 'The similarity in the areas of looper infestation and smelter smoke dispersion might be more than coincidental'. He also noted that:

> There were few or no large firs and spruces [remaining] in the most affected

areas, a larger number of hemlocks, and a preponderance of cedars. This became more obvious some distance away from the smelter. Thus, the relative susceptibilities to fluoride of the predominant coniferous species appeared to be related to their presence in polluted areas.

This would appear to be strong evidence that tree mortality was due to fluoride exposure alone. But the epizootic that developed makes this conclusion implausible. There is a literature suggesting that there may be a relationship between fluoride pollution and increased insect activity (e.g. Alstad *et al.*, 1982; Hughes, 1988), but the size of this attack is still without precedent.

Insect attacks

The Canadian Forestry Service and the Forest Insect and Disease Survey described insect outbreaks that occurred in the Kitimat area over a number of years and they were summarized by Reid Collins (1976). The forest was attacked in 1960 by a massive outbreak of the saddleback looper, causing extensive and indiscriminate defoliation of all species, but especially showing preference for the Pacific silver fir and western hemlock. This attack lasted for 2 years and was more or less confined to an area about 37 km north and 12 km south of the smelter (Fig. 5.2). In some areas, as much as two-thirds of the hemlocks were destroyed. Silver (1961) stated:

> With few exceptions the undergrowth, including Devil's club, elderberry, and other deciduous bushes was defoliated. Most of the coniferous reproduction, regardless of species, was completely stripped . . . This species (saddleback looper) feeds from the forest floor up, and feeding was stratified to the extent that when defoliation was heavy in the upper third of the crowns of intermediate trees, feeding was also heavy on the lower and mid-crowns of co-dominant and dominant trees.

During the same year, a spruce bud worm (*Choristoneura orae*) attack was consolidated and defoliated the young growth of many balsam firs and continued to defoliate firs for 4 years. Eventually, the area of damage extended about 42 km north and 13 km south of the smelter, was often 5–10 km wide and extended as much as 10 km along the western tributaries of the Kitimat River.

In 1961, conditions were favourable for secondary incursions of the balsam bark beetles (*Pseudohylesinus grandis* and *Pseudohylesinus nebulosus*), which colonized the Pacific silver-fir trees weakened by the primary insect attacks. The bark beetles continued to destroy weakened trees until 1969. As a result of the combined insect attacks and fluoride stress, there was an enormous mortality of mature and overmature trees, which had a striking effect on the landscape (Plate 25).

The Reid Collins study

In their study of the impacts of fluoride and insects on the forests which began during the summer of 1974, Reid Collins (1976) determined the probable distribution of smelter emissions by combining knowledge of the physical properties of the gases, fume distribution and dispersion patterns, flow patterns in the area and topography. This was supplemented by aerial and ground surveys, by the results of foliar fluoride analyses carried out during the years in question and by evaluating the extent, abundance and general condition of lichens.

Sixty-four one-fifth acre (0.08 ha) sample plots were established in a grid arrangement covering an area from 18 km north to 8 km south of the smelter and in a band up to 6.5 km in width. Each plot was selected to be representative of the important species of trees and habitats in the area. Sixteen control plots were established in an area known to be free of smelter fumes. The diameters of all trees in the plots were measured, and increment cores were taken from selected trees in each plot to provide a chronological record of growth and of wood density. Foliar fluoride analyses were also made. The data that resulted were used to segregate the area into three zones of effects, 'inner', 'outer' and

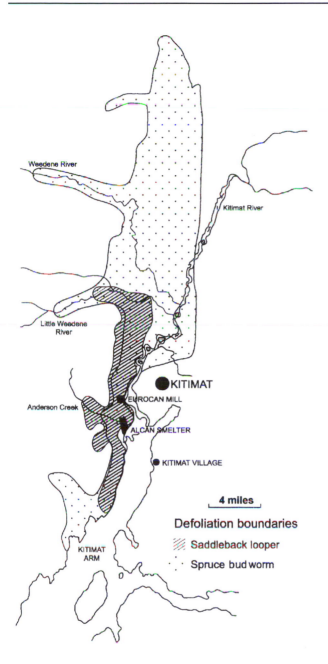

Weedene River

Kitimat River

Little Weedene
River

● KITIMAT

EUROCAN MILL

Anderson Creek

ALCAN SMELTER

● KITIMAT VILLAGE

| 4 miles |

Defoliation boundaries

▨ Saddleback looper

⋰ Spruce bud worm

KITIMAT
ARM

Fig. 5.2. The extent of saddleback
looper (*Ectropus crepuscularia*)
and spruce bud worm (*Cristoneura
orae*) defoliation at Kitimat, BC,
in 1961. (Redrawn from Reid
Collins and Associates Ltd (1976)
*Fluoride Emissions and Forest
Growth.* Aluminium Company
of Canada, Kitimat, B.C.)

'surround', going from greatest to least effects on the forest, and representing 1866, 5197 and 6856 ha in the three zones, respectively.

By 1974, the growth rate of hemlock was reduced by 28.1% in the inner zone, by 19.0% in the outer zone and by 2.2% in the surround zone (Fig. 5.3). This translated to losses of 2690 m³/year in the inner and surround zones due to airborne fluorides between 1954 and 1974 (or a total of 53,800 m³) (Reid Collins, 1976). In a 1979 remeasurement (Bunce, 1985), the outer and surround zones exhibited a 2.8% and 13.6% increase in growth, and hemlock foliage contained concentrations of 29 and 22 mg/kg (dry weight), respectively. The growth stimulation in the surround zone was

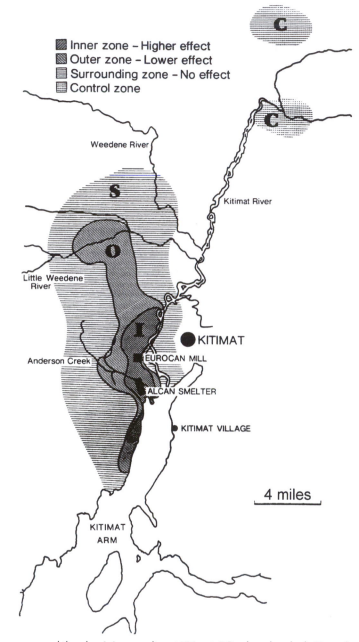

Fig. 5.3. The area around the aluminium smelter at Kitimat, BC, where hemlock (*Tsuga heterophylla*) showed growth reductions in 1974. (Redrawn from Bunce, 1985. Data used from Bunce, H.W.F., *Journal of Air Pollution Control Association* 35, 46–48, 1985, with permission of the Air and Waste Management Association.)

commented upon in Chapter 4. The inner zone continued to show a growth reduction, although lower than in the original study, and this was related to a hemlock foliar fluoride concentration of 87 mg/kg. During the period between the original measurements and the remeasurement (1974–1979), fluoride emissions were reduced by 64% (Table 5.4). The Canadian Forestry Service estimated that the loss due to insects was 283,100 m³ by 1961 and about 2,265,000 m³ when the insect attacks had subsided.

A complementary study of lichens (Reid Collins, 1981) concluded that:

> smelter emissions have caused a depletion in lichen population on the western side of the Kitimat valley . . . This depletion ranges from severe to slight over the area surveyed. It is also concluded that a decline in the lichen population occurred over the three year period 1976–1979 in both the fume and control areas. This decline has occurred despite the reduction in F emissions.

However, there were sources of SO_2 in the area and their effects on the lichens were not taken into account.

How does one reconcile emissions from an aluminium smelter sited in a forested area with such massive outbreaks of two primary insects? Admittedly, there have been a number of reports demonstrating increased insect or mite activity in the presence of fluoride, sulphur dioxide, particles or other airborne contaminants (Alstad et al., 1982), but nothing of this magnitude has ever been reported. Clearly, the evidence suggests that something related to the smelter was responsible for what was deemed the largest recorded outbreak of the saddleback looper in British Columbia. There have been no recurrences of insect attacks of this immensity, despite the interim period of more than 30 years. On the other hand, emissions from the smelter, especially of particulate materials, are now only a fraction of what they were up to 1974, and the incidence and severity of foliar injury were reduced greatly after 1974. Let us summarize some basic facts and examine them individually.

1.   The attacks of primary and secondary insect invaders began in 1960 and ended in 1969. The pattern of injury appeared to be closely correlated with the dispersion of gases and particles from the smelter (Silver, 1961; Alcan Surveillance Committee, 1979).

2.   Silver (1961) described foliar destruction by the insect larvae as progressing from the understorey upward. Fluoride injury of trees generally begins in the upper crown and proceeds downward. That is not to say that insect-weakened trees were not killed by fluoride emissions or that fluoride-weakened trees were not killed by secondary insects.

3.   There were no documented descriptions of fluoride-induced foliar injury until 1971, 11 years after the insect attacks had begun. This makes it difficult to relate emissions, foliar injury and insect outbreaks.

4.   There was a marked decline in the incidence and severity of fluoride-induced foliar injury after 1974, coincident with reduced emissions. Despite this reduction in emissions, Hocking et al. (1981) reported that 1975 year-class needles of hemlock collected in 1976 from an area about 1.5 km north-west of the smelter contained more than 300 mg/kg fluoride on two seasons' growth, which is equivalent to c. 190 mg/kg on a first-season collection.

So what were the inciting factors that caused the calamity at Kitimat? In his field report of 1977, quoted in Alcan Surveillance Committee (1979), Weinstein set forth several possibilities to explain the severity and sharply delineated areas of insect damage:

1.   Fluoride accumulation by tree foliage resulted in trees with reduced vigour, less able to withstand insect colonization. Although this is an attractive theory, and might be valid for secondary invasions by bark beetles, there does not appear to be evidence that primary foliar-feeding insects prefer to attack trees weakened by fluoride.

2.   Uptake of fluoride by tree foliage altered primary or secondary cellular metabolism, rendering the tree more attractive to insects for feeding or oviposition. It is well known that some forest insect pests are attracted by volatile terpenes that emanate from host trees. Renwick and Potter (1981) have shown that exposure of balsam fir (Abies balsamea) to sulphur dioxide can cause increased emissions of terpenes, especially ß-pinene, a compound related to attraction

of spruce bud-worm moths to their hosts (Städler, 1974). If the smelter fumes, which are a mixture of HF, SO_2, carbon monoxide, SF_6, a variety of polycyclic and other hydrocarbons and various particulate materials, have a similar effect on volatile emissions, it is possible that they play a role in attracting insects and in their successful colonization of host trees.

3.   Fluoride, particulate materials or other constituents in the emissions are toxic to parasitic insects, mainly Hymenoptera, which are among the natural control mechanisms for many insects. The toxicity of ingested fluoride to some insects has been documented (Chapter 3), as have the toxic effects of some particles on the survival of parasitic wasps (Bartlett, 1951).

4.   Smelter fumes may be toxic to predacious insects through ingestion of fluoride present in their prey (but see Chapters 2 and 3).

5.   The smelter emissions, which contained an immense amount of particulate matter, created a 'blanket effect', resulting in a somewhat narrower range of temperature extremes, which could have been favourable to the development of the destructive insects. In addition to particulate materials, the emissions were high in CO_2, which could affect ambient temperatures or increase the metabolic activity of the trees.

6.   The adult moths, which are weak fliers, were carried into the Kitimat area by strong winds and were dispersed in the valley in the same pattern as were smelter emissions.

7.   During and after construction of the smelter, there was reportedly an overabundance of electric light-bulbs in areas near the smelter, which emitted wavelengths attractive to the moths, concentrating them at the smelter site at night for distribution in the daylight hours by the same winds that dispersed the smelter emissions.

There is no direct evidence to relate any of these hypotheses to actual events. Over the years, there have been many improvements in the smelter operations, which have resulted in reduced fluoride and particulate emissions, especially through methods of emission control. These changes, along with alterations in forest composition and structure arising from the initial insect attacks, are important factors in any recurrence of the earlier disaster.

The Kitimat experience appears to have other unique features not reported to exist near other fluoride sources. For example, even after the great reduction in emissions around 1974, the fluoride content of foliage collected in many areas north of the smelter continued to be very high, but there was little or no foliar injury on coniferous or broad-leaved species. For many years, it was not uncommon for needles of silver fir, Sitka spruce or hemlock to contain 100 mg/kg dry weight or more, with little or no evidence of foliar injury.

Because the Kitimat situation was not studied during the early years of fluoride injury and insect attack, no definitive answers with respect of causation are possible. The remaining questions are: (i) Why has there been no recurrence at Kitimat in the intervening 30 years? (ii) Why have similar conditions not resulted in explosive situations in other forested ecosystems, such as Long Harbor, Newfoundland (Sidhu, 1979, 1982; Thompson et al., 1979)?

Cubatão, Brazil

Background

The environmental débâcle that occurred in areas near the huge industrial complex at Cubatão in south-eastern Brazil was conceived with the construction of a large industrial complex in the 1950s. The subsequent problems were the result not only of fluoride emissions but also of many other pollutants and it is very difficult to disentangle the effects of any one pollutant. One thing that is clear, however, is that the major effect of fluoride was on the vegetation; no evidence has been established of effects on public health.

Cubatão is located in the state of São Paulo in an area called the Serra do Mar and lies at the foot of a tropical submontane rain forest, the Serra do Mar, often referred to locally as the Atlantic Forest. The city is located between São Paulo, Brazil's largest city, and Santos, its largest port, making the area ripe for major industrialization. Beginning in the early 1950s, an oil refinery and several petrochemical plants were constructed. During the next two decades, several fertilizer plants and a steel smelter were established. Within this zone, 21 uncontrolled industrial installations were located (with about 250 emission sources), all of them at sea level. The military junta made it clear that they cared less about pollution than about industrializing. During the daylight hours, prevailing winds carried the complex mixture of pollutants up narrow valleys and deposited it along the steep, heavily forested mountain scarps of the Atlantic Forest (in the Serra do Mar mountains), which range in elevations up to 800 to 1100 m. During the winter, there are layers of thermal inversions of different depths and intensities. Effects on the forest vegetation were exacerbated by high rainfall and humidity, high temperatures and the topographic aspects of the area. The industries emitted pollutants without any control for over three decades, earning the sobriquet 'the valley of death'. Estimates of the emissions of the major pollutants are shown in Table 5.5.

Table 5.5. Air-pollutant emissions from the industrial complex at Cubatão in 1984 and 1994 (t/year). (From Alonso, C. and Godinho, R. (1992) A evolucão do ar Cubatão. *Quimica Nova* 15, 126–136; and Klockow, D. and Targa, H. (1993) Air pollution and vegetation damage in the tropics. The German/Brazilian Interdisciplinary Project 'Serra do Mar' in the industrial area of Cubatão/ São Paulo, Brazil. In: Junk, W.J. and Bianchi, H.K. (eds) *Studies on Human Impacts on Forests and Floodplains in the Tropics*. GKSS, Geesthacht, Germany, pp. 13–18.)

Pollutant	1984	1994
Particulate matter	114,600	31,000
Hydrocarbons	32,800	4,000
Sulphur dioxide	28,100	18,000
Nitrogen oxides	22,300	17,000
Ammonia	3,100	75
Fluorides	1,000	74

Effects on human health

The intense and varied pollution in Cubatão had an important health effect on the local inhabitants but the evidence points principally to the main effects being due to particulate matter. Within an area of 162 km², Cubatão supported 107,000 inhabitants in the 1980s. Unfortunately, the residential areas were intermixed with industrial plants, and kindergartens were interspersed within the area, generally within 500 m of each child's residence. Children were therefore exposed to a wide mixture of gaseous and particulate pollutants. In 1983 and 1985, Hofmeister and Fischer (cited in Spektor *et al.*, 1991) reported results of lung-function indices, using spirometric techniques, which showed low values for schools located near styrene and petrochemical plants and near a heavily travelled highway and a refinery. In 1988 (Spektor *et al.*, 1991), lung-function indices in kindergarten children were compared with the PM_{10} (particulate matter between 2.5 and 10 µg) or thoracic aerosol mass concentration. At all stations, the PM_{10} values were well above the US annual mean standard of 50 µg/m³, and were as high as 240 ± 122 µg/m³, although the latter was in the Vila Parisi (VP) area, which is non-residential (and the area where most of the fertilizer factories were concentrated). When each of the child's respiratory indices was regressed on the average PM_{10} of the previous month, statistically significant negative slopes were attained, strongly suggesting that fine particulates were the cause of impaired lung function.

As public concern mounted over the reputation of Cubatão as one of the most highly polluted areas in the world, allegations were being heard that there was an unusually high occurrence of anencephaly. This is a condition in which the embryological closure of the neural tube is never completed, leaving the embryo to develop without the upper portion of its skull; these embryos sometimes continue to develop into the fetal stage and may even survive to be born alive, but with upper cranium and scalp missing and the brain open to the outer world. According to Monteleone-Neto *et al.* (1985) and Monteleone-Neto and Castilla (1994), this allegation was vigorously denied in 1980 by local politicians as both *Newsweek* and *Time* magazines disseminated it worldwide. Local citizens founded the Victims of Pollution and Harmful Conditions Association. Investigations were then begun into the quality of air, water, soil and human-effects end-points, such as respiratory diseases (Spektor *et al.*, 1991) and birth defects (Monteleone-Neto *et al.*, 1985). After an intensive study, Monteleone-Neto and Castilla (1994) concluded that:

> The observations reported here did not yield evidence for an elevated birth prevalence rate of anencephaly, neural tube defects, or any other type of major congenital malformation in Cubatão. Furthermore, there is no indication of a high frequency for other adverse pregnancy outcomes such as low birth weight, miscarriages, or perinatal deaths in this community.

The authors concluded that negative results are the rule, not the exception, in studies of alleged disease clusters and because they have such a limited impact, most investigators leave them unpublished. The authors insisted (we believe correctly) that negative results should be published and widely disseminated because unproved allegations soon become 'fact' and affect the future and welfare of the stigmatized community. Cubatão is an excellent example of a stigmatized community and it is still commonly believed to be the 'home' of anencephaly.

At the same time as the controversy over anencephaly, mutagenic effects of particulate matter were also of growing concern after ambient-air genotoxics had been shown to be derived from fuel combustion, waste incineration and industrial processes and formed in the atmosphere through photochemical reactions (Sato *et al.*, 1995). As a result, the mutagenicity of the organic fractions of particulate matter from three sites, two in the São Paulo urban area and one in the Cubatão industrial area, were toxic using the *Salmonella* mutagenicity assay. All samples collected in São Paulo and Cubatão showed mutagenicity with strain TA98. Addition of S9 did not alter the mutagenic response, suggesting that direct-acting frame-shift mutagens were present in the samples from each area. Surprisingly, the mutagenicity rate associated with the two sites in São Paulo was much higher than that in Cubatão, and similar to results for other large urbanized areas of the world.

Effects on vegetation

The cocktail of pollutants resulted in forest degradation, which was most evident on the exposed scarps up to about 800 m elevation. Trees in some areas exhibited severe injury in their upper branches, resulting in a high rate of death, reduction in species diversity, elimination of several tree species and epiphytes and disturbances in ecological function over an area of about 40 km^2 (SMA, 1990; Domingos *et al.*, 1998, 2000, 2002; Klumpp *et al.*, 1996c, 1998). In 1985, this degradation of the forests, especially in the Mogi Valley, resulted in a series of severe landslides, which became a threat to human habitation as well as to industries. This valley is downwind of fertilizer and steel industries, with the major phytotoxic emission being fluorides. The primary and secondary forest, which covered about 53% and 30%, respectively, of the land area in 1962, had nearly disappeared by 1997 (Table 5.6). Species and family diversity in the Mogi Valley, the greatest source of fluoride pollution, was less than 50% of that found in the Pilões Valley, a relatively unpolluted area (Domingos *et al.*, 2000).

In response to the landslide, retaining walls were constructed between the hillsides and industries and, perhaps of greatest consequence, pollution control and refforestation measures were initiated (SMA, 1990; Klumpp *et al.*, 1995). Pollution-control measures resulted in a greatly diminished air-pollution load between 1985 and 1992, but, according to Klumpp *et al.* (1995)

and based on atmospheric fluoride concentrations, even in 1998 it was still far from satisfactory (Table 5.7). After mitigation began in 1984, emissions of all air pollutants were reduced, with fluoride emissions decreasing from about 1000 to 74 t (Alonso and Godinho, 1992; Klockow and Targa, 1998). But fluoride emissions appeared to be somewhat higher in 1998 than in 1992 (Lopes, 1999).

Plant bioindicators and biomonitors

In 1989, an extremely fruitful German–Brazilian research programme was initiated to study the biology, soil science, chemistry and meteorology of the problems at Cubatão (Klockow and Targa, 1993), and most of the

foregoing discussion resulted from that collaboration. It is important to emphasize that, although fluoride was implicated as a major cause of the decline of forests and the subsequent catastrophic landslides in the Mogi Valley, there were almost certainly significant effects caused by the complex mixture of pollutants, which included sulphur dioxide, ozone, peroxyacetylnitrates (PANs) and nitrogen oxides. Visible symptoms were reported for the first three of these pollutants. The programme demonstrated the importance of some of these other pollutants.

The studies of vegetation in the German–Brazilian collaboration included the use of bioindicators in the Serra do Mar (Klumpp *et al.*, 1994, 1995, 1996b, 1997). In an initial study, Klumpp *et al.* (1994) investigated the types of phytotoxic atmospheric

Table 5.6. Changes in vegetation cover determined by aerial photography (in % of total area). (From Domingos, M., Lopes, M.I.M.S. and Struffaldi-De Vuono, Y. (2000) Nutrient cycling disturbance in Atlantic Forest sites affected by air pollution coming from the industrial complex of Cubatão, Southeast Brazil. *Revista Brasileira Botanica* 23, 77–85.)

Vegetation class	1962	1977	1989
Primary forest	53.4	0.26	0
Secondary forest	29.8	0.28	0
Early secondary forest	4.4	0.06	0
Forest affected by pollution	0	43.6	10.3
Forest severely affected by pollution	0	30.6	54.8
Low forest affected by pollution	0	12.1	20.4
Bushy and herbaceous vegetation	1.8	1.9	4.1
Others (agricultural, urban and industrial areas)	10.6	11.2	10.5

Table 5.7. Periodic measurements of gaseous and particulate fluoride concentrations ($\mu g/m^3$) in the atmosphere at the CETESB monitoring site in Vila Parisi, Cubatão. (From Lopes C.F.F. (1999) *Fluoreto na atmosferica de Cubatão*. Technical Report, CETESB, São Paolo, Brazil.)

Month-year	Gaseous fluoride		Particulate fluoride		No. of days of measurement
	Monthly mean	Maximum daily mean	Monthly mean	Maximum daily mean	
9-85	5.38	10.30	0.98	2.33	12
8-87	2.38	5.20	0.41	1.18	10
7-92	0.50	1.77	0.12	0.33	31
8-92	0.73	1.97	0.18	0.54	23
9-92	1.33	5.56	0.43	1.45	30
7-98	1.07	1.97	0.24	0.67	13
8-98	1.63	3.58	0.31	1.32	25
9-98	1.37	2.29	0.18	0.47	28

contamination and their spatial and temporal distribution in the Cubatão industrial area. The wide diversity of industry and the potential for phytotoxic pollutants gave added impetus to the establishment of a programme for testing potential species as bioindicators for fluoride (see Chapter 6), ozone, PANs, other photochemical oxidants and smog components (aldehydes and hydrocarbons). There was also a need to determine the relative loads of heavy metals and sulphur compounds, and biomonitor species were included for this purpose. Because the air mass at Cubatão contained a mixture of pollutants and because it was possible that the primary (or secondary) emissions were spatially separate, we shall examine the total programme (Table 5.8).

Twelve sites were chosen in the Cubatão and Serra do Mar region, locating them in four areas, each with a different pollution composition and load and at different distances from sources and different elevations (Fig. 5.4 and Table 5.9). The VM sites were located in the Mogi Valley and were under the direct influence of fertilizer and steel industries. The PP sites were placed at the end of the Mogi Valley on a plateau near Paranapiacaba, about 10 km from the same sources as the VM sites. The CM sites were placed along the Caminho do Mar, a road connecting São Paulo with the north-western part of the Cubatão industrial zone and near the petrochemical installations. The RP sites were located about 4–5 km south-west and were selected because the forest vegetation in the area appeared to be unaffected by pollution. The IB site was placed at the São Paulo Botanical Garden.

Lolium foliage was extremely effective in accumulating fluoride, ranging between 319 and 677 mg/kg dry wt in the Mogi Valley and with much lower amounts elsewhere.

Bel W3 tobacco plants at the CM sites, especially those at the higher elevations, had a considerable amount of ozone marking, but the amount was about the same as at the São Paulo Botanical Garden. An intermediate amount of injury was found on plants at the PP sites near Paranapiacaba. However, the VM plants had little injury, suggesting that ozone was not a major factor in causing the severe plant damage in that part of the valley. Results with *Urtica urens*, which is sensitive to ozone and PANs, were also greatest at the CM sites and the Botanical Garden, with intermediate amounts of injury at the VM and PP sites. Little injury was seen at the RP sites. *Gladiolus* injury, caused by exposure to fluoride, was almost exclusively confined to the VM sites, influenced by emissions from fertilizer and steel industries (Tables 5.9 and 5.10). A small amount of injury appeared at the two lowest-elevation CM sites, and practically none at the PP, IB and RP sites. *Hemerocallis* was found to be an acceptable bioindicator for fluoride with results paralleling the injury found for *Gladiolus*. The authors (Klumpp et al., 1994, 1996a) concluded that, in the Mogi Valley (VM sites), gaseous fluoride was probably the most important phytotoxic component in the blend of pollutants in the atmosphere and confirmed that fluoride injury was largely responsible for the forest damage and the severe landslides in that area. A few years after mitigation began, the concentration of fluoride in the atmosphere

Table 5.8.   Plant species used in the biomonitoring programme at Cubatão.

Species, variety	Sensitive to
Nicotiana tabacum cv. 'Bel W3'	Ozone
Urtica urens	PANs (and ozone)
Petunia hybr. cv. 'Mirage'	Components of photochemical smog (e.g. aldehydes, hydrocarbons and PANs)
Gladiolus hybr. cv. 'White Friendship' and other cvs	Hydrogen fluoride
Hemerocallis hybr. cv. 'Red Moon'	Hydrogen fluoride
Lolium multiflorum italicum cv. 'Lema'	Accumulation of fluoride, sulphur, heavy metals, etc.

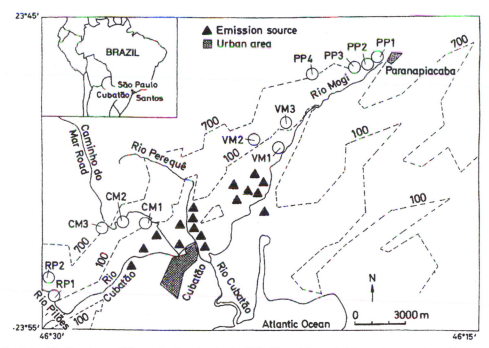

Fig. 5.4. The locations of biomonitoring sites in the Cubatão and Serra do Mar region used by a German–Brazilian team to investigate pollution effects. Major emission sources are shown. VM sites were under the direct influence of fertilizer and steel industries; PP sites were about 10 km from the same sources as the VM sites; CM sites were along the Caminho do Mar road; and the RP sites were selected because the forest vegetation in the area appeared to be unaffected by pollution. (Redrawn and modified from Klumpp, A., Klumpp, G. and Domingos, M. (1994) Plants as bioindicators of air pollution at the Serra do Mar near the industrial complex at Cubatão, Brazil. *Environmental Pollution* 85, 109–116, with permission from Elsevier.)

Table 5.9. Foliar fluoride concentrations (mg/kg dry wt) in *Gladiolus* and *Lolium* foliage at 12 experimental sites in September/October 1990 (from Klumpp *et al.*, 1994). (Reprinted from *Environmental Pollution* 85, 109–116, 1994, with permission from Elsevier.)

Location	Elevation (m)	Site	Lolium	Gladiolus
Pilões Valley	40	RP1	16	7
	150	RP2	14	4
Caminho do Mar	80	CM1	131	21
	450	CM2	165	21
	740	CM3	55	14
Mogi Valley	20	VM1	677	80
	140	VM2	576	57
	250	VM3	319	33
Paranpiacaba	860	PP1	79	19
	800	PP2	37	8
	800	PP3	119	32
	760	PP4	–	–
São Paulo	800	IB	52	10

Table 5.10. The % leaf injury and fluoride content of five varieties of *Gladiolus* grown at Pilões and Mogi Valley sites, Brazil. (From Klumpp, A., Modesto, I.F., Domingos, M. and Klumpp, G. (1997) Susceptibility of various Gladiolus cultivars to fluoride pollution and their sustainability for bioindication. *Pesquisa Agropecuária Brasileira* 32, 239–247.)

Gladiolus cultivar	% leaf injury		Fluoride content (mg/kg dry wt)	
	PB – Pilões	VM – Mogi Valley	PB – Pilões	VM – Mogi Valley
'White Friendship'	0.6	20.3	6.6	224
'Peter Pears'	1.3	18.3	6.1	266
'Gold Field'	0.8	15.5	6.5	230
'Fidelio'	1.1	12.7	6.7	190

and the amount accumulated in *Lolium* exceeded the standards recommended by Verein Deutscher Ingenieure (VDI) in Germany (VDI, 1987, 1989) and an excessive amount of residual fluoride still remained in foliar tissues and in soil solution a decade later (Klumpp *et al.*, 1996a).

Fluoride and Brimstone

Background

Our last case involves a natural source of fluoride and we have included it as a reminder that, even though we can control and reduce industrial emissions, there is still a major source that will continue to cause environmental problems in many parts of the world. Volcanoes are commonly identified with brimstone and other sulphurous constituents but many volcanic eruptions, as well as fumaroles and geothermal springs, not only evolve sulphur- and chlorine-containing compounds (e.g. SO_2, H_2S, HCl, etc.) but also fluoride-containing gases, such as HF and/or SiF_4. In fact, bison and mule deer inhabiting areas around the hot springs in Yellowstone National Park in Wyoming, USA, often have mild fluorosis and display mottled teeth. These fluoride volcanoes are found throughout the world, but Icelandic volcanoes, and its most notorious one, Mt Hekla, will be discussed in greatest detail here. Dozens of others are known and have been studied, including Mt Vesuvius and Etna in Italy, Mt Lonquimay in Chile and a number of

volcanic centres in New Zealand, the most famous of which is Mt Taupo. A notable non-fluoride-emitting volcano is Mt St Helens in the state of Washington, USA.

The Icelandic volcanoes

In general, fluoride-emitting volcanoes have been responsible for disseminating tephra (ash and debris) containing HF and SiF_4 over wide areas, resulting in phytotoxicity and fluoride contamination of vegetation (Thorarinsson and Sigvaldason, 1973; Fridriksson, 1983). The fluoride compounds (as well as other noxious gases) are carried on tephra principally by adsorption to the particle surfaces. Thus, the direct asphyxiating effects on animals and humans, destruction of vegetation and incidence of fluorosis are closely associated with the dispersion of tephra, especially where the tephra layer is thin (< 1–2 mm) and does not interfere with grazing. It is especially dangerous when deposited during the growing season. Damage to grasslands, forests and agriculture has been associated with deposition of tephra over distances up to 500 km. Increased injury with distance is due to the much greater absorptive surface area of small particles, which are dispersed over the greatest distances. Thus, following many Icelandic eruptions, damage has been reported from areas far remote from the original eruption (Thorarinsson, 1979).

Mt Hekla and other Icelandic volcanoes have erupted many times (Fig. 5.5). The first

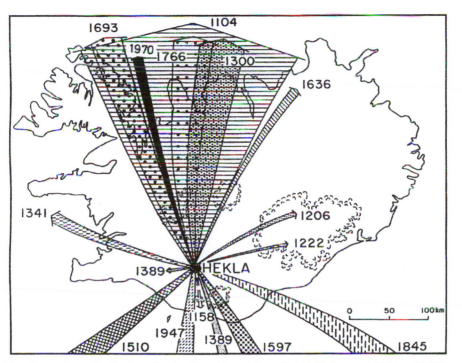

Fig. 5.5. Approximate distribution of tephra from 16 eruptions of Mt Hekla between 1104 and 1970 AD. (Redrawn from Fridriksson, S. (1983) Fluoride problems following volcanic eruptions. In: Shupe, J.L., Peterson, H.B. and Leone, N.C. (eds) *Fluorides. Effects on Vegetation, Animals and Humans*, Paragon Press, Salt Lake City, Utah, pp. 339–344.)

recorded eruption of Hekla occurred in AD 1104. The tephra layer that fell within the 0.2 cm isopach (an isoline designating the depth of tephra) covered an area of 55,000 km². Fine particles from this eruption have been found in peat bogs as far away as Norway and Sweden (cited by Thorarinsson, 1979). The total amount of tephra that fell has been estimated to be 2.0×10^9 m³. In 1693, Hekla erupted for the 11th time, with a total tephra volume estimated at 300 million m³. In Iceland alone, the tephra covered 22,000 km², within which many meadows and cultivated fields were devastated.

Sheep are the most common livestock raised in Iceland and therefore have borne the brunt of the fluoride toxicosis. In this situation they appear to be as vulnerable to fluoride concentrations in forage as dairy cattle, with a threshold of 30–40 mg/kg dry wt. In Iceland, a fluoride content that exceeded 250 mg/kg dry wt of grass killed

sheep within 2–3 days (Sigurdsson and Pálsson, 1957, cited in Thorarinsson, 1979). The contemporary report of the Hekla eruption of 1693 contained the first description of dental lesions in sheep and was written by a farmer and chronicler in 1694:

> In the following autumn and winter people noticed that on the teeth of grazing sheep were yellow spots and some black ones; in some animals the teeth were all black; the teeth fell out in some cases, but small, round-pointed teeth came up afresh, like the teeth of a dog or a catfish; where the spots came the tooth turned soft so it could be shaved like wood. In some animals the flesh peeled away from around the front teeth and molars . . . People thought that this was due to the sand-fall from Hekla.
>
> (Thorarinsson, 1967)

Another contemporary reported that young cattle and horses also had yellow and some black spots on teeth, with some teeth being all blackened. In teeth that formed after being shed by horses and sheep, there were bluish black streaks, often referred to as 'ash-teeth' (Thorarinsson, 1967). The molars of some sheep and horses developed outgrowths, or spurs, which made it impossible for the animals to graze and chew. The condition became known as *gaddur*. Special pliers were used to break off these outgrowths.

Fridriksson (1983) quoted a local clergyman and historian, Jon Steingrimsson, who described the aftermath of an eruption of Laki, a 30 km long fissure that appeared in central Iceland in 1783:

> In horses, sheep and cattle the effects of the perilous fumes from the fires resulted either in death or various diseases; the horses lost all muscle, sometimes the hide decayed along the back. The hair of the tail and mane rotted and fell out if pulled. Joints became enlarged, especially above the hooves. The head puffed up with severe weakness in the jaws and the animals could not graze, swallow or chew. Sheep suffered even more. Practically all extremities were swollen with lumps, especially on the jaws where they tore out of the skin. Large bone lumps grew on the ribs, hipbones and legs. The legs bowed or became extremely fragile. Lungs, liver and heart were edemic and sometimes withered, the insides were decayed and soggy. The meat from these animals was foul smelling, bitter tasting and often poisonous, although people would try to make use of it. The cattle suffered similar hardships. Ossified lumps grew on the jaws and shoulder bones, and in some cases the leg bones would split. Hip joints and other joints moved out of place or became calcified. Part of the tail fell off and then the hooves would loosen in parts or drop off.

Reverend Steingrimsson prayed so intensely that he later claimed responsibility for stopping the lava flow before it engulfed his church.

Roholm (1937), in describing the effects of the 1845 eruption of Hekla, reported:

> The disease attacked animals which ate the grass contaminated with the fallen ash, which means sheep especially, as cattle and horses were kept stabled as much as possible and so escaped. Many animals

died acutely, in the course of days or weeks. Chronic changes developed in the course of months and were observable in the year after the eruption. There were emaciation, increased diuresis, decrease of milk yield, loss of strength, and impairment in the use of the limbs. The bones, especially those of the extremities and the jaws were thickened by growths which could be cut with a knife. No mention is made of spontaneous fractures. It was characteristic that young animals especially were attacked . . . The symptoms disappeared when the animals were taken indoors and fed on hay mown before the eruption. Later symptoms, which could be traced up to ten or twelve years after an eruption, were a variety of dental affections attacking only the permanent teeth of animals which had not shed their milk teeth prior to the eruption. The incisors, which in some cases were smaller and more pointed than normally, were studded with yellow and black spots; they decayed quickly (ash-tooth). The molars had the greater changes, however, being deformed so that the row became irregular, making cud-chewing difficult. Sharp prominences would gnaw holes in the opposite jaw. This dental disease has the ancient Icelandic name of *gaddur* or *gaddjaxl* (*gaddur*, spike; *jaxl*, jaw-tooth).

Roholm also examined bones of adult sheep collected after the 1845 eruption. The teeth all appeared to be normal but the bones were coated to varying degrees with a kind of porous and brittle osseous tissue. In extreme cases, the bone was twice the normal diameter.

Perhaps the best documented of the Icelandic eruptions is that of Hekla in 1970 (Thorarinsson and Sigvaldason, 1973; Thorarinsson, 1979; Fridriksson, 1983). Tephra released from the eruption covered 12,000 km^2 and was deposited 200 km away at a rate of 10 t/ha (Fig. 5.6). This Hekla eruption released an estimated 30,000 t of soluble fluoride adsorbed to particle surfaces. (Assuming that the average aluminium smelter releases less than 0.5 t/day, this would be the equivalent of emissions for 160 years!) In southern Iceland, early in the eruption, the tephra contained 2000 mg/kg fluoride while, in the north, it was 1400 mg/kg (Oskarsson, 1980). Within 2 weeks, the

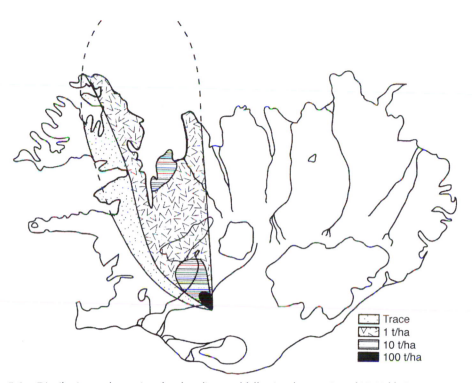

Fig. 5.6. Distribution and quantity of tephra dispersed following the eruption of Mt Hekla in 1970. (Redrawn from Fridriksson, S. (1983) Fluoride problems following volcanic eruptions. In: Shupe, J.L., Peterson, H.B. and Leone, N.C. (eds) *Fluorides. Effects on Vegetation, Animals and Humans.* Paragon Press, Salt Lake City, Utah, pp. 339–344.)

fluoride content had dropped by 90% and, a week later, it was only 1% of the original. The fluoride content of young grass was 4300 mg/kg (dry wt) but it declined rapidly, although, by the end of May, the concentration was still about 200 mg/kg (Fig. 5.7). Coniferous trees were especially vulnerable and dicotyledonous plants exhibited marginal chlorosis. Mosses and lichens were badly affected and eliminated in some areas (Fridriksson, 1983). About 3% of the sheep were affected and loss of lambs was 8–9%. Lamb and sheep bones developed normally but had three to four times as much fluoride as normal animals. Some dental lesions were found in young sheep (Georgsson and Petursson, 1972).

Tephra from the 1980 eruption of Hekla contained 1500–2000 mg F/kg. About 3 days afterwards, vegetation about 150 km north of Hekla contained more than 1000 mg F/kg (dry wt), but it decreased

rapidly by leaching during precipitation events, so that, 21 days later, the fluoride content was near to background level (Fridriksson, 1983).

These effects reported in relation to Iceland were all associated with fluoride in tephra, but there are a few situations where plants are exposed solely to gaseous fluorides. This is the case at high elevations on Mt Etna, where Garrec *et al.* (1977a) reported elevated fluoride levels in vegetation collected in 1976 of between 113 and 295 mg/kg. The species were unnamed. Mt Etna releases about 30 t F/day, most of which is HF. Although fluoride is by far the most phytotoxic of the volcanic gases, a much greater volume of sulphurous (SO_2, H_2SO_4, H_2S, etc.) and chloride-containing compounds is also ejected. The possibility of interactive effects of two or more pollutants cannot be dismissed, although the kinds and degree of injury to plants resemble

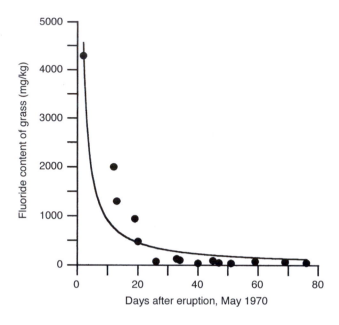

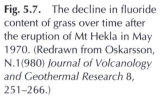

Fig. 5.7. The decline in fluoride content of grass over time after the eruption of Mt Hekla in May 1970. (Redrawn from Oskarsson, N.1(980) *Journal of Volcanology and Geothermal Research* 8, 251–266.)

those induced by fluoride. Garrec *et al.* (1977a) and Thoraninsson (1979) acknowledged this likelihood and suggested that the proportion of sulphur dioxide present with fluoride may reduce the amount of fluoride accumulated by plants (Mandl *et al.*, 1975), in part, perhaps, by stomatal closure induced by sulphur dioxide.

6

Monitoring and Identifying Effects of Fluoride in the Field

Introduction

One of the greatest challenges in fluoride toxicology is predicting the effects of fluoride exposure on plants or animals in natural and managed ecosystems. Although there are adequate methods for measuring fluoride in the atmosphere, they are generally expensive, lack sensitivity and are labour-intensive. Besides, even if the concentration of fluoride in the atmosphere is known, on its own this information is not usually sufficient to predict effects in an ecological context. Where plants are concerned, length and intermittence of exposure, temperature, light intensity, relative humidity and other biological factors play an important part in determining the effects. For animals, such as livestock or insects, knowledge of the total dietary intake is the key requirement. This leads to a self-evident but key question that should always be asked when monitoring is being planned: why is it being undertaken and what are the goals? In some cases monitoring may be a statutory requirement and a way of ensuring that regulations are not being breached. From an industry perspective it may be defensive, providing evidence for the regulator or public that there is no significant impact. Or it may be used to provide an early warning in order to prevent problems, such as fluorosis, which takes time to develop.

However, in too many cases the authors have seen monitoring programmes that do not seem to have an obvious purpose, that are not well planned and that in some cases are excessive. We have seen cases where data were being collected but never examined or analysed, and the scientific literature has many examples of 'monitoring' in which the end result is data that cannot be interpreted in terms of effects or used to prevent environmental damage. Therefore, in this chapter we discuss the methods that have been used to monitor the concentrations of fluoride in the environment and the effects on plants and animals. The intention is to help users to choose the most appropriate methods for their purposes and to avoid some of the pitfalls. The chapter is divided into four main parts: measuring atmospheric fluoride concentrations; biological indicators (bioindicators); biological monitors or accumulators (biomonitors); and field surveys.

Measuring Atmospheric Fluoride Concentrations

Volumetric instruments

The main advantage of using atmospheric fluoride concentrations for monitoring is that, with due care, the data are an absolute measure that is free from much of the uncertainty and problems of interpretation associated with biological methods. Reliable records of atmospheric fluoride concentrations are an invaluable aid to interpreting visible injury shown by vegetation. Several instruments based on spectroscopy are currently available that give real-time measures of the concentrations of HF and SiF_4, but they are expensive and they do not have the sensitivity required for most out-plant monitoring. Their main use is for in-plant surveillance of industrial processes and determining sources of emission (e.g. LaCosse et al., 1999). For the range of concentrations that occur off industrial property and in the threshold range where effects occur, air samples have to be collected for periods of minutes to days, using alkaline absorbents, and then analysed, usually with a fluoride-specific ion electrode. Several commercial instruments are available in which the air is passed through an alkali-impregnated filter paper, through paper tape on a spool (Weinstein and Mandl, 1971) or through a liquid absorbent. An example of the data from a paper-tape sampler was shown in Fig. 2.3 (Chapter 2). In this case the data were very useful in demonstrating the frequency distribution of fluoride concentrations and the fact that the range is greater nearer the source. This kind of information cannot be revealed by bioindicators, although it may be possible using an appropriate biomonitor.

Monitoring using the rate of deposition of HF on alkaline media

The cost and power requirements of instruments mean that in most circumstances only one or two can be deployed, so they cannot show the geographical pattern of atmospheric fluoride concentrations, which is an important aid in field surveys. An alternative is to measure deposition of fluoride on an alkaline surface and to calibrate the system against a volumetric monitor (Davison and Blakemore, 1980). This technique may have had its origins in the use of natural populations of Spanish moss (Tillandsia spp.) as collectors of atmospheric fluoride or, by analogy, with the long-established method for monitoring SO_2 using lead peroxide candles, but the first use of lime-impregnated papers was in the 1950s, when it was found that there was a correlation between the fluoride deposited on the paper and the HF concentration (Miller et al., 1953; Robinson, 1957). These correlations have been confirmed many times (Adams, 1961; Wilson et al., 1967; Israel, 1974a,b; Blakemore, 1978; Lynch et al., 1978; Sidhu, 1979; Davison and Blakemore, 1980). The basic technique is to impregnate a paper with an alkali, often a calcium salt or sodium formate, expose it to the air in a housing that protects it from the rain and then analyse the deposited fluoride after a period from a week to a month. Variants of the method are widely used in many countries, especially by the aluminium industry. Some have used the papers hanging in louvred boxes (Davison and Blakemore, 1980), others cement them in inverted Petri dishes (Lynch et al., 1978; Clark, 1982) or enclose them in envelopes (Israel, 1974a) and others wrap the papers around cylinders to form candles (Wilson et al., 1967).

Most users regard the data as a relative measure of the fluoride concentration and report some form of 'fluoridation index' or 'fluoridation rate', and there is a tendency to regard the system as a 'black box', even though the physics and chemistry of deposition are well known. A few authors have calibrated their systems, notably Robinson (1957), Adams (1961), Mukai and Ishida (1970), Israel (1974a), Lynch et al. (1978) Sidhu (1979) and Davison and Blakemore (1980). The latter authors reviewed the publications up to 1980 and showed that the rates of deposition differed from about 25 to 73 µg F/dm^2/week at an atmospheric concentration of 1 µg HF/m^3. The differences were due to the fact that the different methods used to protect the papers from rain had different effects on the turbulence. The systems that exposed the papers to wind speeds above about 3 m/s had consistent, high rates of deposition, around 65–70 µg/dm^2/week per 1 µg F/m^3. The two studies that were done indoors with low wind speeds had rates of 25 and 33 µg/dm²/week. Davison and Blakemore (1980) concluded that comparability could be obtained by standardizing the housing used to shelter the papers. With adequate control of the turbulence by sheltering the surfaces and after calibrating the individual system, the technique gives a useful estimate of the HF concentration. Figure 6.1 shows data of Davison and Blakemore (1980) collected in England near an aluminium smelter, using paper suspended in a louvred box. Also shown are the data of Lynch *et al.* (1978) collected by CaO plated into upturned dishes exposed near a phosphate fertilizer factory in the USA. The individual regressions are: Davison and Blakemore, HF = 8.85 × deposited F (µg/cm²/day), $r^2 = 0.749$, $n = 98$, and Lynch *et al.*, HF = 5.81 × deposited F (µg/cm²/day) + 0.411, $r^2 = 0.611$, $n = 18$. Considering the differences in the techniques, sources and

locations, the two data sets are remarkably compatible.

An important question about the use of alkaline papers is whether they collect only HF or whether they collect a significant fraction of the particulate fluorides as well. Particles land on the exposed paper surfaces but the rate of deposition depends on the particle size and the wind speed (Table 2.1, Chapter 2). Larger particles (> 10 µm) have a higher deposition velocity than HF, especially at higher wind speeds, so, if they form a significant fraction of the total fluoride, this will be reflected in the total amount deposited. However, particles as large as that only occur very close to certain sources and the louvred housing or upturned Petri dish that is used to shelter alkaline papers reduces the turbulence significantly, so interference from large particles is not a serious problem in most circumstances. For smaller particles (< 5 µm),

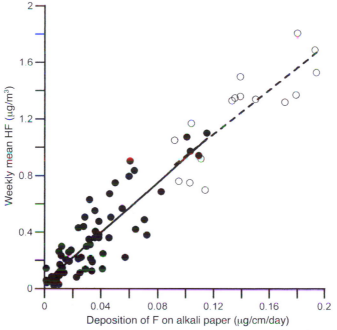

Fig. 6.1. The regression of weekly mean atmospheric HF concentration (µg/m³) on the rate of fluoride deposition on alkaline media. ● = data of Davison and Blakemore (1980) from 2 years of continuous monitoring near an aluminium smelter: HF = 8.85 × deposited F, $r^2 = 0.749$, $n = 98$. ○ = data from Lynch *et al.* (1978): HF = 5.81 × deposited F + 0.411, $r^2 = 0.611$, $n = 18$. (From Davison, A.W. and Blakemore, J. (1980) Rate of deposition and resistance to deposition of fluorides on alkali impregnated papers. *Environmental Pollution Series B* 1, 305–319, with permission from Elsevier.)

the rate of deposition on a paper in a housing is lower than that of HF, so the contribution to the total deposit is small. This interpretation is supported by field data (Davison and Blakemore, 1980).

Alkaline papers can be deployed at any place where the site is clear of major obstructions, such as trees or buildings, and where they are out of the reach of inquisitive animals, including humans. Knowledge of the sources of emissions – stacks, vents and doors or settling ponds – can be used to optimize the locations. As fluoride concentrations decrease in a non-linear way with distance, it may be best to site the samplers more densely close to the source or in the locations where stack fallout is expected. An initial survey of the fluoride content of vegetation (i.e. a biomonitor survey – see later) can be used to assist in choosing sites. Clearly, the method does not give an exact estimate of the HF concentration but it is sufficiently reliable to paint a broad, comparative picture, which can be used to support other investigations and interpret vegetation injury. For example, Fig. 2.2 (Chapter 2) shows how alkaline plates were used by Sidhu (1979) to monitor the HF concentrations near a phosphate plant. This demonstrated how far downwind the HF concentration was elevated above background. A second example, Fig. 6.2, shows the pattern of estimated HF concentrations around an aluminium smelter in Europe. The map was used to predict the locations where there might be the greatest effects on vegetation and therefore where to focus field surveys. Any symptoms recorded outside the lowest contour were regarded as probably not being due to HF and needing further investigation.

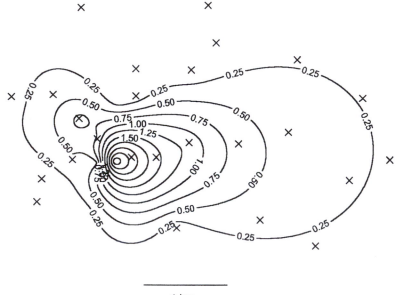

1 km

Fig. 6.2. A contour map of the monthly mean HF concentrations around an aluminium smelter, drawn using data from alkali-impregnated papers. The sampling sites are shown as X. The data were converted to µg/m³, using the regression equation of Davison and Blakemore (1980), and mapped, using SURFER software. The prevailing wind was from the west. The main aim of the exercise was to determine the area where a vegetation survey should be concentrated and to assist interpretation of visible symptoms shown by the plants. (The data were converted to µg/m³ using the regression equation of Davison and Blakemore, 1980)

Precipitation and bulk deposition

Several researchers have found that the fluoride content of precipitation is higher near sources and sometimes it correlates with the fluoride content of vegetation (see Chapter 2). Therefore, one possibility for monitoring is to record total or wet deposition. However, as explained in Chapter 2, the problem with this approach is that the amount of fluoride collected in a deposition gauge depends upon the geometry of the gauge and the wind speed, so the data are not absolute measurements. Also, a deposition gauge does not measure the amount deposited on vegetation and it is impossible to estimate the latter on a routine basis, so deposition measurements are not recommended.

Bioindicators and Biomonitors – Definitions

There is such a plethora of definitions of these two terms that they confuse students and practitioners alike. A bioindicator plant has been defined as a sensitive species that responds in a characteristic and predictable manner to the conditions that occur in a particular region or habitat (Mellanby, 2000). These responses are generally visible changes in leaves, flowers or fruits, and include formation of chlorotic (yellow) or necrotic lesions, other pigment changes, such as the production of red pigments (anthocyanosis), numerous kinds of leaf distortions, fruit deformities, reduced growth, alteration in plant form and other effects. A bioindicator has also been defined as a plant that exhibits a specific array of symptoms when exposed to a particular phytotoxicant (Feder, 1978; Feder and Manning, 1979; Manning and Feder, 1980), which is similar to Mellanby's (2000) definition. One can compare a bioindicator species with Rich's statement that 'crops are like canaries' (Rich, 1964) – bioindicators that have been used since Roman times as indicators of carbon monoxide in coal mines. The canary, which

became unconscious, could be revived in clean air and used over and over again. Another use of bioindicators is the measurement of changes in metabolic, physiological, histological and genetic characteristics resulting from exposure to fluoride, although this approach is time-consuming and not pollutant-specific. Some investigators refer to 'sentinel organisms', which can be defined as 'the most sensitive organisms to fluoride in each plant or animal category'. Thus, sentinel organisms are included as the quintessential bioindicators. In recent years, the term biomarker has been introduced as 'systems that generally include a subsystem of a whole organism to identify a specific endpoint' (Huggett *et al.*, 1992). The US Environmental Protection Agency (EPA) definition of a biomarker is 'any assessment of pollutants or biological effects of pollutants which enter the organism's organs or tissues' (Fowle and Sexton, 1992). From the latter definition, one might conclude that the biomarker could include both bioindicators and biomonitors. Jamil (2001) said that bioindicator refers to 'the presence, absence or abundance of certain species indicating information about environmental quality', and biomonitor to 'the measurement among individuals of their molecular, biochemical, or physiological parameters'. Peakall and Shugart (1991) discussed 'biomarkers of exposure, biomarkers of effect' and 'biomarkers of susceptibility'. We prefer to use biomarker to describe the more subtle changes in organisms, such as alterations in enzyme activity or metabolic processes due to chronic or acute exposures to a toxicant, but which may or may not demonstrate toxicity at the organismic level.

The second category is the biomonitor, which is often defined as 'a plant that is tolerant to fluoride exposure and accumulates it in foliar or other tissues'. Leaf analysis can provide an estimate of the length and intensity of exposure and a rough approximation of the concentration of fluoride encountered – provided that the dynamics of fluoride uptake and loss are known (Davison, 1982, 1983, 1987), and that is an uncommon situation. However, Burton (1986) defined biomonitor in its broadest sense as 'the measurement of growth and effects on metabolism and of concentrations of contaminants in living organisms', thus including both bioindicators and biomonitors. Using biomonitors is also time-consuming, but, as part of a continuing and multifaceted programme, it can be invaluable. The presence and amount of fluoride in plant tissues can also help to verify that visible plant injuries are caused by fluoride and not any of a host of edaphic, climatic and biological factors that can mimic fluoride injury, provided that the data are interpreted with due care.

Bioindicators

Having conducted field studies for a number of years and in many ecological situations, ranging from tropical to cool, temperate rainforests, from high to low deserts and from semi-tropical to temperate forests, it is the authors' experience that at least one species of plant in each area has stood out as a sensitive bioindicator. It is only in heavily industrialized areas that sensitive indicators tend to be totally absent or greatly restricted in their distribution. But suitability as a bioindicator varies with the location; the most sensitive species in one area may be considered as tolerant in another. A good example of this is *Eucalyptus globulus*, which is regarded as being of intermediate tolerance in its native Australia (Doley, 1986). However, it is grown as a timber tree in much cooler and more humid conditions in western Europe and there it appears to be less tolerant, possibly because of the lack of water stress and greater stomatal conductance over the whole of the year. For unknown reasons, an occasional species may be sensitive in some locations and relatively insensitive in others. For example, *Hypericum perforatum* (St John's wort) is usually a bioindicator of choice because it is widely distributed in the USA, Canada and Europe. In nearly all areas, the species is sensitive to fluoride, expressing

symptoms as orange-red tip necrosis of young leaves. However, in at least one area of the Pacific coastal USA (Tacoma, Washington State), *H. perforatum* has been exposed to HF for many years and is relatively tolerant. North of Tacoma in Bellingham, Washington State, also a Pacific coastal city, the species is once again very sensitive. There is also similar variation in parts of Europe. Why is this? Perhaps some populations are more tolerant because of heritable differences in stomatal conductance, leaf structure or water use, or perhaps it is due to local soil conditions characteristic of the area. Determining the causes of this kind of variation needs an experimental approach to determine the contribution made by genetic and environmental factors but it has rarely been done. Whatever the reason, several of the species listed in Table 6.1 may also exhibit differences in sensitivity in some locations.

Many authors have provided information on reactions of plants to fluoride and those species suitable as plant bioindicators. Thus, for North American species, the reader is referred to Treshow and Pack (1970), Weinstein (1977) and Weinstein *et al.* (1998); for South America, Arndt *et al.* (1995) and Weinstein and Hansen (1988); and for Australia and New Zealand, Doley (1986). For Europe, several authors have reported on relative sensitivities, notably Borsdorf (1960), Bolay and Bovay (1965), Bossavy (1965), Guderian *et al.* (1969) and Dässler *et al.* (1972). But all of these lists should be used with great caution. Comparing the rankings is difficult, in part because few species are common to all lists, but also because the sensitivity classes are subjective and based on different criteria. For example, Borsdorf (1960) divided species into four classes: *hochempfindlich* (highly sensitive), *empfindlich* (sensitive), *wenig empfindlich* (slightly sensitive) and *unempfindlich* (non-sensitive). The highly sensitive category included plants that showed obvious injury at the furthest distance from the source that he investigated, while the non-sensitive (tolerant) plants showed little or no injury even when adjacent to the source. The criteria for

Table 6.1. Relative sensitivities of higher plants to atmospheric fluorides, based on foliar symptoms.

Latin binomial	Common name	Response
USA, UK, Europe[a]		
Abies alba	Silver fir	S
Abies balsamea	Balsam fir	S
Abies concolor	Colorado fir	S
Acer campestre	Hedge maple	I
Acer ginnala	Amur maple	I
*Acer negundo	Box elder or Manitoba maple	S
Acer palmatum (and cvs)	Japanese maple	S–I
Acer pennsylvanicum	Striped maple	I
Acer platanoides	Norway maple	I
Acer saccharinum	Swamp or silver maple	I
Acer saccharum	Sugar maple	I
Achillea spp.	Yarrow	T
Agave spp.	Agave	I
*Allium ursinum	Wild garlic	S
*Allium spp.	Onion	S–I
Allyssum saxatile	Golden tuft	T
Alnus rubra	Red alder	I–T
Alnus rugosa	Speckled alder	I–T
Althea rosea	Hollyhock	T
Amaranthus retroflexus	Pigweed	T
Ambrosia artemisifolia	Common ragweed	T
Ambrosia trifida	Giant or great ragweed	T
*Amelanchier canadensis	Service-berry, saskatoon-berry	I
Anthemis cotula	Burdock	T
Apium graveolens	Celery	T
Arctium spp.	Burdock	T
Arctostaphylos uva-ursi	Bear-berry or kinnikinick	T
Asclepias syriacus	Milkweed	I
Asparagus officinalis	Asparagus	T
Aster spp.	Aster	I
Avena sativa	Oat (mature plant)	T
Avena sativa	Oat (young)	S
Berberis julianae	Wintergreen barberry	T
Berberis nervosa, Berberis repens	Oregon grape	S
Berberis thunbergii	Japanese barberry	I
Berberis verruculosa	Warty barberry	T
*Berberis vulgaris	Common barberry	S
Beta vulgaris	Beet	T
Betula lutea	Yellow birch	T
Betula nigra	Black birch	T
Betula papyrifera	White birch	T
Betula pendula	European birch	T
Betula populifolia	Grey birch	T
Brassica kaber	Common mustard	T
Brassica oleracea (and cvs)	Cabbage, kale, Brussels sprouts	T
Bromus tectorum	Downy brome grass	I
Buxus sempervirens	Boxwood	I
Callistyphus chinensis	Rainha-margarida or China aster	I–T
Camellia japonica	Camellia	T
Campanula spp.	Bellflower	T
Capsicum frutescens	Pepper	T

continued

Table 6.1. *Continued.*

Latin binomial	Common name	Response
Carya pecan, Carya glabra, Carya ovata	Pecan, pignut, shag-bark hickory	T
Celastrus scandens	Bitter-sweet	T
Celtis occidentalis	Hackberry	I
**Cercis canadensis*	Redbud	S
Chaenomeles lagenaria	Flowering quince	T
Chamaecyparis spp.	False cypress	T
Chenopodium album	Lamb's quarters	I
Chenopodium spp.	Nettle-leaf goosefoot	I
Chrysanthemum spp.	Chrysanthemum	T
Cichorium intybus	Chicory	T
Cirsium arvense	Canada thistle	T
Citrus spp.	Orange, lemon, grapefruit, tangerine	I
Clematis spp.	Clematis	T
Coffea arabica	Coffee	T
**Convallaria majalis*	Lily of the valley	S
Convolvulus arvensis	Bindweed	T
Cornus canadensis	Bunch-berry	T
Cornus florida	Flowering dogwood	T
Cornus kousa	Korean dogwood	T
Cornus mas	Cornelian cherry	T
Cornus stolonifera	Red-osier	T
Cotinus coggryia	Smoke-tree	S–I
Cotoneaster spp.	Cotoneaster	T
Crataegus crus-galli, Crataegus oxyacantha	Cockspur, scarlet hawthorn	T
**Crocus* spp.	Crocus	S
Cryptomeria japonica	Japanese cypress	T
Cucumis sativus	Cucumber	T
Cucurbita spp.	Squash, pumpkin, water melon, etc.	T
Dahlia spp.	Dahlia	T
Datura stramonium	Datura, jimson weed	T
Daucus carota	Carrot	T
Deutzia gracilis	Slender deutzia	T
Deutzia scabra	Fuzzy deutzia	T
Dianthus barbatus	Sweet william	T
Dianthus spp.	Carnation, pink	T
Digitaria sanguinalis	Crab-grass	S
Echinacea purpurea	Purple cone-flower	I–T
Echinocloa crusgalli	Barnyard grass	I
Eleagnus andustifolia	Russian olive	T
Epilobium angustifolium	Fireweed	I
Euonymus alatus	Winged euonymus, burning bush	T
Euonymus fortunei	Winter creeper	T
Euphorbia spp.	Spurge, poinsettia	T
Fagopyrum esculentum	Buckwheat	S
Ficus spp.	Fig	T
Forsythia spp.	Forsythia	T
Fragaria spp.	Strawberry	T
Fraxinus americana	White ash	S
Fraxinus pennsylvanica	Green ash	I
Freesia spp.	Freesia	S
Fuchsia spp.	Fuchsia	T
Galium spp.	Bedstraw	T
Ginkgo biloba	Maidenhair	T

Table 6.1. *Continued.*

Latin binomial	Common name	Response
*Gladiolus hortus	Gladiolus	S
Gleditzia triacanthos (and cvs)	Honey locust	T
Glycine max	Soybean, soya	T
Gossypium hirsutum	Cotton	T
Hedera helix	English ivy	T
Helianthuus annuus	Sunflower	T
*Hemerocallis fulva (and Hemerocallis flava)	Orange (and yellow) day lily	S
Heuchera sanguinea	Coral-bell	T
Hordeum vulgare	Barley (mature plant)	T
Hordeum vulgare	Barley (young)	S
*Hosta spp.	Plantain lily	S
Hybiscus syriacus	Rose of Sharon	T
Hydrangea macrophylla, Hydrangea paniculata	Blue, pee-gee hydrangea	T
*Hypericum perforatum	St John's wort, goatweed	S
Ilex aquifolium, Ilex verticillata, Ilex opaca	English, inkberry, American holly	T
Impatiens spp.	Impatiens	T
Ipomoea batatas	Sweet potato	I
Iris germanica	German iris, flag	S
Ixora spp.	Ixora	T
Juglans cinerea, Juglans nigra, Juglans regia	Butter-nut, black, English walnut	I
Juniperus spp.	Red cedar, juniper	T
Kalmia angustifolia	Sheep laurel	S–I
Kalmia latifolia	Mountain laurel	T
Kerria japonica	Kerria	T
Kochia scoparia	Summer cypress	T
Koelreuteria paniculata	Golden-rain tree	I
Kolkwitzia amabilis	Beauty bush	T
Laburnum anagyroides	Golden-chain tree	I–T
Lactuca sativa, Lactuca scariola	Common lettuce, prickly lettuce	T
Larix laricina	American larch	S–I
*Larix occidentalis	Western larch	S
Laurus spp.	Laurel	T
Ligustrum spp.	Privet	T
*Lilium spp.	Lily	S
Liquidambar styraciflua	Sweet gum	I
Lobelia erinus	Lobelia	T
Lolium perenne	Perennial ryegrass	T
Lonicera spp.	Honeysuckle	T
Lotus corniculatus	Bird's-foot trefoil	T
Lupinus spp.	Lupin	T
Lythrum salicaria	Purple loosestrife	T
Magnolia grandiflora	Evergreen, saucer magnolia	T
Magnolia stellata	Star magnolia	T
*Maianthemum canadense	False lily of the valley	S
Malus domestica	Apple	T
Medicago sativa	Lucerne	T
Melilotus alba	Sweet white clover	T
Melilotus officinalis	Yellow clover	T
Morus alba	Mulberry	T
Morus spp.	Mulberry	I
Narcissus spp.	Narcissus, daffodil	S–I
Nicotiana tabacum	Tobacco	T

continued

Table 6.1.   *Continued.*

Latin binomial	Common name	Response
Oryza sativa	Rice (mature plants)	T
Oryza sativa	Rice (young plants)	S
Ostrya virginiana	American hop hornbeam	T
Oxydendrum arboreum	Sourwood, sorrel tree	T
Pachysandra terminalis	Japanese spurge	T
Paeonia spp.	Peony	T
Panicum miliaceum	Millet	I
Pelargonium spp.	Geranium	T
Petunia hydrida	Garden petunia	T
Phaseolus vulgaris	Bush or climbing bean	T
Philadelphus spp.	Mock orange	T
Phleum pratense	Timothy	T
Phlox spp.	Phlox	T
Picea engelmanni, Picea abies, Picea mariana	Engelmann, Norway, black spruce	I–T
Picea glauca, Picea pungens	White and Colorado spruce	I
Pieris japonica, Pieris floribunda	Japanese, mountain andromeda	T
Pinus monticola, Pinus banksiana, Pinus nigra	Western white, jack, Corsican pine	I
Pinus strobus, Pinus mugo, Pinus contorta, Pinus taeda, Pinus ponderosa	White, mugo, lodgepole, loblolly, ponderosa	S
Pinus sylvestris	Scots pine (young needles)	S
Pinus taeda, Pinus ponderosa, Pinus sylvestris	Loblolly, ponderosa, Scots pine (old needles)	T
Pisum sativum	Garden pea	T
Plantago major, Plantago lanceolata	Common plantain, English plantain	T
Platanus acerifolia	London plane-tree	I
Platanus occidentalis	Sycamore, button wood, plane-tree	I
Poa annua	Annual bluegrass	I
Polygonatum biflorum	Solomon's seal	S
Polygonum persicaria	Smart-weed, lady's-thumb	I
Populus balsamifera	Balsam poplar	I–T
Populus deltoides	Eastern cottonwood	T
Populus grandidentata	Big-tooth aspen	T
Populus nigra var. *italica*	Lombardy poplar	I–T
Populus tremuloides	Trembling (or quaking) aspen	S
Populus tricocarpa	Black cottonwood	T
Portulaca grandiflora	Portulaca, moss-rose	T
Portulaca oleracea	Purslane	T
Potentilla fruticosa	Shrubby cinque-foil	T
Primula spp.	Primrose	T
Prunis domestica var. *bradshaw*	Bradshaw plum	I–T
Prunus domestica var. *italica*	Italian prune	S
Prunus armeniaca cv. 'Chinese'	Chinese apricot	S
Prunus armeniaca cv. 'Tilton'	Tilton apricot	I
Prunus avium cvs	Sweet cherry	I
Prunus pennsylvanica	Bird cherry	I
Prunus persica	Peach (foliage)	I
Prunus persica	Peach (fruit)	S
Prunus serotina	Black cherry	I
Prunus serrulata	Flowering cherry	S–I
Prunus virginiana	Choke cherry	I
Pseudotsuga menziesii	Douglar fir	S
Pueraria thunbergiana	Kudzu	T

Table 6.1. *Continued.*

Latin binomial	Common name	Response
Pyracantha coccinea	Firethorn	T
Pyrus communis	Pear	T
Quercus alba	White oak	T
Quercus coccinea	Red oak	T
Quercus palustris	Pin oak	T
Quercus phellos	Willow oak	T
Quercus stellata	Posr oak	I
Quercus velutina	Black oak	T
Quercus virginiana	Live oak	T
Rhododendron catawbiense	Catawba rhododendron	T
Rhododendron maximum	Rosebay rhododendron	T
Rhododendron molle	Deciduous azalea	I
Rhododendron obtusum cv. *'Hinodegiri*	Evergreen azalea	T
Rhus radicans	Poison ivy	S–I
Rhus typhina, Rhus glabra, Rhus copallina	Stag-horn, smooth, winged sumac	I
Ribes spp.	Currant	S–I
Ricinus communis	Castor	T
Robinia pseudo-acacia	Black locust	T
Rosa rugosa	Rugose rose	T
Rosa spp.	Hybrid tea, rambling, multiflora	I
Rubus spp.	Blackberry, raspberry, salmon-berry	T
Rudbeckia spp.	Cone-flower, black-eyed Susan	I–T
Rumex spp.	Dock	T
Saccharum officinarum	Sugar cane	S
Salvia spp.	Salvia, sage	T
Sambucus canadensis	American elderberry	I–T
Secale cereale	Rye (mature plants)	T
Secale cereale	Rye (young plants)	S
Sedum spp.	Stonecrop	T
Silene cucubalus	Bladder campion	T
Skimmia japonica	Skimmia	I
Smilax spp.	Cat brier, greenbrier, thorny vine	S
Solanum melongena	Aubergine	T
Solanum nigrum	Black nightshade	Y
Solanum pseudo-capsicum	Jerusalem cherry	S
Solanum tuberosum	White potato	T
Solidago spp.	Golden rod	T
Sorbaria sorbifolia	False spirea	T
Sorbus aucuparia	European mountain ash, rowan	I
Sorghum halepense	Johnson grass	S–I
Sorghum vulgare (and cvs)	Sorghum, milo	S–I
Spinacea oleracea	Spinach	T
Spiraea bumalda 'Anthony Waterer'	Anthony Waterer spirea	T
Spiraea prunifolia	Bridal wreath	T
Spiraea vanhouttei	Vanhoutte spirea	T
Stellaria media	Chickweed	S–I
Symphoricarpos orbiculatus	Coral-berry	T
Symphoricarpos orbiculatus	Snowberry	T
Syringa vulgaris	Lilac	I
Tagetes spp. (and cvs)	Marigold	I
Tanacetum vulgare	Tansy	T

continued

Table 6.1. *Continued.*

Latin binomial	Common name	Response
Taraxcum officinale	Dandelion	T
Taxodium distichum	Bald cypress	I
Thuja occidentalis	Arbor vitae (all forms)	T
Thuja plicata	Western red cedar	T
Tilia spp.	Linden	I
Trifolium repens	White clover	T
Trifolium sativum	Red clover	T
Tsuga canadensis	Canadian hemlock	T
Tsuga caroliniana	Carolina hemlock	T
Tsuga heterophylla	Western hemlock	T
*Typha angustifolia	Cat's-tail	S
Ulmus alata	Winged elm	T
Ulmus americana	American elm	T
Ulmus parvifolia	Chinese elm	T
Ulmus pumila	Siberian elm	T
Vaccinium angustifolium	Low-bush blueberry	S
Vaccinium corymbosum	High-bush blueberry	S
Vaccinium hirsutum	Hairy huckleberry	T
Verbena hybrida	Verbena	T
Viburnum dentatum	Arrowwood	T
Viburnum opulus (and cvs)	Cranberry bush	T
Viburnum prunifolium	Black haw	T
Viburnum rhytidophyllum	Leather-leaf viburnum	T
Vicia faba, Vicia sativa	Broad bean, vetch	T
Vinca major	Periwinkle	T
Vinca minor	Vinca, periwinkle	T
Vitis labruscana	Concord grape	I
Vitis vinifera	European grape	S
*Vitis spp.	Wild grape	S–T
Xanthium spp.	Cockle-bur	T
Yucca filamentosa	Adam's-needle yucca	T
Zea mays	Field corn (maize)	I
*Zea mays	Sweetcorn (many cvs) (maize)	S
Zinnia spp.	Zinnia	T
South America (Brazil)[b]		
Agave sisalana	Sisal agave	S–I
Ageratum conyzoides	Agerato, menastro or ageratum	T
Albizzia molucanna	Albisia or albizzia, mimosa	I
Aloe vera	Barbosa or aloe	T
Amaranthus spp.	Bredo or amaranth	I
Anacardium occidentale	Caju or cashew	I
Anadenanthera colubrina	Angico	I
Ananas commosus	Abacaxi or pineapple	T
Anatherium bicorne	Sapé	I
Arachis hypogaea	Amendoim or groundnut	T
Araucaria angustifolia, Araucaria heterophylla	Pineiro-do-parana or pineiro-brasileiro, parana pine or Norfolk Island pine	T
Aristida pallens	Barba-da-bode or goat's-beard grass	S–I
Artocarpus heterophyllus	Jaca or jackfruit	T
*Astrocaryum tucuma	Tucum or tucuma	S–I
Bambusa vulgaris	Bambu or feathery bamboo	I
Bombax sienopetalum	Paineira	I
Bougainvillea spectabilis	Três marías, bouganvílea, bougainvillaea	I–T

Table 6.1. *Continued.*

Latin binomial	Common name	Response
Brassaia actiniphylla	Cheflera or schefflera	I–T
Brassica spp.	Mostara or wild mustard	T
Buddleia brasiliensis	Verbasco-do-brasil or butterfly bush	T
Cactus spp.	Cacto or cactus	T
Caesalpina echinata	Pau-brasil or Brazil-wood	I–T
Cajanus cajan	Andu, endu, guandu or pigeon-pea	T
Caladium bicolor	Caládio or caladium	I–T
Calliandra spp.	Jurema or powder-puff	I
Callistemon spp.	Bottlle-brush	T
Camellia japonica	Camélia or camellia	T
Canna generalis	Albara or canna lily	S–I
Canna indica	Cana-de-india or Indian shot	S–I
Capsella bursa-pastoris	Bolsa de pastor or shepherd's purse	T
Capsicum annuum	Pimentão or pepper	I
Caryota urens	Banda-de-sargento or duckfoot palm	S–I
Casuarina equisetifolia	She-oak or pineiro	T
Cecropia spp.	Umbaúba, embaúba, imbaúba or cecropia	I
Chamaerops humulis	Palma-de-leque or fan palm	T
Chenopodium ambrosioides	Mastruço or American worm-seed	I
Chrysalidocarpus lutescens	Areca or yellow palm	S
Chrysanthemum frutescens	Margarida or chrysanthemum	I–T
Cleome spinosa	Catainga-de-nedro or cleome	T
Cocos nucifera	Coco or coconut palm	S–I
Codiacum variegatum	Cróton or croton	I–T
Coffea arabica	Café or coffee	T
Coleus blumeri	Coléos or coleus	I
Coreopsis lanceolata	Coreópsis or coreopsis	I–T
Cosmos spp.	Cósmes or cosmos	T
Crescentia cujete	Coité, cuité or calabash tree	T
Cynodon dactylon	Capim-do-burro or Bermuda grass	T
Cydonia oblanga	Marmela or quince	I–T
Cyperus papyrus	Papiro or papyrus	S–I
Delonix regia	Flamboiã or royal poinciana	I–T
Dracaena marginata	Dracena	T
Elaeis guineensis	Dendé-do-para, caiaué or dende palm	S–I
Eriobotrya japonica	Ameixeira or loquat	S–I
Eucalyptus citriodora	Eucalipto-laranja or lemon-scented gum	S
Eucalyptus globulus	Eucalipto or bluegum	I
Eucalyptus grandis	Eucalipto or rose gum	S–T
Eucalyptus saligna	Eucalipto or Sydney bluegum	S–I
Eugenia uniflora	Pitanga or Surinam cherry	T
Euphorbia pulcherrima	Papagaio or poinsettia	T
Euterpe oleracea	Açai or assai palm	S–I
Ficus benjamina, Ficus dolliaria, Ficus elastica	Ficus-benjamim or wooping fig, gamelleira, ficus-japones or rubber plant	T
Fragaria chiloensis, Fragaria vesca	Morango or strawberry	T
Fuchsia triphylla	Brinco-de-princeza or fuchsia	T
Gardenia jasminoides	Dama-da-noite, jasmin-do-cabo or gardenia	I–T
Genipa americana	Genipapo or genipap	T

continued

Table 6.1.   *Continued.*

Latin binomial	Common name	Response
*Gladiolus communis	Lagrima-de-Sta Rita or tears of Sta Rita	S
Grevillea robusta	Grevilha or silk oak	T
Helianthus tuberosus	Topinambo or Jerusalem artichoke	T
Heliconia behai	Banana-do-mato	I
Hyacinthus orientalis	Hyacinth	S
Hydrangea macrophylla	Hortência or hydrangea	T
*Hypericum teretiusculum	Arruda-do-campo or St John's wort	S
*Iris spp. (young leaves)	Lírio or iris	S
Jacaranda ovalifolia	Caroba-guaca or jacaranda	T
Jatropha curcas	Pinhão branco or Barbados nut	T
Kalanchoe pinnata	Folha-da-costa or air plant	T
Lagerstroemia indica	Reseda or crape myrtle	I
Lantana camara	Cambará, camará or lantana	T
Lathyrus spp.	Comanda or sweet pea	I–T
Ligustrum lucidum	Privet	T
Lilium candidum	Lírio-dos-postas or lily	S–I
Lonicera caprifolium	Madressilva or honeysuckle	T
Magnolia grandiflora	Magnólia-branca or bull bay	I
Magnolia × soulangeana	Magnólia or saucer magnolia	I
Mangifera indica	Manga or mango	I
*Manihot utillissima, Manihot esculenta	Mandioca, macaxeira or manioc, cassava	S–I
Mauritia flexuosa	Buriti or ita palm	I
Maxililiana maripa	Inajá or cucurite palm	I
Melaleuca leucadendron	Óleo-de-cajeput or river tea tree	T
Melia azedarach	Cinamoma or chinaberry tree	T
Melinis minutiflora	Capim-gordura	T
Mentha spp.	Hortela or mint	T
Mimosa pudica	Sensitivo or sensitive mimosa	T
Monstera deliciosa	Sete-chagas or cut-leaf philodendron	I
*Moquilia tomentosum	Oiti	S
Musa paradisiaca cv. 'Sapientum'	Banana	I
Myrica gale	Alecrim-do-norte, sweet gale	T
Nasturtium officinale	Agrião or watercress	T
Nectandra spp.	Canella	T
Nerium oleander	Espirradeira or oleander	T
Nicotiana sanderae	Wild tobacco	T
Olea europeae	Oliveira, oliva or olive	T
Opuntia spp.	Palmatória or prickly pear	T
*Orbignya barbosiana	Babaçu or babassu palm	S–I
Oxalis spp.	Azedinha or sorrel	I
Pandanus utilis	Pandano or screw pine	S
Pandorea spp.	Sete-léguas or Australian bower vine	I
Paspalum notatum	Grama-batatais or Bahia grass	T
Passiflora alata	Maracujá or passion-flower	I
Pennisetum clandestinum	Capim-cucuiu or kikuyu grass	T
Persea americana	Abacatei or avocado	S
Petrea subserrata	São-Miguel or purple wreath	S–I
Phaseolus limemsis	Feijão-de-lima or lima bean	T
Phaseolus vulgaris	Feijão or common bean	T
Philodendron imbe	Cipó-imbé or philodendron	I
Phoenix dactylifera	Tamareira, palmareira-de-igreja or date-palm	S–I
Phyllanthus acidus	Groselha or Otaheite gooseberry	I

Table 6.1. *Continued.*

Latin binomial	Common name	Response
Pimpinella anisum	Anis or anise	I
Pinus caribaea	Pinho or Caribbean pine	I
Plumeria drastica	Janaúba or frangipani	I
Poa spp.	Capim-sempre verde or spear-grass	I
Posoqueria latifolia	Paptera or needle-flowered tree	I
Prosopis spp.	Algarobo or mesquite	T
Psidium guajava	Goiaba or guava	S
Rhododendron indicum	Azaleá or azalea	S–I
Ricinus communis	Mamona or castor bean	T
Rosmarinus officinalis	Alecrim or rosemary	T
Roystonea regia	Palmeira-imperial or royal palm	S
Ruta graveolens	Arruda or rue	S
Sabal mexicana	Palmeira-mexicana or sabal palm, palmetto	T
Saccharum officinarum	Cana-de-açúcar or sugarcane	S
Sanseviera zeylanica	Espada-de-ogum or snake plant	T
Schinus terebinthifolius	Aroeira pimenteira or Christmas berry tree	I–T
Solanum panicum	Jujurubeba or nightshade	I–T
Spathodea campanulata	Espatodéia or South African tulip-tree	I
Syragrus campestris	Arirí palm	S
Syzygium jambos	Jamalão or rose-apple	I
Tabebuia avellandae	Ipê-amarelo or trumpet-tree	I
Taxodium distichum	Pineiro-do-brejo or bald cypress	T
Terminalia catappa	Amendoeira, Castanhoda or Indian almond	I–T
Thea sinensis	Chá or tea	T
Theobroma cacao	Cacau or cocoa	I
Tipuana tipu	Tipuana or rosewood	T
Tropaeolum majus	Chagas or nasturtium	I–T
Typha spp.	Tabua or cat's-tail	S
Vanillosmopsis erythropappa	Candeia	I
Verbena camaedryfolia	Camaradinha or verbena	T
Yucca filimentosa	Ávore-da-pureza or Adam's-needle yucca	I
Australia/New Zealand (NZ)[c]		
Acacia aulacocarpa	Hickory wattle	S
Acacia fimbriata	Brisbane wattle	S
Acacia pulchella	Prickly moses	S
Angophora costata	Rough-bark apple	S
Angophora floribunda	Smooth-bark apple	S
Banksia serrata	Saw-leaved banksia	S
Bursaria spinosa	Blackthorn	S
Callistemon linearis	Narrow-leaved bottlebrush	S
Callistemon verminalis	Weeping red bottlebrush	S
Coprosma grandiflora	Large-leaved coprosa (NZ)	S
Corymbia calophylla	Marri	S
Corymbia citriodora	Lemon-scented gum	S
Corymbia ficifolia	Red-flowering gum	S
Corymbia gummifera	Red bloodwood	S
Corymbia intermedia	Pink bloodwood	S
Corymbia torrelliana	Cadagi	S

continued

Table 6.1. *Continued.*

Latin binomial	Common name	Response
Eucalyptus amygdalina	Narrow-leaved peppermint	S
Eucalyptus baxteri	Brown stringy-bark	S
Eucalyptus globulus	Tasmanian bluegum	S
Eucalyptus grandis	Flooded gum	S
Eucalyptus haemastoma	Scribbly gum	S
Eucalyptus marginata	Jarrah	S
Eucalyyptus microcorys	Tallow-wood	S
Eucalyptus nitida	Shiny-leaved mallee	S
Eucalyptus pilularis	Black butt	S
Eucalyptus punctata	Grey gum	S
Eucalyptus robusta	Swamp mahogany	S
Eucalyptus rudis	Western Australian flooded gum	S
Eucalyptus saligna	Sydney bluegum	S
Eucalyptus viminalis	Manna gum	S
Griselina littoralis	Broadleaf (NZ)	S
Petalostigma pubescens	Quinine tree	S
Phormium tenax	Native flax (NZ)	S
Pittosporum eugenoides	Lemonwood (NZ)	S
Pittosporum undulatum	Pittosporum	S
Pseudopanax arborens	Five-finger (NZ)	S
Pseudowintera colorata	Horopito (NZ)	S
Rubus squarrosus	Bush lawyer (NZ)	S
Xanthorrhea australis	Grass tree	S
Xanthorrhea johnsonii	Grass tree	S
Xanthorrhea preissii	Grass tree	S

[a]From Guderian *et al.* (1969); Treshow and Pack (1970); NAS (1971); Weinstein (1977); Weinstein *et al.* (1998); A. Davison, personal communication, and others. [b]From Weinstein and Hansen (1988); Arndt *et al.* (1995). [c]From Doley (1986) and D. Doley, personal communication (2002). *Species particularly suitable for use as bioindicators.

the two middle classes were not defined in the paper. On the other hand, Bossavy (1965) divided species into only two classes, *sensibles* (sensitive) and *résistantes* (resistant), based on the presence or absence of traces of necrosis in June. The result is that *Iris germanica* was classified by Borsdorf (1960) as highly sensitive and by Bossavy (1965) as sensitive. Similarly, hawthorn (*Crataegus monogyna*) was described as slightly sensitive by Borsdorf but as resistant by Bossavy. The authors' field experience is that this species is very tolerant because they have never seen it injured, even in places where the mean HF was > 3 µg/m³ for several weeks. Guderian *et al.* (1969) did not have a resistant or non-sensitive category, but only had *sehr empfindlich* (very sensitive), *empfindlich* (sensitive) and *weniger empfindlich* (less sensitive). The lack of a non-sensitive category might be misinterpreted as meaning that all species have a degree of sensitivity to ambient fluoride concentrations, which is not true. As Borsdorf (1960) observed, there are many species that do not show injury even at the highest ambient concentrations. In fact, the majority of species possess a considerable degree of resistance. Furthermore, there can be an added degree of confusion when a relatively narrow group of plants, such as legumes, grasses or grains, is ranked according to the relative susceptibilities of the species, as in Guderian *et al.* (1969). The most 'susceptible' plants in these groups may be assigned as being much more tolerant in broader lists. The different means of determining susceptibility and tolerance account, at least in part, for the anomalous positions of some species in the many tabulations that have been published, although the lists may be generally similar.

In order to gain consistency in classifying species, it would seem logical to use atmospheric fluoride concentrations as the baseline for comparison. The most sensitive category would include those species that are affected by the shortest durations of exposure to the lowest concentrations that are known to cause injury, while the most resistant would be those that remain uninjured by the highest concentrations encountered in the field. However, even this approach can cause problems if the range of concentrations is not sufficiently great. Guderian *et al.* (1969) fumigated a large number of species in Mylar greenhouses with 0.85 µg HF/m^3 for 16 days or 1.1 µg HF/m^3 for 49 days. These concentrations are three to four times higher than the minimum required to induce injury in the most sensitive species. Both of these exposures produced only 'very slight chlorosis' in white clover (*Trifolium repens*) and yet this species was placed in the same sensitivity class as *Gladiolus*, a species that can show severe necrosis at 1.1 µg/m^3 and which is injured by concentrations as low as 0.25–0.3 µg/m^3. Borsdorf (1960) classified white clover as insensitive and this is confirmed by the authors, who have rarely seen it injured, even in locations where the HF concentration exceeded 2 µg/m^3 for periods of weeks. Therefore, any fumigation can be misleading if it does not include the lowest threshold concentrations and the most sensitive species to act as boundary markers.

Mimicking symptoms

Unfortunately, there are many climatic, edaphic and biological factors that cause symptoms that are very similar to those cause by fluoride (see Chapter 4). They include such conditions as drought, cold, heat, sun scald, nutrient deficiencies, nutrient excesses, non-nutrient element excesses (such as cobalt, nickel, cadmium, aluminium and others), toxic organic compounds (such as many pesticides and herbicides), insect injuries (including leaf-miners and borers, mites, etc.) and pathogens (such as mosaic and yellows viruses, mycoplasma, wilt fungi and needle fungi).

So how can one be relatively certain that the symptom being observed is indeed due to fluoride toxicity? Obviously, there must be a known source in the area and the source strength must be sufficient to be toxic to plants. It is important, therefore, to have some knowledge of atmospheric fluoride concentrations over the time-scale needed to cause injury. The geographical pattern of symptoms should be approximately similar to the pattern of dispersion of the pollutant, being greatest in a downwind direction and generally decreasing with distance, although this ideal situation is rare because it is unusual to find the same species so well distributed around a source. Equally important, injury to plants must follow known plant sensitivities. For example, necrotic lesions that appear on a species that is known to be very sensitive and not on other, less sensitive species supports the conclusion that the lesions were caused by fluoride. If, on the other hand, there are symptoms on both a sensitive and a tolerant species or just on the tolerant species, another interpretation must be sought. The absence of an elevated fluoride concentration in leaves that are injured can help in the interpretation, but the presence of elevated fluoride in the leaves of injured species does not necessarily prove that it was the cause. Diagnosis and interpretation are rarely easy and may not even be possible in every case. There is no substitute for the experience of observing injury in as many different species, climates and situations as possible and repeating observations over several seasons or years.

Indicator species should be reviewed periodically to verify that they are consistent and reliable. Their selection should have some scientific rationale, e.g. response to controlled fumigations or use in other field situations. The bioindicator should produce consistent effects within each survey area. We have tabulated some of the most sensitive plant bioindicators found in several different countries and climates in Table 6.1.

Most of the species designated as suitable as bioindicators have been observed by the authors under field conditions and in a number of cases the atmospheric fluoride concentrations were known, but all of the reservations expressed above apply to the list. The legal warning *caveat emptor*, let the buyer beware, is very appropriate here.

Using native species *in situ*

The best bioindicator for airborne HF, as with other pollutants, is usually the most susceptible species and its signature is usually the development of visible foliar symptoms. As mentioned earlier, injury to leaves takes many forms, depending upon the species, the fluoride concentration in the air and other factors, but it is usually expressed as chlorosis, anthocyanosis, necrosis, distortion or their combinations (Plates 6–18). Most routine bioindicator surveys have used vascular plants, but lichens and mosses may be appropriate in some locations. They are well known to be among the most valuable bioindicators of several pollutants, especially SO_2, and they have been used for fluoride but, unfortunately, there has been less research in relation to fluoride and very few controlled fumigations have been conducted to establish dose–response relationships. None of the field studies of lichens have reported the atmospheric fluoride concentrations, so the scientific basis for using this group is much weaker than for vascular plants. Furthermore, in the field, SO_2 is often emitted from the same processes as fluoride, so it may be very difficult to assign injury to one pollutant (Chapter 4). Nevertheless, some lichens appear to be sensitive receptors to airborne fluoride, so they are worth consideration provided the expertise is readily available. Where fluoride is present in the atmosphere, the symptoms often occur as colour and morphological changes (Chapter 4; Gilbert, 1971, 1973; LeBlanc *et al.*, 1971; Nash, 1971; Perkins *et al.*, 1980; Perkins, 1992). Correct interpretation of the symptoms depends on having a lichenologist with experience of the effects of airborne pollutants. One of the most significant limitations of using lichens is that, once the thallus is injured or killed, recovery is extremely slow, so symptoms are a record of a past event or events that may have happened years before in response to occasional accidental emissions or to concentrations that no longer occur because of changes in the industrial process.

For both bioindication and biomonitoring purposes, it has been a practice to transplant native lichens and mosses to urban or industrial environments (Schönbeck, 1969; LeBlanc *et al.*, 1971; Palomäki *et al.*, 1992). LeBlanc *et al.* (1971) described the effects of fluoride pollution from an aluminium smelter on transplanted discs of *Parmelia sulcata* and was able to detect injury after 4–12 months of exposure at distances up to 15 km from the source. However, the atmospheric concentrations were not measured, so dose–response could not be estimated. In general, the algal symbiont appeared to be more sensitive than the fungal partner.

Mosses are another group of non-vascular organisms that are used as bioindicators and biomonitors of fluoride, although the latter use has been more common. As indicated in Chapter 4, there has not been enough research on the effects of fluoride or on the relative sensitivity of mosses to make them suitable for routine use as bioindicators.

Cultivated vascular species *in situ* as biomonitors

A few cultivated species, such as stone fruits, are sensitive to low concentrations of HF; the symptoms are characteristic and the economic effects can be severe, so they may be important biomonitors. The best-known example, described in Chapter 4, is suture red spot or soft suture of peach (Benson, 1959; Plate 24). Relatively high concentrations of fluoride can cause injury to apricot, cherry, apple and pear fruits (Bolay and Bovay, 1965; Bonte, 1982). In cherry, a similar syndrome has been termed 'snub nose' or 'shrivel tip' by Treshow and Pack (1970).

This group of plants is ideally used for repeated field surveys to indicate year-to-year changes in the impact of the pollutant. Individual trees or plots can be easily marked and rerecorded each year.

Transplants of cultivated vascular species

Where suitable bioindicators are absent, it may be possible to transplant a range of suitable species of known sensitivity. Sometimes this involves planting in self-watering pots and in other cases whole gardens are created, with the plants grown in standard soil mixes. In general, the following parameters should be considered when siting bioindicator transplants for the detection of fluoride: type of source, characteristics of dispersion and topography. There are several good examples of the use of transplants, including the German–Brazilian study described in Chapter 5, but the most extensive study was undoubtedly that carried out in The Netherlands on a national scale (Posthumus, 1982). This started in 1973 with a regional study but by 1976 there was an effects network consisting of 40 experimental fields spread all over the country. The monitoring was performed only during the vegetative period from April to November and it included both bioindicators and biomonitors. The fields were visited weekly. Plants were grown in standard soil in plastic containers fitted with an automatic watering system. Acute effects were measured as visible symptoms and chronic effects were recorded by comparing the growth of sensitive plants after long-term exposure. Bioindicators were included for fluoride, SO_2 and O_3, and rye-grass was included as a biomonitor. The tulip variety 'Blue Parrot' was used as a bioindicator of fluoride in spring and Gladiolus 'Snow Princess' was used in summer. A computer-generated map of the Gladiolus leaf injury observed in 1977 (Fig. 6.3) shows clearly the greater degree of necrosis in the south-west of the country in the vicinity of Rotterdam. The programme finished in the 1980s but recently a new

project, called EuroBionet, was started to provide a European network for the assessment of air quality using bioindicators (EuroBionet, 1999). The purpose is to raise the awareness of the urban population to problems related to air quality, which is one of the most important uses of bioindicators, because it gives the public something they can see for themselves and it is relevant to their environment. The partner cities are Barcelona, Copenhagen, Düsseldorf, Edinburgh, Klagenfurt, Lyon, Nancy, Sheffield, Valencia, Verona and Ditzingen.

Animal bioindicators

Mammals are the most important animal bioindicators because scoring of dental lesions gives a direct, unequivocal estimate of the effects of contamination and it can be repeated non-destructively. Insects and other invertebrates cannot usually be used as bioindicators because they do not show easily assessed symptoms. An exception is the use of bee colonies (Bromenshenk et al., 1985, 1996; Mayer et al., 1988).

Mammals, as already described, display typical markings and lesions of the teeth and bones that are specific to fluoride, so routine examination of cattle and sheep plays a vital part in detecting the onset of fluorosis and preventing economic loss to farmers. Usually routine examinations are supported by monitoring the dietary intake of the animals (see later). Teeth of cattle and sheep are easily examined, but it needs a trained observer (NAS, 1974) and a crew to assemble and hold the animals, so, although it is very effective and non-invasive, it can be expensive. With livestock, usually a score is assigned to the incisors of a representative sample of the herd, but molars are not usually examined because it is difficult to examine them in a live animal. Table 6.2 shows an example of biomonitoring using the prevalence of dental lesions in cattle and buffalo from the vicinity of a phosphate plant in India. Other examples include Grunder et al. (1980) and Riet Correa et al. (1986). The main limitation is that the

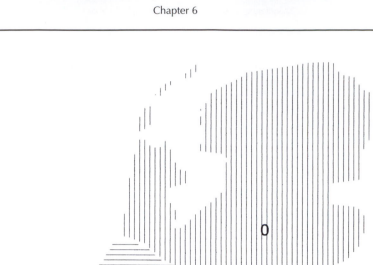

Fig. 6.3. Map of fluoride injury (cm of leaf necrosis) shown by *Gladiolus* cv. 'Snow Princess' in The Netherlands at the end of August 1977. Data from 40 monitoring stations. (Redrawn from Posthumus, A.C. (1982) Biological indicators of air pollution. In: Unsworth, M.H. and Ormrod, D.P. (eds) *Effects of Gaseous Air Pollution in Agriculture and Horticulture.* Butterworths, London, pp. 27–42; with permission of Elsevier.)

Table 6.2. The prevalence of dental lesions in cattle and buffalo grazing near a phosphate plant in India. The fluoride content of fodder and water samples ranged from 401–875 mg F/kg and 0.36 to 2.46 mg F/l, respectively. (Reprinted from Patra, R.C., Dwidedi, S.K., Bhardwaj, B. and Swarup, D. (2000) Industrial fluorosis in cattle and buffalo around Udaipur, India. *The Science of the Total Environment* 253, 145–150, with permission from Elsevier.)

	Cattle			Buffalo
Age:	< 1 year	1–3 years	> 3 years	> 1 year
Number of animals	23	43	75	25
Number with dental lesions	0	12	24	18
% of total	0	28	32	72

effects on teeth only occur if the fluoride is excessive during the period of development. The occurrence of bone fluorosis can be determined post-mortem, but this is used more as a check of diagnosis than for routine monitoring.

For wild mammals, monitoring the diet is difficult (Chapter 2), but using the teeth and bones as bioindicators is a viable option. There are several publications that demonstrate the relationship between the condition of the teeth of wild mammals and fluoride exposure. For example, Kierdorf *et al.* (1993) examined the mandibles of roe deer (*Capreolus capreolus*) collected by hunters in the Saxonian Ore Mountains for age estimation. One of the strengths of this study was that it was possible to locate the exact areas where the animals were killed because of the meticulous record keeping of the hunters. The teeth were scored and the mandibles assigned to one of seven fluorosis categories and then the data were mapped to show the distribution of fluorosis over an area of about 95 × 60 km. This showed very clearly that the most severe dental fluorosis was in the southern municipal districts, which lay about 10–30 km north of a number of brown-coal powerplants in the Czech Republic. The fluorosis decreased with distance from the south. Other investigations of wild mammals include those by Shupe *et al.* (1984), Suttie *et al.* (1987) and Vikøren and Stuve (1996a).

Some of the common British small mammals – voles, shrews and mice – are very well researched (Andrews *et al.*, 1989; Cooke *et al.*, 1990, 1996; Boulton *et al.*, 1994a,b,c, 1995, 1997, 1999). Dental lesions have been described in controlled laboratory experiments and recorded in the field, so this makes the more sensitive species strong candidates for use as bioindicators (Boulton *et al.*, 1995, 1997). The paper by Boulton *et al.* (1999) is particularly valuable because it gives a detailed account of applying tooth-lesion scoring to biological monitoring of

fluoride. The authors point out that the incisors and molars of short-tailed field voles (*Microtus agrestis*) grow through the life of the animal, so they are very sensitive to fluoride. Because they can be trapped and examined in the field, biomonitoring using voles has the great attraction of being non-destructive. The relationships between tooth-lesion scores and tooth fluoride concentration for two species of small mammal are shown in Fig. 6.4, indicating the potential for using an assessment of dental lesions to give a direct measure of the effect of fluoride and to estimate the fluoride burden. The scoring system used by Boulton *et al.* (1999) is shown in Table 6.3, and Plate 26 illustrates the teeth and scores for the South African mammal, *Mastomys natalensis*.

Biomonitors

Plants, animals, algae or any other suitable group of organisms can be used as biomonitors–collectors of fluoride. The attraction is that they can be used at little cost, and they are especially valuable where there is no power available. Perhaps the

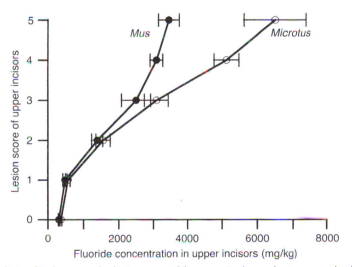

Fig. 6.4. The relationship between the lesion score of the upper incisors of two mammals, the short-tailed field vole (*Microtus agrestis*) and the white mouse (*Mus musculus*), and the fluoride content of the incisors (mg F/kg). (Drawn using data from the *Journal of Applied Toxicology*, vol. 15, Boulton *et al.*, Fluoride accumulation and toxicity in laboratory populations of wild small mammals and white mice, pp. 423–431, 1995. © John Wiley & Sons Limited. Used with permission.)

Table 6.3. Scoring system for classification of fluoride-induced lesions in the upper incisors of the short-tailed field vole (*Microtus agrestis*). (From Boulton, I.C., Cooke, J.A. and Johnson, M.S. (1999) Lesion scoring in field vole teeth: application to the biological monitoring of environmental fluoride contamination. *Environmental Monitoring and Assessment* 55, 409–422.)

Lesion score	Classification and appearance
1	Normal: glossy, smooth enamel, with uniform orange-yellow colour; incisor pair tightly aligned with gradual curvature; cutting edges sharp and even
2	Altered appearance: enamel smooth and glossy, with faint 'flecks' and/or slight reduced depth of pigmentation; tooth curvature normal, cutting edges less regular but still sharp
3	Slight fluorosis: narrow, faint concentric enamel 'banding'; slight 'mottling' or 'chalking' of enamel; cutting edges with slightly increased rate of erosion
4	Moderate fluorosis: pronounced banding, mottling and chalking of the enamel surfaces; erosion or 'blunting' of cutting edges, enhanced curvature of incisors
5	Marked fluorosis: enamel virtually devoid of pigmentation, with pitted, scarred and roughened surfaces (enamel hypoplasia); frequent enamel 'scabbing' with exposure of underlying dentine; curvature increased; cutting tips eroded almost to stubs
6	Severe fluorosis: enamel devoid of pigmentation; hypoplasia very common, surfaces heavily pitted, scarred and scabbed; increased tooth curvature, incisors 'splayed' outwards towards tips; cutting edges eroded to blunt stubs

greatest advantage is the capacity of organisms to absorb and integrate doses of fluoride under different environmental and climatological conditions (Gilbert, 1968; LeBlanc and DeSloover, 1970; Oshima, 1974; Treshow, 1975; Weinstein and Laurence, 1989; Weinstein *et al.*, 1990). However, the amount of useful information that they supply and the cost-effectiveness vary greatly. Unless the use of biomonitors is focused and applied with specific goals in mind, surveys run the risk of showing only that the fluoride concentration is higher in one place than in another or higher in one organism than in another.

Plant biomonitors

Vascular plants are usually the first choice because there is almost always one species that is widespread and common enough to provide a useful picture of the pattern of fluoride contamination. The best plant biomonitors are species that are relatively tolerant to fluoride because there is often an inverse relationship between the degree of sensitivity and fluoride accumulation. Acute fluoride exposure may affect fluoride absorption and, in addition, necrotic tissue

accumulates greater amounts of fluoride than adjacent, healthy tissue, so necrosis alters the relationship between exposure and leaf fluoride content. In more tolerant species, accumulation continues without interruption. One of the main limitations of using leaf concentrations is the problem of determining the proportion of the fluoride that is deposited externally on the leaf surfaces and the fraction that is in the parts of the tissues where it can exert a toxic effect. Washing tissues with water or various solvents does not remove a constant, well-defined fraction of the fluoride, so that does not provide an answer, although the method used by emission spectroscopists of washing in a non-ionic detergent containing a chelating agent, such as ethylene-diamine tetra-acetic acid (EDTA), has been adopted in some laboratories. A second problem is uncertainty about the residence time or turnover of fluoride that is absorbed into the tissues. As indicated in Chapter 2, leaf analysis gives a snapshot of the fluoride accumulated over days, weeks or even years, depending on the species, but there is no way of knowing what a particular concentration represents without a study of the dynamics. The analytical problems that were discussed in Chapter 1 are also a consideration and the normal variance caused by sampling and analysis means that the confidence limits can be large (Davison et al., 1979).

Despite the problems involved in using plants as collectors, there are many excellent examples where a simple approach has yielded valuable information. One was the use of the street tree Chinese tallow (*Sapium sebiferum*) to help in an investigation of an accidental release of HF at Corpus Christi, Texas, described in Chapter 1. It was essential to try to establish the area that was affected by the release, so leaves were sampled from 72 locations up to 1 month after the incident and analysed for fluoride content. The data were mapped (Fig. 6.5) and they showed very clearly the plume dispersion path and the areas that were potentially affected. More comprehensive uses of biomonitors have been reported by several authors, including Sidhu (1979, 1981) and Aluminiumindustries Miljøsekretariat (AMS, 1994). Sidhu (1979, 1981) used foliar analysis to define the seasonal and spatial patterns of fluoride in the vicinity of a phosphate factory. He was also able to relate foliar fluoride to visible injury and concentrations in the air. The Norwegian survey (AMS, 1994) was probably one of the most

comprehensive and valuable studies of the effects of emissions from aluminium smelters because it included studies of vegetation, soil, fresh water, animals, the marine environment and human health. Biomonitoring using foliar analysis was particularly effective at demonstrating the way in which emissions travelled very long distances along valleys, and the effects of reducing emissions over time (Fig. 6.6).

Undoubtedly the most important use of plant biomonitoring is in the prevention of fluorosis in livestock. The dietary tolerances of common farm animals are well known (Chapter 3) and in many parts of the world regular monitoring of the fluoride content of forage is used as a measure of air quality and as the basis for regulation of industrial emissions. In general, most industries and regulators use criteria based on those outlined by the National Academy of Sciences (NAS, 1974) and Suttie (1969, 1977). Essentially, most guidelines state that the annual or long-term average fluoride content of the diet (forage + water + supplements) must be lower than 40 mg/kg. Some authorities recommend a lower threshold of 35 or even 30 mg/kg, and some specify thresholds for shorter periods. It is common practice to

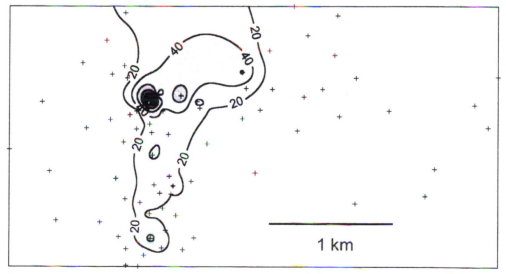

Fig. 6.5. The fluoride concentration (mg F/kg dry wt) in leaves of Chinese tallow tree (*Sapium sebiferum*) collected from 72 locations around the source of an accidental release of HF in Corpus Christi, Texas, USA (L.H. Weinstein, unpublished).

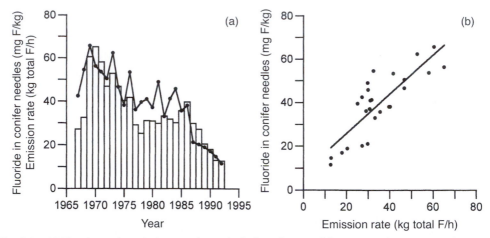

Fig. 6.6. (a) The change in emission rate (kg total F/h) from the Sunndal (Norway) aluminium smelter between 1967 and 1992, and the fluoride content of conifer needles (mg F/kg). (b) The same data plotted as a regression of the fluoride content of conifer needles on the emission rate: needle [F] = 0.898 × emission rate + 7.9, r^2 = 0.679. (Redrawn from AMS, 1994.)

collect and analyse forage and hay samples once a month and then to calculate a growing season or annual mean. Although this is accepted as standard practice, the high variability of forage fluoride concentrations over time means that the annual average estimated from monthly samples is subject to a high degree of uncertainty (Davison *et al.*, 1979). Furthermore, sampling and analytical protocols are very rarely stipulated, even by regulatory agencies, and inadequacies in either of these can lead to further errors. The difficulty of obtaining an estimate of the fluoride intake of a herd of cattle with an accuracy better than ±20% should not be underestimated. After reviewing the subject, Davison *et al.* (1979) made the following recommendations:

1.   At least 25 approximately equal-sized subsamples should be cut from the sampling area and bulked to form a composite sample.
2.   Ideally, subsamples should be collected randomly but, if that is not possible, they should be collected at approximately equal intervals on a W-shaped path that fills the area being sampled.
3.   All sites where livestock are grazing should be sampled at least once each week. In order to reduce the workload, the weekly samples collected during the month can be

combined to produce a single sample for analysis.
4.   Forage is only one source of fluoride in the diet, so Suttie's (1969) recommendations for sampling hay, silage, mineral supplements and water must be followed.

Because of the problems associated with estimating the fluoride content of forage, in some European countries, notably Germany, it is common practice to use standardized grass cultures. In some places these are a legal requirement. Sites are set up in appropriate locations with small, self-watering grass cultures (Scholl, 1971; Arndt *et al.*, 1987; Fig. 6.7). A prescribed variety of grass is germinated and grown in a low-fluoride area, after which pots are exposed for standard intervals in different locations. This technique has the advantage of minimizing some of the sources of variability, such as differences in species composition, water stress and soil splash. In Chapter 2 we showed how van der Eerden (1991) was able to use a similar system to produce a simple model of fluoride deposition. Unlike deposition gauges, it gives an estimate of the total deposition to a living plant sward. However, the simplicity is also a weakness because it is so artificial. It does not give a realistic picture of the fluoride

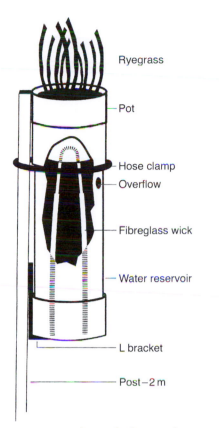

Ryegrass

Pot

Hose clamp

Overflow

Fibreglass wick

Water reservoir

L bracket

Post–2 m

Fig. 6.7. Diagram of a standard grass-culture system in which a prescribed variety of grass is germinated and grown in self-watering pots in a low-fluoride area, after which they are exposed for standard intervals in different locations. (Redrawn from Arndt, U., Nobel, W. and Schweizer, B. (1987) *Bioindikatoren*, Eugen Ulmer GmbH and Co., Stuttgart.)

phytosociological method to produce an index of atmospheric purity (IAP) over a 250 km² area. This showed very clearly the zone of influence of the emissions around the industrial centre, though the authors assigned all of the effects to fluoride and did not comment on the possible effects of the SO₂ from nearby pulpmills. The main drawback to the method, apart from the previously mentioned problem of knowing when the lichens were affected by the pollution, was the demand for highly trained field staff. However, it is notable that they also measured the fluoride content of the lichen *P. sulcata* and that there was a striking inverse relationship between it and the more elaborate IAP. This suggests that, once the relationship was established, the simpler biomonitoring technique could be used as an alternative, or at least as an initial step to define the potential problem areas.

Two other notable studies are those by Perkins *et al.* (1980) and Davies (1982). The former group made a comprehensive study of the effects of an aluminium smelter at Anglesey (Wales), an area that was completely rural. Because they made their observations from start-up, they were able to follow the increase in damage to lichens, but they also used the lichens as biomonitors to map the pattern of contamination and follow the increases in fluoride content of the thalli over time. Interestingly, this suggested that corticolous (growing on bark) lichens accumulated fluoride more quickly than saxicolous (growing on rocks) species. They thought this might be due to greater exposure of the corticolous habitats. Davies (1982) worked in the brick-fields of Bedfordshire (England), where fluoride emissions were estimated at about 1400 t/year. Emissions of SO₂ were known to be high enough to have affected the lichens and the area was impoverished in terms of lichens, but it was not known if the fluoride had any effect on them. The main problem in trying to assess the effect of fluoride in a situation such as this is that the emissions were from several sources with tall stacks. The latter would be expected to lead to relatively low (but still potentially toxic) concentrations occurring over a large but difficult-to-define area.

concentration in pastures or meadows, so it should not be used to indicate the potential fluoride intake by grazing animals. It is also more expensive than a forage-monitoring programme.

Lichens and, to a lesser extent, mosses have been used very successfully as biomonitors. One of the first studies was that of LeBlanc *et al.* (1972) around the aluminium smelters and other industries in Arvida (Quebec, Canada). At that time, industrial emissions caused a great deal of damage to the epiphytes and many species were eliminated from the most heavily polluted zones. The authors used a complex

Davies (1982) examined one of the few wide-spread lichens, *Xanthoria parietina*, and analysed the fluoride content of the thalli collected from 60 sites in an area of about 58×38 km. This simple and rapid procedure enabled him to produce a map that showed the main area of deposition, which he was able to relate to the visible appearance of the lichens.

Mosses have been less used than lichens but, as their fluoride concentrations are often reported as being much higher than those of vascular plants, it might be concluded that they could be used as efficient collectors. For example, Roberts *et al.* (1979) and Sidhu (1979) reported that fluoride accumulation in the moss *Polytrichum commune* was about ten times greater than in needles of balsam fir (*Abies balsamea*) in the inner fume zone near a phosphorus plant in Newfoundland. However, as indicated in Chapter 2, caution must be used in comparing the fluoride contents of different taxa. The reason is that the ratio between the absorbing surface area and dry weight of a moss differs from that of a conifer needle by a factor greater than ten times: that is, the moss has over ten times the surface area to absorb fluoride per gram of tissue than *Abies*. That fact alone can account for the apparent difference in the amount 'accumulated'. This supports the present authors' view that organisms should only be used as biomonitors if the dynamics of fluoride uptake and loss is well known.

Animal biomonitors

There have been many studies in which animals or animal tissues have been used as collectors of fluoride. These have generally been conducted in order to relate fluoride accumulation to distance and direction from the source or as part of an evaluation of the ecological or economic effects on the species or ecosystem. Working with mobile animals, especially those that have large home ranges, presents an obvious problem, but there are several other difficulties that limit the potential for using animals as biomonitors. Some will be obvious from the discussion in Chapter 2 about the transfer of fluoride from the air to animals. For example, in adult butterflies, almost all of the total body fluoride is surface deposits, the rate of deposition is non-linear with time and fluoride is lost postfumigation. This means that two individual insects taken in the field might have the same fluoride content but very different histories of exposure. Also, whole-body analysis can be misleading unless the gut contents are removed, especially in detritus feeders. An example is the earthworm, in which the fluoride content is almost entirely due to the soil contained within the gut (Walton, 1986).

In vertebrates, fluoride accumulates in calcified tissues, which makes them an attractive proposition for biomonitoring, but the fact that the fluoride content increases non-linearly with time means that the age of the animals must be known – something that is not always possible. Where cattle are concerned, the fluoride intake is sometimes monitored by analysis of the fluoride content of tail bones. Animals are given a local anaesthetic, the tail is shaved and a single bone is removed by a veterinary surgeon (Plate 27). As the animals do not show any signs of distress, this can be an effective and unequivocal way of tracking fluoride accumulation in a herd, but it is invasive and expensive (Fig. 6.8). With small mammals, the population turnover rate may be so rapid that the mean bone fluoride of the population changes significantly as young are recruited and older animals are replaced, making it difficult to sample adequately without decimating the population. On the other hand, larger animals, such as deer that are hunted, may provide an excellent source of aged bone samples so that changes in fluoride load can be followed over time at little cost. Species that are at the higher levels of the food-chain, particularly top predators, are not usually considered for biomonitoring because of the small numbers and conservation issues. Overall, it is difficult to avoid the conclusion that using animal biomonitors presents real problems that should not be underestimated. In many cases, it is probably more productive to use animals as bioindicators (e.g. tooth examination of mammals) and plants as biomonitors.

Three good examples of animal biomonitoring are provided by: (i) the Norwegian smelter study (AMS, 1994); (ii) Walton (1984, 1985, 1986, 1987a,b, 1988); and (iii) research on bees (Mayer *et al.*, 1988; Bromenshenk, 1994). The first included records of the fluoride content of several animal tissues, including deer, elk and hare bones, geese and gulls' eggs. They helped to build an overall picture of the extent of contamination and of which groups were being affected. The authors were able to determine the ages of deer and elk and therefore to define in sound statistical terms the upper limits for the natural background fluoride content. That information and data on dental fluorosis allowed them to define a critical bone fluoride content for the most sensitive age group of

red deer (*Cervus elaphus*). Table 6.4 shows the risk of dental fluorosis associated with different bone fluoride concentrations in young (1.5 years) deer, and the % of animals falling into each category in the vicinity of three aluminium smelters.

Walton's work (1984, 1985, 1986, 1987a,b, 1988) provides an interesting contrast to the Norwegian study. He trapped and/or collected earthworms, woodlice, small mammals and fox bones in a systematic study of fluoride contamination around the Anglesey aluminium smelter, starting shortly after it began emissions. The research provided an essential database of the range of fluoride concentrations in different groups of animals, which was lacking, and it showed clearly the changes in fluoride accumulation with distance from the smelter. However, the effects on the animals in terms of population numbers were not

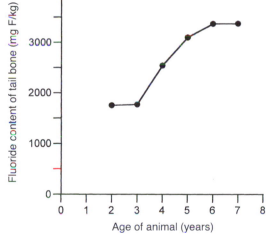

Fig. 6.8. The fluoride content (mg F/kg) in the tail bones of a herd of beef cattle. The age of each individual was known from tags and branding. In this herd, the fluoride content appeared to level off at around 6–7 years old, which may have been due to reduction in emissions. Some individuals had mild (cosmetic only) dental fluorosis but none showed any other symptoms. (From A. Davison, unpublished data.

Table 6.4. The risk of dental fluorosis associated with different bone fluoride concentrations in young (1.5 years) red deer, and the % of animals falling into each category in the vicinity of three aluminium smelters in Norway (data from AMS, 1994).

Bone fluoride (mg F/kg)	Response	% of animals in each risk category in three municipalities		
		Årdal	Sunndal	Kvinnerad
< 1000	None	62	90	96
1000–2000	Risk of developing dental damage	12	10	4
> 2000	Causes dental damage with score of 3 or > 3	26	0	0

studied, so it is still not known whether the smelter had any effect on the animals or which groups might be at greatest risk.

Although insects present difficulties in relation to biomonitoring fluoride, honey-bees (*Apis mellifera*) have been successfully used for more than 20 years in the USA and Europe as biomonitors for several contaminants of terrestrial ecosystems (Bromenshenk, 1994). Users consider that, among their virtues as biomonitors, honey-bees provide a rapid, inexpensive and elegant method to evaluate the range and approximate intensity of pollution impacts within the range of individual hives. Also, there is a considerable literature on the toxicity of fluoride to honey-bees, much of it conducted in the laboratory under controlled conditions, but also with many field observations on bee mortality in fluoride-contaminated areas. A major advantage of using honey-bees is the large numbers of individuals that are mobilized and that return to a fixed location, where data, such as fluoride ingestion and pollen contamination, can be readily accessed. Field protocols for the use of honey-bees have been outlined by Bromenshenk (1994) and his group has used the technique successfully in several areas. In the Puget Sound area in north-western Washington State, Bromenshenk et al. (1985) recruited 64 bee-keepers, who volunteered to collect samples and make measurements at 72 sites that spanned an area of about 7500 km². The fluoride concentrations in forager bees were mapped, using a kriging algorithm. In a similar but more tightly controlled study, Bromenshenk et al. (1996) established bee colonies at 61 locations at the Idaho National Engineering Environmental Laboratory (INEL) and other areas in the upper Snake River plain. The 3-year study was designed to accomplish the following tasks: map fluoride distribution from several industrial and natural sources; compare the levels of fluoride in bees to those found in other environmental samples; evaluate the use of bees as monitors in a semi-arid desert; and determine the source strength of the potential fluoride sources in the study. Fluoride analyses of forager bees revealed that the major fluoride source in the area was a phosphate rock-processing plant south of the INEL, with bee concentrations of up to about 80 mg/kg. Bees from the INEL had a range of about 10–30 mg fluoride/kg, but the values differed from year to year, with much higher concentrations found at the INEL in the third year from an apparent fluoride release. Although the concentration of fluoride in bees was often high near the phosphate rock-processing plant, much higher concentrations have been reported near aluminium smelters, e.g. Dewey (1973), 221 mg/kg; Bromenshenk et al. (1985), 182 mg/kg; Mayer et al. (1988), 358 mg/kg.

There are, however, some questions remaining about the use of bees and interpretation of the fluoride content. As discussed in Chapter 3, it is still not clear how much of the fluoride is surface deposition and whether the rate of surface deposition is non-linear with time, so the fluoride content of a bee cannot easily be related to atmospheric concentrations. In the study carried out by Mayer et al. (1988) described in Chapter 3, total fluoride concentrations were used to monitor the exposure of the bees (Fig. 6.9). The fluoride concentrations of bees from the three sites in the vicinity of the smelter differed significantly, but the differences were not as great as might have been expected. For example, compare the difference between the sites that were 0.8 km downwind and 6.4 km east of the smelter (Fig. 6.9); in August and September they were very similar. There was a distinct seasonality, with an increase from July to September but a fall in October. The authors did not comment on the reasons for the peak but they thought that the drop in October was due to decreased foraging for pollen in the autumn. These data demonstrate that a great deal remains to be learned about the dynamics of fluoride accumulation by insects.

Use of Bioindicators and Biomonitors in the Field

Many different approaches have been devised to summarize and display the results of field studies of bioindicators and biomonitors, but there are few cases in which the cost-effectiveness can be compared. However, one case is provided by Weinstein *et al.* (1990). They surveyed the extent of fluoride injury to *Gladiolus* over a reach of over 100 km in the Rhône Valley of Switzerland and superimposed the results on a map of the area. This study was accomplished in about 2 days by car, estimating the length of tip burn on gladiolus leaves using binoculars and plotting each data point on a map according to injury classes (Fig. 6.10); it was a simple and relatively unequivocal display of the occurrence and

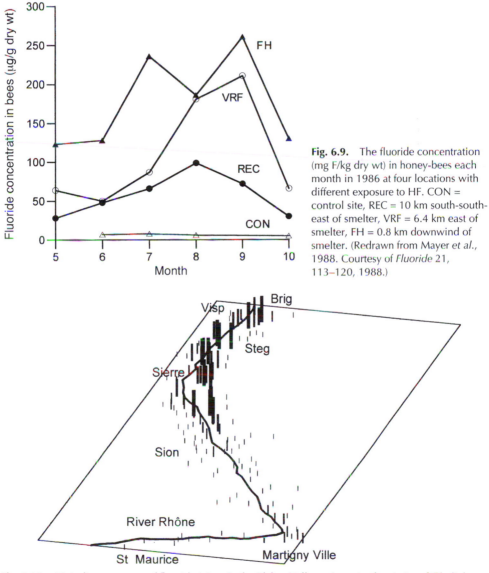

Fig. 6.9. The fluoride concentration (mg F/kg dry wt) in honey-bees each month in 1986 at four locations with different exposure to HF. CON = control site, REC = 10 km south-south-east of smelter, VRF = 6.4 km east of smelter, FH = 0.8 km downwind of smelter. (Redrawn from Mayer *et al.*, 1988. Courtesy of *Fluoride* 21, 113–120, 1988.)

Fig. 6.10. Bioindicator survey of fluoride injury in the Rhône Valley using mixed varieties of *Gladiolus* growing in gardens. The vertical bars are proportional to the % of leaf area injured. (Data from Weinstein *et al.* (1990). Weinstein, L.H., Laurence, J.A., Mandl, R.H. and Wälti, K. (1990) Use of native and cultivated plants as bioindicators of pollution damage. In: Wang, W., Gorsuch, J.W. and Lower, S.R., *Plants for Toxicity Assessment*, ASTM, Philadelphia, PA, pp. 117–126.)

relative intensity of fluoride toxicity to a sensitive bioindicator. During this same season, current-year and 1-year-old needles of Scots pine and leaves of apricot and grapes were collected every 2 km and at elevations of 1000, 500 and 0 metres on the north and south sides of the valley at each location. The foliage was then analysed for fluoride content and the results displayed using mapping software (Fig. 6.11). Obviously, the analytical approach was more accurate overall, but it was not that much more informative than the *Gladiolus* survey for an experienced observer, and the differences in cost were enormous.

We can summarize some of the pros and cons of bioindicators and biomonitors as follows:

Bioindicators
1. Response not instantaneous but may take several days depending on the concentration.

2. Provide an immediate estimate of the degree of injury at the time of the survey.
3. Semi-quantitative but can often estimate approximate dose.
4. Relatively inexpensive if it does not need a trained taxonomist (e.g. lichens).
5. Characteristic symptoms induced in some species but symptoms can be confused with different causes of injury.
6. Symptom expression may be modified by climatic and edaphic factors.

Biomonitors
1. Provide quantitative measure of fluoride accumulation by appropriate organisms.
2. Repeated analyses may reveal topics that need further basic research.
3. Data may be difficult to relate to emission strength or exposure unless the dynamics have been studied. Some accumulated fluoride may be lost by weathering.

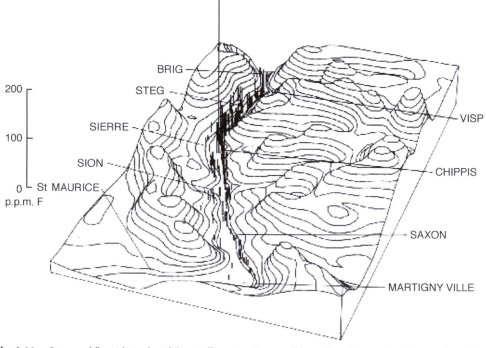

Fig. 6.11.   Survey of fluoride in the Rhône Valley using 1-year-old needles of Scots pine (*Pinus sylvestris*) as biomonitors. The vertical bars are proportional to the fluoride content of the needles. (Data from Weinstein *et al.* (1990). Weinstein, L.H., Laurence, J.A., Mandl, R.H. and Wälti, K. (1990) Use of native and cultivated plants as bioindicators of pollution damage. In: Wang, W., Gorsuch, J.W. and Lower, S.R. *Plants for Toxicity Assessment*, ASTM, Philadelphia, PA, pp. 117–126.)

4. Sample collection for analysis may be slow and analyses can be expensive, which argues for having clearly defined goals and protocols.

Elements of Field Surveys of Vegetation

General considerations

Many industries and some regulators have monitoring programmes that involve recording atmospheric fluoride and the fluoride content of vegetation, but a key additional element is often a regular survey of the health of vegetation. This is because of the high sensitivity of some bioindicators and the importance of crops and native plants in the landscape, so the last section of this chapter is a brief discussion of the practicalities of field surveys. As discussed earlier, improvements in engineering and the arrival of dry scrubbing have greatly reduced fluoride emissions from industry, but, if emissions are so well controlled, why should fluoride problems still exist today? First, this is because of its extreme toxicity to living organisms, especially to plants. Secondly, fluoride emissions depend on factors other than scrubbing efficiency, such as collection efficiency, and this is dependent not only on the quality of the control equipment, but also on its maintenance and housekeeping practices.

The type and scope of a field survey depends on the goals. Surveys may be conducted to:

1. Provide background information before an industry is sited. This is particularly valuable where there are already industries in the area, which might be emitting air pollutants or affecting the vegetation in other ways (e.g. by altering groundwater). An example is the Alcan aluminium smelter at Lynemouth, England. An early survey showed that coal burning in domestic heating systems and emissions from burning mine waste were sufficient to injure sensitive species and raise grass fluoride levels close to those that cause fluorosis (Davison *et al.*, 1973).

2. Evaluate the adequacy of planned control and abatement procedures by industry or governmental agencies.
3. Assess injury or damage to vegetation as a result of an unexpected release of fluoride or of complaints or for purposes of arbitration or litigation.
4. Give warning of a deterioration in equipment or provide confirmation of improvements in emissions after the installation of new equipment. In one case known to the authors, a field survey revealed a source of emissions that was unknown to the management and that had not been detected by other monitoring.

Field surveys are tedious for the surveyor, and there is a tendency to regard them as not making a useful contribution to science. The latter point can be true, but diligent observation can reveal unanswered questions that need further research. For example, if there had been regular surveys in the early days of the Kitimat smelter (Chapter 5), they might have provided a better insight into the causes of the events that followed and led to new information about plant–insect–pollution interactions. The US National Research Council (NRC Committee, 2000) made a statement that is appropriate in this context:

> Conducting field surveys may be denigrated by many scientists as monotonous and mundane and not really scientific research. But the collection of data (or observations) over a period of time can provide answers to scientific questions as well as can laboratory research because, after all, the latter also entails careful planning and repetitive data collection before there is any scientific advancement.

Thus, the authors consider that field surveys should be an essential part of environmental management practices.

Basic requirements

Although Skelly *et al.* (1979) suggested that field surveys 'should be performed by a multidisciplinary team of qualified individuals such as plant pathologists, ecologist-foresters, entomologists and soil scientists', this ideal is rarely achieved and most field surveys are conducted by one or two people. Considering the complex nature of field evaluation, the individual(s) responsible for the survey should be skilled observers and be capable of recognizing pollutant-induced symptoms, nutrient stresses, plant diseases and common insect pests, as well as being familiar with agronomy, horticulture, ecology, plant identification, the characteristics of emissions and their dispersion, air-monitoring technology and statistics. This does not mean that the surveyor must be an expert in each field, but he/she must certainly be expert in recognizing at least the first four categories; the essential requirement is first-hand experience of authentic fluoride injury in a range of species and circumstances. Therefore training is vital. When unusual problems arise, such as when a nutritional problem is suspected, a specialist may be required to participate in subsequent surveys.

It should be apparent that no single method can be prescribed that is suitable for all the situations encountered in the field. Therefore the discussion that follows will suggest general techniques common to all surveys. With these in mind, it should be possible to design an approach for surveillance that considers the nature of the problem and the extent of the information available. The recommended first step is to obtain an emission inventory to determine the amount and forms of the fluoride emitted. As already indicated, the proportion of gaseous to particulate fluoride varies greatly between industries and even in the same industry. For example, not only are there large differences between the proportion of gases and particles emitted by aluminium smelters using the prebake and Söderberg processes, but there are differences even between smelters using the same process, depending on the design, the operating conditions during smelting, the source of the alumina and the type and operation of the air-cleaning equipment. All of the sources of emission should be identified: main stacks, fugitive sources, dumps of fluoridated materials, etc. Any periodicity in the process

should be noted, especially if it relates to the plant growth cycle. Secondly, an estimate of the temporal and spatial distribution of the pollutant should be available for the area, based upon topographical and meteorological data and/or atmospheric monitoring. A wind rose is essential but it should be borne in mind that large buildings can produce local patterns to the wind, so local knowledge may be necessary. If an indication of the spatial pattern of the fluoride concentrations is not available, it should be possible to identify the areas with high fluoride by means of a preliminary survey of the fluoride content of vegetation. Grass or a widespread species is suitable, as in the example cited earlier in this chapter (Fig. 6.5). Thirdly, the surveyor should have a large-scale map of the area and a satellite positioning system to locate and relocate sites. A global positioning system (GPS) receiver is particularly valuable in wooded terrain, where it is difficult to locate sites exactly from a map alone. Fourthly, a preliminary inventory should be made of the plant species present, their distribution and abundance and their economic and aesthetic significance. Fifthly, the surveyor should make use of local knowledge to identify past disturbances and stress at each site. These can influence plant response to fluoride or they may injure plants or cause similar symptoms. Significant disturbances include: construction (which can alter the water-table); drought (which reduces fluoride uptake but which may affect plant growth for years after the event); fire (which may damage epiphytes); flooding (which can cause chlorosis and anthocyanosis, as well as damaging roots); animal grazing/pest outbreaks that produce visible symptoms; and the occurrence of plant pathogens. A good example of the effects of disturbance is provided by the Anglesey aluminium smelter in Wales. In order to build the smelter, a major new road was constructed and it involved cutting through bedrock. Shortly afterwards, many mature trees died on one side of the cutting because the water-table had been drastically altered. However, the deaths were firmly blamed on fluoride emissions from the smelter, even though the deaths happened before there were any significant emissions. Thousands of saplings of the same, fluoride-resistant species (sycamore, *Acer pseudoplatanus*) replaced the dead trees, but still the idea persisted that pollution was the cause of the tree deaths.

The main problem encountered is often one of finding sites that are accessible and in the most appropriate locations, but sites must be selected that will not be altered over the coming years by land development. The sites should be sufficiently numerous and distributed so that effects due to fluoride can be distinguished from other local effects; and they should contain a reasonable number of indicator species at each site, preferably of the same species. If the latter is not possible, then species may be used whose ranges overlap at some sites (Weinstein and McCune, 1970). The number of sites will vary, depending on the emission strength of the source and the degree of injury observed on plants from the preliminary survey. In our experience, a total of 20–25 sites is usually sufficient to evaluate and monitor a field problem, with the majority of sites located in the general downwind direction, assuming that these are characteristic of the area. Sources that are located in valleys may require sites that extend over long distances (see Figs 6.10 and 6.11), so it is important not to underestimate the distance a plume can travel. Thirty years ago it was often necessary to examine vegetation for distances > 10 km from major sources, but the reductions in emission rates mean that in the better-controlled industries the zone of interest may be < 2 km.

Regardless of the number of field-survey sites established, they should be arranged in a systematic manner if possible. The design chosen will depend on the final objectives of the survey and the topography of the area. Clearly, the number and locations of sites should not be so extensive that the survey becomes impractical for reasons of time, available funds and analytical resources. Carlson and Dewey (1971) established sites along transects radiating from the source. Krupa and Kohut (1976) (discussed by Skelly *et al.*, 1979) established

sites at varying distances around a point source along eight radial gradients each 45° apart.

Finally, what should be recorded? Based on Skelly *et al.* (1979), the following are recommended:

1. At each site, photograph a general view of the site, healthy specimens and the whole range of symptoms shown by each species.
2. Devise an objective, quantitative method for recording the symptoms, such as the % of the length of pine needles that is necrotic or the % of the leaf area that is chlorotic. Descriptions such as 'severe' or 'mild' should not be used because they are highly subjective and do not have a scientific basis.
3. Assess the % of leaves that are injured and the % of individuals of each species that are injured. Note and investigate cases where the symptom expression varies greatly within a site. Are there any obvious causes, such as shelter or variation in water availability?
4. Collect enough injured and uninjured leaves of the same age for chemical analysis. Washing leaves before analysis is a matter of judgement, but, if a nutritional imbalance is suspected (e.g. iron deficiency), leaves must be washed before analysis. It should be clear from previous statements that interpreting the fluoride concentration is not straightforward. Analysis does not provide a panacea that will reveal whether symptoms are caused by fluoride.
5. The timing of the survey and the number of surveys will depend upon the occurrence and degree of injury shown by different species. But, if only a single survey is to be conducted, it should take place late in the growing season and before natural senescence becomes a confounding factor. This situation might be suitable for temperate climates, but surveys in tropical and subtropical areas require special considerations. For example, in these areas, it is impossible to determine when it is late in the growing season. Often, the differences are the dry or rainy seasons. One other consequential factor is that the length of plant exposure in these areas is 12 months as

compared with the usual 6-month exposures in temperate areas.

6. The surveyor should collect as much pertinent information as is possible within the time allotted to the survey. Some individuals prefer keeping a notebook, or log, of their observations. Others prefer using printed data sheets that are methodical and include the salient information derived from the survey. A sample data sheet, modified from Skelly *et al.* (1979), is given in Table 6.5.

7. Finally, it is important to remember that it is not always possible to identify the causes of symptoms, especially after a single survey. Often the causes are only apparent after several years and when plants have been observed under a range of different environmental conditions.

Table 6.5. Example of a data sheet. (Modified from Skelly *et al.* (1979), *Handbook of Methodology for the Assessment of Air Pollution Effects on Vegetation*, Air Pollution Control Association, Pittsburgh, PA, pp. 12-1 to 12-30; courtesy of the Air & Waste Management Association.)

Surveyor's name:	Title:

Affiliation and address:
Telephone, fax and email:
General location of survey:
Map sheets:
Name and address of client:
Description of problem:
Suspected pollutants, other than fluoride:

Site Recording Sheet	Site number:
GPS location:	Date:

General topography and landscape features:
Altitude, aspect, slope:
Soil type, drainage:
Direction and distance from source(s):

Plant species injured:	Symptoms:	% leaf injured

Uninjured species:

Remarks, observations:

7

Environmental Standards to Protect Humans, Other Animals and Plants

Introduction

Standards, criteria, guidelines, goals, objectives, limits, lowest observable effect levels, no observable effect concentrations, maximum acceptable concentrations and critical levels are all part of the vocabulary associated with increasing global demands to reduce the risk of adverse effects of pollutants on humans, other animals, plants and ecosystems. These legal or quasi-legal devices may be enacted by national, regional and local governments, by multi-state compacts or by international groups with varying powers, such as the World Health Organization, the European Union (EU) or the United Nations Economic Commission for Europe (UN ECE). The use of the terms is not consistent between organizations. In one country a term might have a very specific meaning but in another it may be a more general term. For example, in the USA, air quality standards refer to time-weighted concentrations. They are prescriptive and enforceable but in other countries the term standard is used loosely, without having any legal definition. The term criteria is used in the USA to refer to descriptions of the relationships between the time-weighted concentrations of pollutants and their effects. The same term does not carry such a meaning elsewhere. Other terms, such as no observable effect concentration, are used specifically in relation to water but not to air. In other words, the nomenclature is a semantic minefield, so in this chapter we shall use 'standard' as a catch-all term, except for instances where text is cited direct from a source and a specific term is used. Environmental standards for fluoride are based on control of emissions at source and the concentrations in air, vegetation and water. In the following discussion, we shall consider each of these in some detail.

Standards Based on Control of Emissions

The earliest approach to pollution abatement was by controlling the rate of emission from the source. Attempts were made to reduce pollution (principally smoke, SO_2 and odours) in Europe from the 13th century, but the first effective legislation was probably the British Alkali Act of 1863. It was followed by various modifications but, in essence, the principles embodied in the Act were the mainstay of pollution control in Britain until 1975. The key statement in the Act was one that specified that to prevent the escape of noxious gases the 'best practicable means' had to be employed. These three words meant that industry could be compelled, in theory at least, to adopt new technology as it became available to reduce emissions. Gradually, practicable means included the maintenance of equipment and any other operation that affected the emissions. The phrase allowed flexibility because the word practicable meant that the industry could negotiate with the regulator on the grounds that the technology was not practicable because it was too expensive. Nevertheless, the Act led to important improvements and all without specifying any actual rate of emission of any pollutant or quantitative air-quality standards.

Nowadays, emission standards to restrict the amount of fluoride that is permitted to be released into the atmosphere are the main basis for abatement and prevention in many countries. Limits are set for several industries, including aluminium smelting, glass manufacturing, phosphate-fertilizer processing and semiconductor manufacturing. One of the crucial requirements of any standard is that it should be possible to monitor the situation and readily detect breaches. This is one of the main advantages of basing control on the rate of emission from the source because modern equipment often allows emission rates to be measured in real time with adequate accuracy, so there can be a continuous record of performance. However, there may be practical problems in monitoring emissions where the design of the building does not allow the use of spectrophotometric methods or where it is difficult to assess the rate of air movement through the building. Estimating fugitive emissions and those from large open areas, such as phosphogypsum settling ponds, is difficult. Furthermore, using the emission rate on its own does not take into account the effects of topography or climate or the presence of sensitive plant and animal receptors, so the same emission rate might have different effects in different regions. MacLean (1982) discussed this and considered the pros and cons of having seasonal and regional standards.

In the USA, as part of the National Emission Standards for Hazardous Air Pollutants (NESHAP), the US Environmental Protection Agency (US EPA, 1997b) established emission values based upon maximum achievable control technology (MACT) for the aluminium industry. In essence, this document establishes fluoride emission standards for each new and existing pot line, paste-production plant and anode-bake furnace associated with the 23 primary aluminium-reduction smelters in the USA (although, at the time of writing, three are not in operation). In 1997, the US nation-wide emissions from all primary aluminium smelters were estimated at 6400 tons (5806 t) per year of total fluoride, of which about 2500 tons (2268 t) was gaseous fluoride. After implementation of the proposed standards, the US EPA estimated that emissions would decrease to 3400 tons (3084 t) of total fluoride per year. This included a 97% reduction in total fluoride emissions from the anode-bake furnaces (from smelters that employ prebake technology) (US EPA, 1997b). The emission limits are outlined in Table 7.1. Note that, in common with many countries, the limits are defined in terms of kg of fluoride per tonne of aluminium. This means that the total emission rate, and therefore the potential impact on the environment, depends on the production capacity of the plant. This appears to have been recognized by the Norwegian study (AMS, 1994) because it did not express the maximum emissions that would not cause damage in terms of kg/t but as kg/h. It concluded that:

> The critical fluoride loads for damage to forest vary considerably with the location of the aluminium plant. As a rough guide . . . new damage to conifers does not appear probable if emissions are kept to below about 10 kg F/hour.

In the past there have been differences in standards in force in different European nations, but with the formation of the EU there has been increasing convergence. Currently in the UK, for example, the operator of a fluoride source, such as an aluminium smelter, must seek authorization, a licence that allows specified emission rates from each of the sources within the operation. The authorization is for 4 years and it includes an agreed improvement programme, so there is a rolling scheme for reduction of pollution. The licence can be altered retrospectively by both parties – for example, by the regulator if it is found that

Table 7.1. Emission limits for total fluorides from existing, new or reconstructed sources (from US EPA, 1997b).

Type of reduction cell or anode-bake furnace	Emission limit (pounds/ short ton of aluminium)	Emission limit (kg/t of aluminium)
Limits for existing sources		
Centreworks prebake 1	1.9	0.95
Centreworks prebake 2	3	1.5
Centreworks prebake 3	2.5	1.25
Sideworks prebake	1.6	0.8
Vertical-stud Söderberg 1	2.2	1.1
Vertical-stud Söderberg 2	2.7	1.35
Horizontal-stud Söderberg	2.7	1.35
Anode-bake furnace	0.2	0.1
Limits for new or reconstructed sources		
All technologies, prebake and Söderberg	1.2	0.6
Anode-bake furnace	0.02	0.01

the emission rate was set too high. In practice, the emission standards that are set within the EU are very similar to those in Table 7.1 because they are a reflection of the available technology and what is practicable. It is envisaged that in the near future further regulation will be introduced by the EU.

Standards Based on Air Quality

Air-quality standards for fluorides have been enacted by several US states and many other countries that house emitting industries (Table 7.2). They refer to the time-weighted concentration of atmospheric fluoride, and they are designed to protect specific receptors, usually plants. In most countries, the standards are specified so that the highest concentrations are acceptable for only the shortest durations – recognizing the non-linear, inverse relationship between atmospheric fluoride, duration of exposure and effects (Chapter 4; McCune, 1969). The standards for the US states and Australia are similar and are based upon the standard designed for The Dalles, Oregon, in 1967 (Weinstein, 1969). Air-quality standards in several other jurisdictions were also based on the case of The Dalles, so the 'Dalles' standard is discussed below to provide a historical context.

In 1969, the Federal District Court of Oregon entered into a consent decree between fruit growers and Harvey Aluminium Company, both in The Dalles, Oregon, to appoint an arbitration panel to investigate and mediate allegations that gaseous fluoride emitted by the smelter was causing damage to the cherry, apricot and peach crops. The growers appointed Moyer D. Thomas of the University of California at Riverside as their arbitrator (he was replaced later by A. Clyde Hill of the University of Utah). Leonard H. Weinstein of the Boyce Thompson Institute was selected by Harvey Aluminium Co. Together, the two arbitrators elected a third, neutral arbitrator, O. Clifton Taylor, also of the University of California at Riverside. The arbitrators were designated not only to investigate the alleged problem in the field and to arbitrate monetary awards, if any, for damage to fruits, but also to establish an air-quality standard specifically adapted for growth conditions and smelter emissions in The Dalles. The arbitrators agreed to submit individual proposals for a standard, from which one would be negotiated. The final version of the standard contained some fractional values for allowable concentrations of fluoride, because they were the mean of the values proposed:

1. Gaseous atmospheric levels of fluoride not to exceed 4.5 p.p.b. (3.7 μg/m^3) on a daily average or for any one 12 h period.

Table 7.2.   Ambient air quality standards and objectives for gaseous fluorides ($\mu g/m^3$) and for fluoride concentrations in vegetation (mg/kg dry weight) in various jurisdictions. (Modified from Streeton, J.A. (1990) *Air Pollution Health Effects and Air Quality Objectives in Victoria*. Environment Protection Authority, Victoria, Australia, with kind permission of the author.)

Jurisdiction	Fluoride concentration in air ($\mu g/m^3$)						Fluoride content (mg/kg)
	90 days	30 days	7 days	24 h	12 h	Other	
Kentucky	–	1	0.8	2.9	3.7	0.4 annual mean	40 – growing season 60 – 2 months 80 – 1 month
Maryland	–	–	–	1.2	–	0.4–72 h	40 – growing season 60 – 2 months 80 – 1 month 35 – field crops[a] 50 – fruit-trees, berries, others[a] 100 – deciduous trees, shrubs[a] 75 – conifers, evergreens[a] 150 – grasses, herbs[a] 40 – ornamentals[a,b]
Montana	–	–	–	–	–	–	35 – growing season 50 – 30 days
New York	–	1	1.7	2.9	3.7	–	40 – growing season 60 – 2 months 80 – 1 month
South Carolina	–	1	1.6	2.9	3.7	–	–
Tennessee	–	1	1.6	2.9	3.7	–	–
Texas	–	1	1.7	2.9	3.7	–	–
Wyoming							30 – annual
Statewide	–	0.4	0.5	1.8	3.0	–	60 – 60 days
Regional[c]	–	1.2	1.8	4.0	10.0	–	80 – 30 days
Australia							
New South Wales	0.50	1	1.7	2.9	3.7	–	–
Special land use	0.25[d]	–	2.0	2.9	–	–	–
Victoria	0.59	–	–	–	–	–	–
Japan	–	–	0.5	1	–	–	–
The Netherlands	–	1	–	2.8	–	0.4 – growing season	–
New Zealand							
General land use	0.5	0.84	1.7	2.9	3.7	–	–
Special land use	0.25	0.4	0.8	1.5	1.8	–	–
Conservation areas	0.1	–	–	–	–	–	–
Norway	–	0.4	–	1	–	–	–
Canada							30
Proposed	0.4	0.4	0.5	1.1	–	–	–
Desirable	–	0.35	0.20	0.40	–	–	–
Adequate	0.20[e]	–	0.55	0.85	–	–	–
Manitoba	–	–	–	–	–	–	–
Desirable	–	–	–	0.40	–	–	–
Acceptable	–	–	–	0.85	–	–	–
Newfoundland	–	0.45	–	0.9	–	–	–
Ontario	–	0.34	–	0.86	–	–	–
(15 Apr.–15 Oct.)							
British Columbia	–	0.35	0.55	0.85	–	–	–
Montreal urban community	–	0.34	–	0.86	–	–	–

[a]Washed leaves.
[b]Current year's needles.
[c]Selected portions of Sweetwater County, Wyoming, USA.
[d]0.20 $\mu g/m^3$ in Western Australia.
[e]70 days.

2. Gaseous atmospheric levels of fluoride not to exceed 3.5 p.p.b. (2.9 µg/m³) on a daily average or for any one 24 h period.
3. Gaseous atmospheric levels of fluoride not to exceed 2 p.p.b. (1.7 µg/m³) on a weekly average.
4. Gaseous atmospheric levels of fluoride not to exceed 1 p.p.b. (0.82 µg/m³) on a monthly average.
5. Gaseous atmospheric levels of fluoride not to exceed 0.6 p.p.b. (0.5 µg/m³) on a growing-season average.
6. Gaseous atmospheric levels of fluoride not to exceed 0.3 p.p.b. (0.5 µg/m³) during the flowering period.

The values in this proposed standard were based in part on the criteria published by McCune (1969), on personal experience of the arbitrators in their own research and on field observations in cherry orchards at The Dalles. Although this standard was established for fruits, especially sweet cherries grown in eastern Oregon, it was endorsed soon by several states in the USA and, as of 1990, was still in use in at least six of the nine states with air quality fluoride standards (Table 7.2).

The strength of air-quality standards is that they are designed to protect the most sensitive receptors, so they will automatically protect all other, less sensitive organisms, including humans and other animals. They can be tailored to suit the local environment – for example, by applying a more stringent standard in conservation areas or during the growing season. The main weakness is in the quantity of good-quality scientific data that are available to use to set the more stringent standards. As indicated in Chapter 4, there have been only a few experiments worldwide that used open-top chambers and HF concentrations at the lowest levels that cause injury, so standards at those levels have a high degree of uncertainty. They involve expert opinion and guesswork. It is worth reflecting on the fact that all of the standards listed in Table 7.2 are based on the same, limited number of scientific publications. The differences are due to interpretation, opinion, regional sensitivities and politics. In addition, there is the problem of surveillance and validation. Instruments that are capable of measuring HF at the lower concentrations are expensive and usually few in number. A standard as low as 0.20 µg HF/m³, expressed to two decimal places, implies an accuracy in surveillance capability that just does not exist. How is it possible to police an area around a major source with only one or even a few instruments? Given that few instruments are usually available and that there are usually few remote locations where power is available, can there be adequate surveillance? It would be interesting to know how many instruments are in use in those countries and states that are listed in Table 7.2 and how much validation actually occurs.

Critical values

The terms critical level and loads are used loosely to define the exposure above which there are adverse effects, but in the original concept critical levels and loads have a different purpose from the standards discussed so far. Whereas air quality standards are usually legally enforceable, critical levels represent a different approach to pollution abatement. The original definition of a critical level was 'the concentration of pollutant in the atmosphere above which direct adverse effects on receptors, such as plants, ecosystems or materials, may occur according to present knowledge' (Nilsson and Grennfelt, 1988). The concept was devised in Europe under the auspices of the UN ECE, which was seeking a sound basis for the development of strategies for the abatement of transboundary air pollutants. The UN ECE has not developed critical levels for fluorides because they are not within its remit of transboundary pollutants, but New Zealand has adopted standards that are referred to as critical levels (NZ Review, 2000, 2002; Table 7.2). However, it is important to appreciate the differences between the UN ECE critical levels and the air quality standards in Table 7.2. In Europe, critical levels are used to map areas of exceedance, which identifies areas at risk and allows strategic planning of abatement. The direct adverse effects are not defined and there can be different levels for different types of effect (physiological effects, visible injury, loss of crop yield). The decision on which adverse effect to adopt for use in abatement is a political decision, not a scientific one. Finally, the caveat 'according to present knowledge' is important because critical levels are not cast in stone, they are revised periodically as new scientific information becomes available. In fact, where ozone is concerned, it has been the need for refinement of the critical levels that has driven much of the research in Europe on that pollutant.

Standards Based on Visible Injury to Vegetation

As hydrogen fluoride causes distinctive symptoms in sensitive plants at very low concentrations, it would seem obvious that plant injury might form the basis for a standard either on its own or in conjunction with other techniques. In the 1960s, there was a great deal of discussion in the USA regarding standards to control the effects of industrial fluoride emissions on crops, forests, other native plants and ornamentals. Earlier, there had been a considerable amount of litigation concerning fluorides and plant life and it continued to be an important subject of discussion. There were cogent arguments for standards of various kinds, the most common of which were those based upon limiting: (i) the degree of foliar injury; (ii) the foliar fluoride content; or (iii) atmospheric concentrations. These discussions were joined by Hill (1969), who reviewed the three approaches to a fluoride standard. He argued that the degree of foliar injury, if it averaged more than an established percentage of the foliar area, should be in violation. In special cases, injury to fruit could be considered. The greatest advantage would be that relatively little time would be required to conduct vegetation surveys in a reasonably large area, and a survey would:

reflect a summation of all of the effects of many complex variables such as the species and varieties present, the chemical forms of the fluoride, and variation in exposure time and concentration. Controls would be based on need rather than speculation.

(Hill, 1969)

Hill concluded that fluoride-induced foliar injury would be the most practical approach to a fluoride standard based upon vegetation effects and that inspections should be made two or three times during the growing season.

However, in a discussion of Hill's paper, Weinstein (1969) felt that this standard would be neither practical nor adequate:

Standards should protect the receptor rather than assess what has already occurred. There is no provision, for example, to evaluate episodes which might occur between vegetation surveys. A standard of this type might therefore result in an intermittent estimation of injury to vegetation after the fact.

This approach would not be helpful to the grower or to industry since little meaningful information would be available for remedial action. Another problem is the need for qualified personnel to conduct the vegetation surveys and the potential for mis-identification and consequent dispute. How would the effects of the weather be taken into account and what would happen in areas where there are few or no bioindicators? Therefore, this approach on its own does not provide unequivocal answers to the problem. Weinstein (1969) suggested a series of time-weighted air quality standards that reasonably reflected the state of knowledge at that time (McCune, 1969), but indicated that these concentrations should be used in conjunction with appropriate atmospheric-sampling stations, vegetation analysis and periodic vegetation surveys. He was suggesting an integrated approach to regulation. As an example, he went on to describe the system established for The Dalles, Oregon, which required the establishment of six air sampling stations and delineated a vegetation-sampling method and sampling periods, combined with periodic vegetation surveys. This approach, he went on, was 'being used to judge the acceptability of not only the industrial emissions but also of the adequacy of the standard itself'.

There are some general problems in respect to establishing standards involving vegetation which were addressed by MacLean (1982). He discussed the need for standards based upon the probability that, within a state or other demographic area, there may be vastly different climatic zones and topography, which could influence the manner in which plants respond to foliar injury or the degree of foliar accumulation. His arguments for tailored air quality standards included: the differences in vegetation types in a region; alterations in sensitivity of plants at different times during development; and the intended use of the plant to be protected. He proposed possible seasonal, regional and forage standards. For example, in relation to a seasonal standard, he said:

> The seasonal variability [in plant suscepti-
> bility] would justify a standard that con-
> siders the life cycle of the receptor(s) at risk
> and the fact that most, if not all, vegetation
> in the temperate zone of the U.S. is suscep-
> tible to fluoride only during a finite
> period each year, and that during the win-
> ter months, foliage is either absent or more
> tolerant of fluoride.

Thus, during the winter months in temperate climates, foliage is either absent or relatively dormant and a higher standard could be tolerated or, at least, if an existing standard was exceeded during that period, this would not be a violation.

Regional standards exist in some countries for limited areas or specific purposes. A regional standard is not necessarily one that provides a variance for higher concentrations in, say, an industrial zone, but it may be a more rigorous standard to protect plants in special land-use areas or for threatened or endangered species, as in Australia and New Zealand. MacLean (1982) discussed the rationale for considering regional standards by citing a hypothetical example. In this, two sources were located in a single state that had an air quality standard. Each industry emitted the same amount of fluoride, but one was located in an area of a coniferous forest that supported recreation and forestry and the second was in a semi-arid rangeland where the principal agricultural activity was raising beef cattle. The existing standard was designed to protect the most sensitive receptors, which were the developing needles of a conifer near one industry. This standard would not take into consideration the much greater tolerance to fluoride of the vegetation in the semi-arid area, and there might be valid arguments for a variance or more lenient standard. On the other hand, although the vegetation may be more tolerant, a more lenient air quality standard could result in greater accumulation of fluoride, violation of a forage standard and an increased risk of fluorosis to cattle and other native herbivores. It would appear that the best arguments for regional standards would be for the protection of particularly sensitive areas of vegetation or for highly industrial areas where little vegetation exists.

In general, special standards, whether regional or seasonal, require greatly increased efforts on the part of regulators than a single fixed standard. MacLean (1982) concluded:

> Land use patterns, vegetation habitats,
> and agricultural operations near real or
> potential sources of fluoride would have to
> be assessed. Further, determination of the
> relative susceptibility of plant species of
> economic, aesthetic, or ecologic importance
> would be required. Policing the standard
> would require comprehensive monitoring
> of fluoride in the air and inland vegetation.
> Superimposed on these requirements is
> the need to evaluate the differential
> susceptibility of the vegetation at risk in
> each area at different times of the year.

He went on to argue that the increased workload and costs to regulatory agencies would be offset by the greater flexibility that would result from the regional and seasonal standards, with the result that the economy would be improved and would allow agriculture and industry to coexist. He also speculated that an outcome of regional standards would be improved guidelines for land-use management — areas with the lowest probability of detrimental effects could be set aside for industrial development. Seasonal standards, in which one set would be enforced during the growing season, with less rigid standards in force for the winter months, would offer flexibility for industrial operations. For example, certain procedures during which there might be a greater release of fluoride, would be scheduled for the period when the risk of an effect might be greatly reduced.

Using the Fluoride Content of Vegetation

The fluoride content of vegetation can be used as a standard to protect both plants and animals. In this section we focus on plants; protection of animals is considered later. A few regulatory agencies have established maximum allowable fluoride concentrations in foliage (Table 7.2). However, there seem to be few advantages to this approach, except perhaps for species where there is a good correlation between the fluoride content and injury, as in the case of *Gladiolus* (Hitchcock *et al.*, 1962). But these are rare and, if a plant like *Gladiolus* were used, it would require establishing special plots and cultural conditions, and it would not be available throughout the whole year. Regardless of what species is selected, there are several problems that make leaf fluoride a dubious choice as the basis for standards. Differences in fluoride accumulation by leaves of different ages and different species require careful standardization of leaf age, time of sampling, species and varieties, a system of collecting representative samples and, not least, specification of the method of fluoride analysis. There are additional potential problems. Much of the fluoride associated with the leaf resides on the surfaces and has little or no physiological activity in the plant. This requires removal of the surface fluoride before analysis and is accomplished by washing, using a method that maximizes surface removal and minimizes leaching from the leaf interior (Chapter 2). Some agencies stipulate washing, but the exact technique determines

the outcome and the same procedure will remove different fractions from different plant species. Others specify a single maximum allowable concentration for conifers and evergreens, but species differ in their rate of accumulation of fluoride, even within a restricted group such as conifers, and the concentration may fluctuate over short periods of time. But perhaps the most telling criticism of this approach is that there is no scientific basis for assuming that using a particular concentration as a limit will actually protect either the selected species or any other species.

Standards to Protect Livestock and Other Domestic Animals

As discussed in Chapters 2 and 3, the main source of fluorides for herbivores is their food, and there is a direct, well-established relationship between the fluoride content of forage and the effects on vertebrates. The dietary tolerances of economically important livestock are well known and universally accepted. Furthermore, the dynamics of fluoride deposition and loss from forage have been investigated in temperate climates and sampling protocols are available. The result is that forage fluoride concentrations have the best scientific basis for use as regulatory standards. They have been adopted officially by several US states but, in addition, many industries and regulators in other countries use forage fluoride as an unofficial guideline to judge the performance of sources. A major advantage of using forage fluoride is that it acts as an early-warning system, so it can be used to prevent fluorosis on a field-by-field and herd-to-herd basis. Because cattle are among the most sensitive of animals, applying a forage standard will also protect less-sensitive wildlife. The main weaknesses are usually the errors introduced by infrequent sampling and the lack of specification of sampling protocols and analytical methods (see Chapter 2).

The most commonly used forage standard in the USA is:

Average for 12 months not to exceed 40 mg/kg (dry wt).
Average for any 2-month period not to exceed 60 mg/kg (dry wt).
Average for any 1-month period not to exceed 80 mg/kg (dry wt).

The 12-month standard concentration has been modified commonly to 30 or 35 mg/kg fluoride, but the 2-month and 1-month allowable levels have been generally accepted in several jurisdictions (Table 7.2). In the USA, one state, Maryland, has not only regulated the amount of fluoride that can be accumulated in forage or pasture crops, but also in baled hay or silage:

Running averages for 12 consecutive months may not exceed 35 mg F/kg.
The average for any 2 consecutive months may not exceed 60 mg F/kg.
The average of samples collected during any 1 month may not exceed 80 mg F/kg.

Water-quality Standards

In general, the standards adopted for water have a stronger scientific basis than those related to air pollution, largely because it is easier to control fluoride concentrations and to test organisms, especially invertebrates. There has also been a substantial amount of research effort on water in relation to dental caries and fluoridation. Some general water quality standards for all uses are shown in Table 7.3 and some for livestock and wildlife are summarized for several jurisdictions in Table 7.4.

The effects of fluoride in drinking water on livestock are comparable to effects on humans (McKee and Wolf, 1963). In Canada,

the total fluoride recommended for dairy cattle, breeding stock and other long-lived animals is 1.0 mg/l for a 30-day average, with a maximum level of 1.5 mg/l. For all other livestock, the 30-day average would be 2.0 mg/l, with a maximum of 4.0 mg/l. If, however, the animals are receiving supplemental fluoride in feed or mineral additive, the 30-day average would be reduced to 1.0 mg/l and a maximum of 2.0 mg/l (Marier, 1977).

Fluoride in municipal sewage effluent is derived mainly from waters discharged by municipalities that have fluoridated water. There are undoubtedly other minor sources because the average fluoride concentration of discharge water from 57 municipal sewage-treatment plants in Ontario averaged 1.0 mg/l (EPS, 1978). The maximum, minimum and mean concentrations in the USA have been reported to be 1.21, 0.74 and 1.03 mg/l (AWWA, 1985). With the natural background concentrations being about one order of magnitude lower, untreated waste water that is released could raise the fluoride concentration significantly and elevate the fluoride levels of wells and of municipal reservoirs, with the exposure of aquatic biota.

Finally, in some jurisdictions there are standards to control the amount of fluoride allowed in water used for various specific purposes, such as irrigating crops and beverage manufacture (Table 7.5). The fluoride present in irrigation waters should have little effect on plants if it is applied solely to soils with a reasonable clay content and a slightly acid to neutral pH (Chapter 2), but long-term application to lighter soils or organic media could lead to undesirable rates of uptake by the crop. Another problem is that, if soluble fluoride is applied to

Table 7.3. Water-quality standards for fluoride in several international jurisdictions (mg/l). Value for the US EPA is maximum contaminant level (MCL), others are maximum acceptable concentration (MAC). (Environmental Health Criteria 227, WHO, 2002b, reproduced courtesy of WHO, Geneva.)

WHO	US EPA	Canada	EEC	South Africa	Taiwan	China	Czech Republic	Britain
1.5	4	1.5	1.5	1	0.8	1	1.5	1.5

WHO, World Health Organization; EEC, European Economic Community.

Table 7.4. A selection of water-quality standards for livestock and wildlife in several jurisdictions. (From Warrington, P.D. (1996) Ambient water quality criteria for fluoride. http://www.gov.bc.ca/wat/wq/ BCguidelines/fluoride.html)

mg F/l	Conditions	Jurisdiction	Date
0.8–1.7	Temperature-dependent livestock and wildlife criteria	British Columbia	1969
1	Livestock drinking water when food contains supplementary fluoride	Canada	1987
1.5	Surface water quality objectives	Manitoba	1979
1.5	Surface water quality objectives	Saskatchewan	1975
		Alberta	1977
2	Livestock drinking water when food contains no fluoride additives	USA	1973
		Canada	1987
		Australia	1974
2	Toxicant level for livestock watering	Australia	1982
	95th percentile stock-watering criteria	Britain	1982
3	99th percentile stock-watering criteria	Britain	1981

Table 7.5. Standards for fluoride in irrigation waters for several jurisdictions. (From Warrington, P.D. (1996) Ambient water quality criteria for fluoride. http://www.gov.bc.ca/wat/wq/BCguidelines/fluoride.html)

Irrigation water – tolerable level (mg/l)	Conditions	Jurisdiction
1	All soils, continuous use, unlimited time period	Canada
		USA
		Australia
1	Toxicant level for irrigation supply	Australia
1.2	95th percentile for spray irrigation of field crops	Britain
1.5	Surface water quality objective	Saskatchewan
		Alberta
10	General irrigation purposes	British Columbia
15	Maximum of generally 20 years' use on fine-textured soils, neutral to alkaline (pH 6.0–8.5)	Ontario
		Canada
		USA
Standards for water used in food production		
0.2–1.0	Carbonated beverage production	USA
1	Carbonated beverage production, brewing, canning, freezing, drying, general food processing, food washing and food-equipment washing	Australia
		USA
1.5	Class 3B waters used for industries except food processing	Manitoba

the foliar surfaces by sprinkler irrigation, sensitive crops may exhibit chlorotic and/or necrotic symptoms of toxicity, as has been reported for sprinkler-irrigated foliage plants in Florida (Poole and Conover, 1973, 1975; Conover and Poole, 1982).

8

Natural Organofluorine Compounds

Introduction

For most of the 20th century, environmental concern about fluorides was focused on inorganic compounds but there are about 30 naturally occurring and many manufactured organofluorides that occur in or are released into the environment. This group of compounds includes some of the most toxic chemicals known, but many are non-toxic, stable, useful and even essential to modern living. This chapter outlines the important properties of organofluorides and then deals with the naturally occurring organofluorides, concentrating on mono-fluoroacetate[1]. The next chapter considers examples of the environmentally important manufactured compounds.

Schofield (1999) surveyed the number of new organofluorine compounds registered by the Chemical Abstracts Service and found that between 1969 and 1989 the total was 610,873. In the SynQuest Laboratory Research Catalogue for 2000, Banks stated that 'More than one million compounds containing one or more carbon-fluorine bonds are known, and since barely more than ten of those occur naturally, organofluorine chemistry is virtually a completely man-made branch of organic chemistry.' The main reason why there are so many organofluorine compounds is because a fluorine atom can normally replace any hydrogen atom in linear or cyclic organic molecules, but the reason for the huge investment in their synthesis and production is because fluorination produces properties that are invaluable in medicine, agriculture, industry and the home.

The important properties of organofluorides are due to the small radius of the fluorine atom, its high electronegativity and the strength of the carbon–fluorine bond. The latter is the strongest single bond found in organic chemistry, having a bond energy of 466 kJ/mol, and consequently it is resistant to attack. Perfluorinated organic compounds (i.e. those containing only carbon and fluorine) are resistant to various aggressive chemicals, to high temperatures and even to X-rays and nuclear radiation. For example, MFA may be distilled from concentrated sulphuric acid and, even when it is refluxed with 10% sodium hydroxide for 1 h, no free fluoride ion is produced. Cleavage requires refluxing in 30% sodium hydroxide for 1 h, after which most of the fluoride ion is released (Saunders, 1972).

Electronegativity is a measure of the attraction for the electrons involved in a chemical bond, and fluorine is the most electronegative of all the elements, which results in it altering the distribution of electrons in a molecule. The effects of these properties on organic compounds can be illustrated by an example involving enzyme inhibition. In the human body, alcohol (ethanol) is oxidized to acetaldehyde by the enzyme alcohol dehydrogenase, using the cofactor nicotinamide adenosine dinucleo-tide (NAD⁺) as an electron acceptor:

$$NAD^+ + CH_3CH_2OH \;\rightleftarrows\; CH_3CH{=}O \;+$$
$$\text{ethanol} \qquad \text{acetaldehyde}$$
$$NADH + H^+$$

Molecules with a structure similar to ethanol may compete for the active site of the enzyme, leading to a reduction in the rate of the oxidation; if trifluoroethanol (CF_3CH_2OH) is added to the mixture, it acts as a competitive inhibitor, reducing the reaction velocity. This is because the high electronegativity of the fluorine atoms holds the electrons in the C–H bonds, preventing them from transferring to the NAD⁺. The significance of this property is that many health disorders are caused by or involve abnormal enzyme activity and regulating activity provides an important basis for therapeutic treatment. Fluorination can be used to design specific enzyme inhibitors, such as the drug 5-fluorouracil. It is used to treat certain cancers because it is structurally similar to uracil, the precursor to thymine, but it inhibits the synthesis of thymine, blocking DNA synthesis.

The strength of the carbon–fluorine bond, which makes fluorinated compounds resistant to metabolic and chemical transformation, is used to good effect to produce non-stick, heat-resistant surfaces and weatherproof materials, such as polytetrafluoroethylene. However, this property is also a potential problem for the environment because defluorination of these compounds may be slowed or even prevented due to the strength of the C–F bonds. As Key et al. (1997) have pointed out, there are many thousands of fluorinated compounds being released into the environment, but very little is known about the pathways of transport or the fate of the carbon–fluorine moieties of the molecules in the environment (see Chapter 9).

Natural Organofluorides of Geological and Biological Origin

Until recently the only organofluorides that were known to occur in nature were those of biological origin, but in the last decade evidence has emerged that around 30 organofluorides can be found in the natural environment (Gribble, 2002) and that some are of geological origin (Table 8.1). However, it must be stressed that there has still been relatively little research on the latter group and in some cases their existence still needs to be confirmed. In the past, research on biological organofluorides was severely limited by the low concentrations involved and the crude analytical techniques that were available. The latter have improved immensely but there are still problems with quantifying the geological compounds, including extracting the compounds without altering their chemical composition and preventing contamination of samples.

The first five compounds in Table 8.1 (CF_4, CF_3Cl, CF_2Cl_2, $CFCl_3$ and CHF_3) are gases with global-warming potentials, and some have an ozone-depleting potential, but when they are trapped in minerals they are not of immediate environmental concern. However, their presence in minerals does raise the question of whether emissions of naturally occurring organofluorides from degassing of minerals or volcanoes might contribute to the atmospheric burden of warming gases. Jordan et al. (2000) concluded that $CFCl_3$, the only chlorofluorocarbon (CFC) detected in volcanic emissions, is not released into the atmosphere in sufficient quantities to contribute to climate change. Harnisch et al. (1996) have suggested that about half of the current atmospheric burden of CF_4 arose from degassing of natural sources in the lithosphere. The rest of the CF_4 (and some C_2F_6) arises mostly from short-term upsets in the process of primary aluminium smelting. Clearly, more research is needed on the natural geological organofluorides.

The biological compounds nucleocidin and 4-fluorothreonine are produced by particular strains of microorganisms isolated from soil and grown in laboratory culture. The amounts produced in nature and the environmental significance of the organisms and the organofluorides are therefore not known, but the

Table 8.1. Examples of naturally occurring organofluorine compounds. Some are chlorofluorocarbons (CFCs), so they are also known by CFC numbers. The empirical formula of a CFC can be determined by adding 90 to the CFC number. The three digits are then the number of carbon, hydrogen and fluorine atoms, respectively. The total number of chlorine atoms is $= 2(C + 1) - H - F$. For a complete list of natural organofluorides, see Gribble (2002).

Compound (common names)	Origins	Authors, comments
Perfluoromethane, CF_4 (CFC-14)	Geological: fluorites, granites, atmosphere?	Harnisch et al. (1996, 2000); Harnisch and Eisenhauer (1998)
Trifluorochloromethane, CF_3Cl (CFC-13, Freon 13)	Geological: fluorite from Wölsendorf, Germany	Harnisch et al. (2000)
Difluorodichloromethane, CF_2Cl_2 (CFC-12, Freon 12)	Geological: fluorites, granites, volcanites, metarhyolite, gneiss	Harnisch et al. (2000)
Trichlorofluoromethane, $CFCl_3$ (CFC-11)	Geological: fluorites, granites, volcanites, metarhyolite, gneiss, volcanoes?	Isidorov et al. (1993a,b); Harnisch et al. (2000); Jordan et al. (2000)
Trifluoromethane, CHF_3 (HFC-23, fluoroform)	Geological: fluorite from Wölsendorf, Germany	Harnisch et al. (2000)
Monofluoroacetic acid, CH_2FCO_2H (occurring as the acetate, sodium salt known as compound 1080)	Biological: *Streptomyces cattleya* Biological: *Acacia georginae*, some *Dichapetalum* spp. *sensu lato*	Sanada et al. (1986) Summarized in Gribble (2002); see also Twigg et al. (1996a)
Trifluoroacetic acid, CF_3CO_2H (occurring as the acetate)	Biological and/or geological	Jordan and Frank (1999); Ellis et al. (2001b, 2002)
Fluoroacetone? C_3H_5FO Fluoroacetaldehyde? C_2H_3FO	Biological: *Acacia georginae*, *Streptomyces cattleya*	Peters and Shorthouse (1967), but fluoroacetone doubtful; fluoroacetaldehyde postulated, see O'Hagan and Harper (1999); Moss et al. (2000)
ω-Substituted fluoro fatty acids	Biological: several isolated from *Dichapetalum toxicarium*, e.g. ω-fluoro-oleic acid	Peters et al. (1960); Ward et al. (1964); Harper et al. (1990); Hamilton and Harper (1997); Gribble (2002)
4-Fluorothreonine	Biological: *Streptomyces cattleya*	Sanada et al. (1986); see also O'Hagan and Harper (1999)
Nucleocidin, 4′-fluoro-5′-O-sulphamoyladenosine	Biological: *Streptomyces calvus*	Morton et al. (1969)

organisms are of potential importance to the chemical industry as a source of bio-engineered organofluorides (O'Hagan and Harper, 1999). In contrast, monofluoroacetic acid (MFA) and trifluoroacetic acid (TFA) are not only natural products, but are also manufactured and of considerable environmental interest. The former is discussed below and in the next chapter along with TFA.

Poisonous plants and monofluoroacetic acid, a historical review

Poisonous plants have been recognized, avoided and used by native people for centuries, if not millennia. Renner (1904a,b, also quoted by Power and Tutin, 1906) discussed the use of *Dichapetalum toxicarium* (ratsbane) (= *Chailletia toxicaria*) as a poison in Sierra Leone:

For some years now, death by poison has been the subject of talk in the colony of Sierra Leone, and one could scarcely credit the statements so often made with respect of this subject. No one, it would appear, dies from natural causes. Poisoning in one form or another is put down as the cause of death, not only among the poor but also among the rich, and yet no one would or could come forward to attest the fact that such a one had been poisoned by such a substance. Some of these deaths which appear to be mysterious have given cause for great alarm and anxiety, and aroused a feeling of dread and bitterness against the Country Doctors to whom they are attributed. The Country Doctors, who have the true knowledge of the cause of the deaths, keep it a secret, attributing it to some mysterious influence of the devil or to witchcraft or to something occult, and thereby make much profit from their knowledge and wield great power over the masses. A peculiar kind of disease, of the origin of which no account could be given, is now and again met with among the people. The European doctor, when called in, is puzzled, for he sees before him a young healthy man or woman, between the ages of twenty-three and forty years struck down with paralysis . . . The individual may recover or die, this depending on the nature and the quantity of the poison administered. These cases are brought forward daily. It has been my fortune lately to discover the cause of one of the mysterious and sudden diseases that affects the natives in the form of paralysis of the lower limbs.

Renner (1904b) continued:

The *Chailletia toxicaria* . . . is known as 'Ratsbane'. In the Colony proper it is called 'Broken Back' . . . It grows plentifully in West Africa and South America [*sic*?] and the fruit, which contains the poisonous substance, is used largely for the destruction of rats and other animals; but beyond this is also used extensively by the people of the Colony and the Hinterland to poison one another. Conversing with some of the older men of the Timnes and the Mendis, I found that with them it is very frequently used against their enemies to poison wellwater or streams which supply hostile villages. Domestic animals poisoned by it are seen to rush around in great excitement as if in severe pain; they vomit, and drag their hind limbs, which ultimately become paralyzed. The forearm and chest muscles become paralyzed also, and they die from paralysis of the respiratory centre.

When settlers started raising cattle, sheep and goats in the new colonies of South Africa and Australia, they encountered plants that often killed animals after ingestion of relatively small amounts of tissues. Some of the plants involved had common names that indicated their toxicity, such as ratsbane, but in other cases there was little indication of the toxicity. Indeed, in Australia, it was observed that native animals could consume *Gastrolobium* species with apparent impunity, so it was thought that they could not be the cause of deaths of domesticated animals (Everist, 1981). Further confusion was caused by the fact that some were leguminous (peas, beans, clovers) species and both botanists and cattlemen knew from their European backgrounds that this group of plants generally provide an excellent source of foodstuff for livestock. New settlers in the British colony in Western Australia were hard-pressed to find adequate forage for their livestock, and many of them, not surprisingly, settled on leguminous shrubs of the genus *Gastrolobium*. Many of these plants were found later to be extremely toxic (Aplin, 1971c). According to Cameron (1977), most poisonous plants of Europe:

are cyanogenic, or contain poisonous, sometimes narcotic, alkaloids. With the exception of foxglove (*Digitalis purpurea*), their poisonous qualities were suggested through their acrid taste and the foetid smell of their sap and bruised leaves. Settlers were well aware of these characteristics and, as might be expected, their attention centered first on number of cyanogenic plants, particularly the blind grasses (*Stypandra imbricata* and *S. grandiflora*) which induce a range of symptoms similar to those induced by *Gastrolobium* and *Oxylobium* and the narcotic Woodbridge poison (*Isotoma hypocrateriformis*), a member of the *Lobelia* family.

Some of the early colonists in Western Australia suspected that the stock losses were caused by a pea-flowered plant with 'sage-colouring' (Drummond, cited in Marchant, 1992). When this was reported to other colonists, they refused to believe that leguminous plants could be responsible. The controversy over the identity of the toxic plants continued for several years, but Drummond (1842, cited in Marchant, 1992; Everist, 1981) became convinced that the grey-leaved, pea-flowered shrub, *Gastrolobium calycinum* (York Road Poison), was the culprit. However, a well-known Prussian botanist, Ludwig Preiss, disputed this claim and asserted that *Gastrolobium* was the best leguminous plant that the colonists could cultivate for forage. He even drank a wine glass of an infusion of leaves to prove his point! Marchant (1992) remarked that as Preiss suffered no ill effects he must have made the infusion from a plant in 'non-toxic mode'. The controversy was resolved experimentally in 1841 (Marchant, 1992, citing Drummond) when, by feeding infusions of *G. calycinum* and leaves of the plant itself to sheep and goats, the animals died within about 3 h after feeding. The report went on to say that after the experiment, 'three valuable dogs were poisoned by getting at the guts of the sheep experimented upon . . .' (Marchant, 1992). It took until 1944 to identify the toxic principle in *Dichapetalum*, 1961 in *Acacia georginae*, 1963 in *Palicourea marcgravii* and 1964 in species of *Gastrolobium*. In all these cases the toxic principle was found to be fluoroacetate, though *D. toxicarium* has also been found to contain ω-fluoro-oleic, ω-fluoropalmitic and possibly other longer-chain fatty acids in addition to fluoroacetate (Table 8.1). Some species of *Gastrolobium* have also been shown to contain di- and trifluoroacetates (F. Wang, personal communication). Estimates of the lethal doses of fluoroacetate for different species are given in Table 8.2.

Identification of the toxic principles in plants

The impetus for isolating and determining the mode of action of the toxic principle(s)

Table 8.2. Lethal doses of sodium fluoroacetate in various animal species. (From Hodge, H.C., Smith, F.A. and Chen, P.S. (1963) Biological effects of organic fluorides. In: Simons, J.H. (ed.) *Fluorine Chemistry*, Vol. III, Academic Press, New York, pp. 1–240.)

Species	Lethal dose, range or minimum lethal dose[a] (mg/kg body weight)
Dog	0.06–0.15
Sheep	0.25–5[a]
Human	2–5
Rat	0.1–15
Frog	150
Toad	1500

in *Dichapetalum* and other fluoroacetate-containing species was their great toxicity. However, progress was slow because of the lack of appropriate biochemical and analytical methods available at the time. Even to this day, fluoroacetate assay can still be problematic, and the results often display considerable variation. Fluoroacetic acid was first synthesized by Swarts in Ghent, Belgium, in 1896 (Marais, 1944), but apparently he did not recognize its toxic properties and it received scant attention from him and other chemists and none from pharmacologists. During research in the USA in the 1940s, sodium fluoroacetate was assigned the number '1080' and this led to the name 'Compound 1080', a tag that has been used ever since. The development and use of 1080 and the potential environmental problems associated with its use are discussed in the next chapter. The identification of MFA as the toxic principle in *Dichapetalum cymosum* was one of the first examples of an organofluoride compound found in nature and it was a brilliant piece of work, especially considering the limited methods that were available at the time. Later, the fluoro-fatty acids, ω-fluoro-oleic and ω-fluoropalmitic acids were identified in *D. toxicarium*. The identification of fluoroacetate in each species provides its own story of detection and analytical skill.

Dichapetalum cymosum

Steyn (1928) reported that the first person to confirm the general belief among farmers that *gifblaar* (Afrikaans for 'poison leaf') was poisonous to stock was Theiler in 1902, who fed the plant to oxen and rabbits with fatal results. Shortly thereafter, Dunphy (1906) concluded 'that the poison acts on the nervous system and has little, if any, effect on the bowels'. Walsh (1909) found that two or three leaves were sufficient to poison an ox, although other accounts claimed that a far greater quantity of old leaves was required 'unless the ox drinks water after eating the plant'. In 1916, Stent (cited by Steyn, 1928) confirmed the toxicity of the plant and it was discussed for some time afterwards (see Rimington, 1935). The toxic plants were generally confined to the Transvaal region. Forty-two years after Theiler's report, Marais (1944) identified the toxic principle as fluoroacetate in his laboratory at Onderstepoort. A year earlier, he had reported on the isolation of a highly toxic substance he called 'potassium cymonate' (Marais, 1943). Marais was not able to characterize the compound because of insufficient material. However, when he finally made the identification, he was surprised that one of the main ingredients was fluorine. Through a series of elegant steps, Marais (1944) clearly demonstrated that potassium cymonate was the salt of MFA ($F.CH_2.COOH$). The concentration of fluoride was found to be 0.015%, which would be the equivalent of about 790 mg/kg of dried plant material (or 790 p.p.m.) of sodium fluoroacetate, assuming that all the fluoride was present in fluoroacetate. *Gifblaar* plants were found to have a marked difference in toxicity according to their stage of development and season. Leaves were most toxic when young and in the spring and autumn (Steyn, 1928). In recent years, ingestion of *D. cymosum* has been responsible for about 3000 head of cattle being fatally poisoned in Africa each year.

Dichapetalum toxicarium

Because the powdered fruit was used to control rats, *D. toxicarium* became known as *ratsbane*. The shrub grows in soils that contain a low concentration of fluoride, but the plant has the unusual capacity to sequester fluoride in high concentrations. In contrast, many species from several different plant families growing nearby all contain less than 2 mg/kg dry weight (Vickery and Vickery, 1972). The mechanism of fluorine uptake and accumulation from soil by organofluorine-accumulating species is not understood and clearly deserves more attention. Peters *et al.* (1954, 1959, 1960) and Peters and Hall (1960) reported that the seeds contained at least two fluorinated long-chain fatty acids. Peters and Hall (1959) found that one of these was a fluoroacid containing 18 carbon atoms, one double bond and one fluorine atom, i.e. a fluoro-octadecanoic acid. This acid was present in by far the greatest quantity (*c.* 80%) and it was identified as the unsaturated 18-carbon fatty acid, ω-fluoro-oleic acid. The second acid was not identified because of difficulty in its isolation and purification, but, later, Ward *et al.* (1964) were able to separate it as a white crystalline solid and subsequently identified it as the saturated 16-carbon fatty acid, ω-fluoro-palmitic acid. Vickery and Vickery (1972) suggested that young leaves may be the main site of fluoroacetate biosynthesis in these plants, but this is yet to be proved. Fluoroacetate concentration is greatest in the small leaves whose petioles are adnate to the peduncles of the inflorescences. These leaves develop with the cyme (up to 1100 µg/g fresh weight) and appear to utilize fluoroacetate as an energy source. Fluoroacetate also appears to be utilized by the embryonic seeds, where the transformation to long-chain fluoro-fatty acids can also occur.

A final feature of the genus *Dichapetalum* is that it is not restricted to Africa, occurring across the tropics in countries as far apart as Borneo, China, the Philippines,

Brazil and Mexico, but the species in these countries are non-toxic or, at least, there have been no reports of toxicity from any countries other than Africa (G.T. Prance, personal communication). These may be worth further study.

Acacia georginae

In 1955, Bell *et al.* considered that except for drought, 'Georgina poisoning' or 'Georgina River disease' was the most important factor limiting animal production in an area of more than 60,000 square kilometres in western Queensland and in the north-east of the Northern Territory of Australia (Fig. 8.1). In recent years cattle producers in Queensland and the Northern Territory estimated that the total loss across 15 properties was about 4500 cattle per year. Some incidents are even more severe. In the 1980s, Gary Haines of Argadargada station in the Northern Territory lost around 1000 head of cattle, valued then at Aus$350,000, and later at the Linda Downs station on the Queensland/Northern

Territory border 17,000 sheep were lost from a herd of 70,000 (K. Gregg, personal communication) (Plate 28). From 1896 onwards, many attempts were made to identify the plant responsible for livestock deaths in the region, but it was only in 1955 that it was finally confirmed to be *A. georginae* (gidyea, gidgee or Georgina gidgee) (Everist, 1981). One of the reasons for this delay was that the toxicity is very varied, and the amount of toxic principle ranges widely, with many gidyea plants being non-toxic. Toxicity depends upon the locality and the season (Bell *et al.*, 1955). This was confirmed by Whittem and Murray (1963), who reported that large tracts of *A. georginae* to the east and north-east of Alice Springs were non-toxic, while in the north of the Northern Territory toxic stands were confined to the Georgina River basin. In the past, the ingestion of the pods, seeds or leaves has accounted for extensive livestock fatalities, with as many as 300–500 animals being recorded in a single incident. Barnes (1958) reported similar problems in the toxic gidyea areas of the

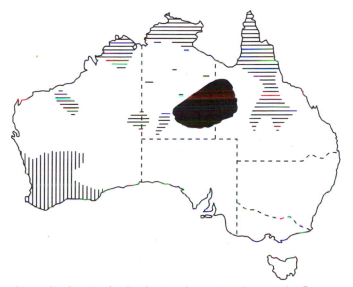

Fig. 8.1. A map of Australia showing the distribution of vegetation that contains fluoroacetate. Vertical lines = 33 species of *Gastrolobium* (*sensu lato*). Horizontal lines = *Gastrolobium grandiflorum*. Solid black area = *Acacia georginae*. (Redrawn and modified from Twigg and King (1991), courtesy of *Oikos*.)

Northern Territory. It is paradoxical that although pods and suckers of *A. georginae* were shown to be toxic, the farmers persisted in considering the plant to be a valuable source of fodder in the Georgina River basin, and, because of its widespread nature, eradication would be both difficult and undesirable (the region mainly consists of low-rainfall pastoral rangelands). In order to devise a strategy that would still utilize the plant as fodder, it was first necessary to identify the toxic principle. This was achieved in 1961, when Murray *et al.* reported the suspected presence of fluoroacetic acid in tissues of *A. georginae* collected from the northern areas of the Northern Territory. They were able to demonstrate the presence of organically bound fluoride and the highly toxic nature of their extracts, but they were unable to obtain a purified compound. During the same year, Oelrichs and McEwan (1961) showed that the toxic principle from *A. georginae* seeds was, in fact, MFA. They reported 25 mg sodium fluoroacetate/kg of air-dried leaves.

After the discovery of fluoroacetate in *A. georginae*, it was postulated that its synthesis might be regulated by the availability of fluoride in the soils, but soil surveys conducted by Smith *et al.* (cited in Whittem and Murray, 1963) and further studies by Murray and Woolley (1968) found no differences in the fluoride content of surface soils with and without toxic gidyea. However, both total and water-soluble fluoride increased with soil depth, and this suggested to Chippendale (cited in Whittem and Murray, 1963) that toxicity may be a function of plants suffering from water stress. At such times, plants may draw their water from lower in the soil profile and hence take greater amounts of fluoride.

Gastrolobium, Oxylobium *and* Nemcia

These three Australian genera are all members of the legume family, the *Fabaceae*. The older literature contains references to toxic species in all three genera, but in the 1990s the relationships between the taxa were revised (see papers by Aplin, 1967,

1968a,b,c, 1969a,b, 1971a,b,c; Crisp and Weston, 1995). In 2001, Chandler *et al.* suggested that all *Oxylobium*, *Nemcia* and other genera should be included in *Gastrolobium* and in 2002 Chandler *et al.* produced a definitive taxonomic revision of the group that confirmed this. They also described several new taxa. However, because almost all of the literature on the toxic effects of these plants uses the older names, in this book we shall continue to use them. Table 8.3 lists the 30–40 species that have been described as being toxic due to fluoroacetate.

The organofluorine-containing species of Western Australia provide a striking example of the way a flora may have a profound effect on the colonization and social history of a huge region, and also on the evolutionary responses of native animals (Mead *et al.*, 1985a). Cameron (1977) stated that these plants 'constituted the major hazard for the nascent pastoral industry, losses from them exceeding combined losses from all other hazards including bushfires, floods, droughts, aboriginal depredations and the attacks of native dogs'. The coast of Western Australia was settled in the 1820s by colonists, who brought with them limited supplies of seeds, plants, feed and livestock. They were in a land where they were dependent on finding fodder for livestock but where the flora was unknown. As the overall number of livestock was low, any losses were very important. The first stock losses due to the toxic plants were reported as early as 1830 (Everist, 1981). The seriousness of this is indicated by a Perth newspaper (cited by Cameron, 1977), which reported that 2315 sheep, goats, cattle and horses were poisoned by these plants between 1833 and 1840, representing as much as half of the total livestock in the colony at that time. Once the identity of the toxic plant was known, steps were taken to eradicate it from the main roads and stock routes. Ironically, this eradication, together with the clearing of habitats and use of fertilizers, has led to some species of *Gastrolobium* now being threatened with extinction and, in 1989, 14 were thought to be under threat (Marchant, 1992).

Table 8.3.   Plant species reported to synthesize organofluorine compounds (data from numerous sources).

Species	Common name	Family	Country
Gastrolobium oxylobioides Benth.	Champion Bay poison	*Fabaceae*	SW Australia
Gastrolobium microcarpum Meissn.	Sandplain poison	*Fabaceae*	SW Australia
Gastrolobium bennettsianum C.A. Gardn.	Cluster poison	*Fabaceae*	SW Australia
Gastrolobium propinquum C.A. Gardn.	Hutt River poison	*Fabaceae*	SW Australia
Gastrolobium rotundifolium Meissn.	Gilbernine poison	*Fabaceae*	SW Australia
Gastrolobium bilobum R. Br.	Heart-leaf poison	*Fabaceae*	SW Australia
Gastrolobium forrestii A.J.	Ewart River poison	*Fabaceae*	SW Australia
Gastrolobium velutinum Lindl.	Stirling Range poison	*Fabaceae*	SW Australia
Gastrolobium villosum Benth.	Crinkle-leaf poison	*Fabaceae*	SW Australia
Gastrolobium ovalifolium Henfr.	Runner poison	*Fabaceae*	SW Australia
Gastrolobium polystachyum Meissn.	Horned poison and Hill River poison	*Fabaceae*	SW Australia
Gastrolobium tomentosum C.A. Gardn.	Woolly poison	*Fabaceae*	SW Australia
Gastrolobium parvifolium Benth.	Berry poison	*Fabaceae*	SW Australia
Gastrolobium glaucum C.A. Gardn.	Spike poison	*Fabaceae*	SW Australia
Gastrolobium hamulosum Meissn.	Hook-point poison	*Fabaceae*	SW Australia
Gastrolobium appressum C.A. Gardn.	Scale-leaf poison	*Fabaceae*	SW Australia
Gastrolobium crassifolium Benth.	Thick-leaf poison	*Fabaceae*	SW Australia
Gastrolobium stenophyllum Turcz.	Narrow-leaf poison	*Fabaceae*	SW Australia
Gastrolobium densifolium C.A. Gardn.	Mallet poison	*Fabaceae*	SW Australia
Gastrolobium grandiflorum F. Muell.	Wallflower poison	*Fabaceae*	N, NE and central Australia
Gastrolobium floribundum S. Moore	Wodjil poison	*Fabaceae*	SW Australia
Gastrolobium laytonii J. White	Breeyla or kite-leaf poison	*Fabaceae*	SW Australia
Gastrolobium calycinum Benth.	York Road poison	*Fabaceae*	SW Australia
Gastrolobium spectabile (Endl.) Crisp	Roe's poison	*Fabaceae*	SW Australia
Gastrolobium graniticum (S. Moore) Crisp	Granite poison	*Fabaceae*	SW Australia
Oxylobium parviflorum		*Fabaceae*	SW Australia
Oxylobium heterophyllum		*Fabaceae*	SW Australia
Oxylobium racemosum		*Fabaceae*	SW Australia
Oxylobium rigidum		*Fabaceae*	SW Australia
Oxylobium tetraconophyllum		*Fabaceae*	SW Australia
Nemcia spathulata		*Fabaceae*	SW Australia
Acacia georginae F.M. Bailey	Gidyea	*Fabaceae* or *Mimosaceae*	Georgina River basin, Australia
Dichapetalum cymosum (Hook.) Engl.		*Dichapetalaceae*	Transvaal, Africa
Dichapetalum toxicarium Baill.	Ratsbane	*Dichapetalaceae*	Sierra Leone, Africa
Dichapetalum braunii		*Dichapetalaceae*	Africa
Dichapetalum flexuosum		*Dichapetalaceae*	Africa
Dichapetalum barteri (Engl.)		*Dichapetalaceae*	Nigeria
Palicourea marcgravii St Hil.	Erva de rato	*Rubiaceae*	Brazil
Arrabidea bilabiata (Sprague) Sandw.	Chibata or gibata	*Bignoniaceae*	Brazil
Spondianthus preussii Engl.		*Anacardiaceae*	Cameroon, Nigeria, Ivory Coast, Africa

Over 120 years after Drummond (cited in Everist, 1981, from Cameron, 1977) commented that 'It is very difficult to analyse unknown vegetable poisons or to explain their effect on animal life' McEwan (1964a,b) identified the toxin in *Gastrolobium grandiflorum* as MFA. The toxic species of *Gastrolobium* are mostly confined to the south-west corner of Western Australia (South-western Vegetation Province), although some species extend into the Northern Vegetation Province in Western Australia and into the Northern Territory and Queensland (Fig. 8.1). *G. grandiflorum* occurs across the top of Australia, including northern Western Australia, northern Queensland and parts of the Northern Territory. Bailey (cited in McEwan, 1964a) referred to *G. grandiflorum* as 'the most poisonous to stock of any of the Queensland flora'. *Gastrolobium brevipes* is found only in a relatively small region of central Australia (Twigg *et al.*, 1999).

Up to 185 mg fluoroacetate/kg of air-dried leaf tissue was found in a sample of *G. grandiflorum* collected in central Queensland (McEwan, 1964b). However, it is not unusual to find over 1000 mg fluoroacetate/kg in the air-dried leaves of many species (Aplin, 1967, 1971c). In fact, eight of the 35 toxic species have over 1000 mg fluoroacetate/kg in their leaves (Twigg, 1994) and 13 of 34 species tested contained more than 400 mg/kg (Aplin, 1971c). Some species are exceptionally toxic and Aplin (1971c) reported that heart-leaf poison (*Gastrolobium bilobum*) has been shown to contain up to 2650 mg/kg in air-dried leaves and over 6000 mg/kg in seeds. However, there is considerable intra- and interstand variation in the fluoroacetate content of these plants (Twigg 1994; Twigg *et al.* 1996a,b, 1999). Although the toxic principle is present throughout the year in *Gastrolobium*, it rises dramatically at the flowering and pod-filling stages and when the plants are growing rapidly. For example, Aplin (1969a) demonstrated, in a stand of Champion Bay poison (*Gastrolobium oxylobioides*), the concentration of fluoroacetate rose from 'none detected' to 1050 mg/kg dry wt over a 5-week period during late spring. Increasing toxicity is related to periods when abundant new growth and/or pods are present. At a concentration of 1000 mg/kg of air-dried plant material, it was calculated that 0.88 oz (*c.* 25 g) would kill a 110 lb. (50 kg) sheep.

But, as suggested above, the concentration of fluoroacetate in *Gastrolobium* species at any given time can be variable and, even within an area of 200–300 ha, the differences in concentration may vary more than tenfold. Young leaves of two individuals of *Nemcia spathulata*, growing about 600 m apart, were shown by Twigg *et al.* (1996b) to contain between 40 and 80 mg/kg on a dry weight basis.

Palicourea, Arrabidea *and* Spondianthus

Much less is documented about the remaining three genera that have been reported to contain fluoroacetate. *P. marcgravii* is a shrub distributed in several states in Amazonian Brazil, where it occurs in moist woodland. It is known in South America as *erva de rato* or rat weed, and it is regarded as being one of the most toxic of plant species (de Oliveira, 1963). MFA was isolated from leaf tissues by de Oliveira (1963), using paper chromatography, and the isolate was toxic to guinea-pigs. Using a different analytical technique, ^{19}F nuclear magnetic resonance (NMR) spectroscopy, Krebs *et al.* (1994) found that their sample of *P. marcgravii* contained a low level of fluoroacetate, 5.4 mg/kg. As there are so few data on the fluoroacetate content of this species and as the concentrations appear to be relatively low, it is not possible to tell if there is the same degree of intraspecific or seasonal variation as in the Australian species.

Arrabidea bilabiata is endemic to the Amazonian area and is found as far south as Argentina. It is extremely toxic, causing rapid death in grazing animals, but the principle was not investigated until 1994. Krebs *et al.* (1994) reported fluoroacetate levels of 3.0 mg/kg in the leaves and 64.1 mg/kg in seeds of a single sample. The possibility of there being other toxins has not been investigated, so it would be premature to conclude that the extreme toxicity of this species is due solely to fluoroacetate.

Fluoroacetate has been reported in *Spondianthus preussii*, but little more is known about the ecological or economic significance of its toxicity (Tatou-Kamgoue *et al.*, 1979; Sere *et al.*, 1982). Other toxins have been recorded in it (Tessier and Paris, 1974), but the relative importance of each toxin is as yet unknown. *S. preussii* is a tree up to 25 m high that is usually classified as being in the *Anacardiaceae* but is sometimes placed in the *Euphorbiaceae*. There are two subspecies recognized, with different geographical distributions. The variety *preussii* grows along the coast of the Gulf of Guinea from Guinea to Angola. while var. *glaber* grows in the same region but it is also found more in the interior of Africa, especially in Central Africa, spreading eastward as far as Tanzania.

Do Organofluorine Compounds Occur in Other Plants?

In the 1950s and 1960s, the occurrence of organofluorine compounds in many of the species discussed above was well established, but it was not known if they occurred in other species at very low or biologically insignificant concentrations. Reckendorfer (1952, 1953) analysed the fluoride content of cherry leaves before and after ether extraction and concluded that about 26% of the total fluoride in the leaves was in the form of organofluorine compounds. He estimated that 104–128 p.p.m. (mg/kg) fluoride was in organic combination, which, if expressed as fluoroacetate, would represent between about 425 and 525 mg/kg. However, the weakness of this interpretation is that differential solvent extraction is not specific and the non-polar solvent would be subject to contamination by fluoride associated with the lipid fraction. Furthermore, the accuracy of fluoride determination was generally too low at that time to produce reliable estimates based on the difference between solvent-extracted and unextracted tissues. Wade *et al.* (1964), following non-polar solvent extraction and paper chromatography, also concluded that traces of fluoro-organic compounds were present in salad and forage crops exposed to fluoride. No estimate of the actual amount of organofluoride was given but they were

considered to be too low to be of concern. These reports did not cause a great stir in the scientific community. However, they were followed by a series of astounding reports from the Miller laboratory at Utah State University (Cheng *et al.*, 1968; Lovelace *et al.*, 1968; Yu *et al.*, 1970, 1971), which indicated the occurrence of large quantities of fluoroacetate and fluorocitrate in forage, crested wheat grass and soybean. They stated that this was the first evidence that the biosynthesis of fluoroacetate and fluorocitrate was a general feature of plants. The reports were received as a bombshell by plant physiologists and by employees of fluoride-emitting industries, because they implied that potentially toxic organo-fluoride concentrations might be widespread in the vicinity of pollution sources. This prompted the question of whether the biosynthesis of organofluorides was restricted to the previously known species or whether it was more general but less prominent in many other species. The answer was important to an understanding of both the biosynthesis of organofluorides and the potential hazards from environmental contamination by inorganic fluoride.

In the first paper by Lovelace *et al.* (1968), the authors collected a pasture mix that contained *Medicago sativa* (lucerne) and *Agropyron cristatum* (crested wheat grass) within 2 miles of a rock-phosphate-processing plant. Horses grazing in this area showed 'severe fluoride injury' from grazing on vegetation that contained up to 1000 mg total fluoride/kg dry weight, with no apparent visible injury to the plants themselves. Other samples were collected from an area of low fluoride. Based principally on paper chromatographic evidence, the authors found 179 mg fluoroacetate/kg dry wt and 896 mg fluoroacetate/kg dry wt in the pasture mix. The material identified as fluorocitrate was tested for its inhibition of aconitate hydratase and it was found to correspond closely to commercial fluorocitrate at comparable concentrations. The authors concluded:

> Considerable amounts of fluorocitrate and fluoroacetate accumulated in plants that

were exposed to fluoride pollution. Because of the known high toxicity of these compounds to animals, consumption of these plants could result in adverse effects greater than those resulting from inorganic fluoride. Blood samples from animals grazing on pastures in an area of fluoride air pollution contained greater citrate content than control animals. Controlled experiments are now in progress to determine the effects of fluoroacetate and fluorocitrate at field levels on animals over a prolonged time.

(Lovelace *et al.*, 1968)

The work was continued by Cheng *et al.* (1968), who reported the presence of fluoro-organic acids in soybean leaves exposed either to experimental atmospheric HF gas or to NaF added to hydroponic culture. These compounds were chromatographically similar to fluoroacetate and fluoro-citrate, and the latter was shown to inhibit pig-heart or soybean aconitate hydratase. Of interest at this point is the report of Treble *et al.* (1962) that plant aconitate hydratase (from a sycamore cell suspension) was about 2000-fold less sensitive to fluorocitrate than the pig-heart enzyme, so soybean aconitate hydratase might have been expected to be more resistant to the inhibitor. Louw *et al.* (1970) reported that the aconitate hydratase from *D. cymosum* was as sensitive to fluoro-citrate as that from *Parinarium capense*, but less sensitive than the enzyme from pig heart. Later, two papers were published (Yu and Miller, 1970; Yu *et al.*, 1971) stating that both field- and greenhouse-fumigated crested wheat grass (*A. cristatum*) contained both fluoroacetate and fluorocitrate. The kidney from a horse suffering from severe fluorosis due to ingestion of high-fluoride forage was also reported to contain the two fluoro-organic compounds. Clearly, if the concentrations reported by the Utah State University group were correct, there would have been an ecological situation of some gravity because of the high toxicity of the two organofluorides. Given the concentrations that they reported, the characteristic symptoms of fluoroacetate poisoning and livestock mortality should have been common and unmistakable. However, no symptoms or poisonings were reported,

which raises the strong suspicion that, at the very least, the concentrations were greatly overestimated.

Other literature on the subject is somewhat contradictory. Peters and Shorthouse (1964) concluded that no organic fluoride was present in grass seedlings or *Camellia* provided with fluoride in the nutrient medium, nor was any detected in commercial tea. This conclusion was changed in 1972 (Peters and Shorthouse, 1972b), when they reported that, based upon chromatographic evidence, they found 100 µg and 30 µg of fluoroacetate and fluorocitrate, respectively, in 24 g of soybean tissue culture. A tissue culture of tea yielded 5–10 mg fluorocitrate/kg of tissue, and commercial tea contained up to 30 mg/kg of leaves. A sample of commercial oatmeal reportedly contained up to 62 mg/kg of fluorocitrate and small, but unspecified, amounts of fluoroacetate (Peters and Shorthouse, 1972a). In contrast, Weinstein *et al.* (1972) were unable to find evidence of organo-fluorides in crop and forage species. They exposed two cultivars of soybean (*Glycine max* cvs 'Lee' and 'Hawkeye') and crested wheat grass (*A. cristatum*) to HF under controlled conditions. One of these was the same cultivar used by Cheng *et al.* (1968) in which they reported the presence of fluoro-acetate and fluorocitrate. 'Lee' soybeans were exposed to a mean HF concentration of about 10.5 µg/m³ for 39 days, while control plants were exposed to a mean concentration of 0.17 µg/m³ for the same length of time. 'Hawkeye' soybean was exposed to 7.3 and 26.6 µg/m³ for 33 and 27 days, respectively. A control group of plants was exposed to filtered air (< 0.1 µg/m³). Crested wheat grass was exposed to about 10.5 µg/m³ or filtered air for 15 days. In addition to these species, three samples of maize, one of tomato, two of hay, two of grass and one of lucerne that had been archived were examined. The source and fluoride content of each sample are shown in Table 8.4. Extracts of each species were analysed by paper chromatography, oxygen-flask combustion and gas chromatography, and in no case was there any evidence for the presence of organofluorine compounds. The authors concluded that,

Table 8.4. Fluoride treatment and concentration of fluoride in tissues of plants studied for the presence of organofluorine compounds (from Weinstein *et al.*, 1972).

Species	Treatment or source	Tissue fluoride (mg F/kg)
Tomato	NaF	621
Maize 1	HF	2300
Maize 2	HF	550
Maize 3	HF	1200
Hay 1	Field collection	130
Hay 2	Field collection	4930
Grass 1	Field collection	828
Grass 2	HF	718
Soybean var. Lee	Field collection	999
Soybean var. Hawkeye	7.3 µg F/m^3 33 days upper leaves	485
	7.3 µg F/m^3 33 days lower leaves	908
	26.6 µg F/m^3 27 days upper leaves	1149
	26.6 µg F/m^3 27 days lower leaves	3278
Crested couch grass	10.5 µg F/m^3 15 days upper leaves	1234
Lucerne	10.5 µg F/m^3 15 days upper leaves	285

although this study did not disprove the presence of organofluorine compounds in the crop and forage plants studied, if they were present, they were below a conservative estimate for the detection of fluoroacetate and fluorocitrate in these experiments of 5 and 2.5 mg/kg, respectively. More recently, traces of fluoroacetate were reported in tea and guar gum by Vartiainen and Gynther (1984) and Twigg *et al.* (1996a), so the situation remains unclear. The overall position at present seems to be the same as stated by Hall (1974), who said:

> Whilst it is possible that small amounts of organically-bound fluorine may be formed in common plants when growth is subjected to unusual ecological conditions, the experiments described here indicate that such formation is not easy to induce artificially. Pollution of the environment and its effects must be under constant surveillance but it seems unlikely that the levels of toxic C–F compounds in pastures, arising from the sources of industrial contamination mentioned earlier, if they are formed at all, constitute a serious hazard.

If there are small amounts of monofluoroactetate or monofluorocitrate formed in common plants, it is important to know the biochemical pathway and why the concentration does not increase as the supply of fluoride increases.

Biosynthesis of Carbon–Fluorine Bonds

Possibly the least understood aspect of biological organofluorine compounds is the fluorination process: the production of carbon–fluorine bonds. O'Hagan and Harper (1999) commented that 'Despite a considerable interest and a variety of speculative suggestions as to the mechanism, no specific details of the biochemistry of fluorination in any organism are known.' Peters *et al.* (1965a,b) and Peters and Shorthouse (1967) studied the synthesis of the carbon–fluorine bond in *A. georginae*, a plant that germinates easily and grows under greenhouse conditions. Peters *et al.* (1965a) found that seedlings of *A. georginae* varied considerably in their capacity to take up fluoride from the growth medium and convert it to the organic form, suggesting that the toxicity of the plants is determined genetically. More organic fluoride was usually found in roots than in the aerial parts of the plant. They postulated that chloroacetic acid could be an intermediate in the pathway of the synthesis of fluoroacetate, but they were not able to demonstrate the presence of any organochlorine compound in *Acacia*, even after inducing high chloride uptake in the plants. A second hypothesis presented, also shown to be incorrect, was that formation of the carbon–fluorine bond is initiated by inhibition of an enzyme, such as succinate dehydrogenase, which might lead to attachment of HF to the double bond of fumarate. Also, they demonstrated that, in young, soil-grown plants of *A. georginae* under normal and fluoride-stressed conditions, the enolase activities, although low, were not significantly different. Thus, there appeared to be no inhibition of enolase *in vivo*. A similar conclusion was arrived at by McCune *et al.* (1964) in experiments with bean and milo maize, where enolase was apparently stimulated after exposure to HF.

Mead and Segal (1972) also proposed pathways for the synthesis of fluoroacetate in higher plants. In one sequence, pyridoxamine phosphate-bound fluoropyruvate equilibrates with pyridoxal phosphate-bound fluoroalanine. Fluoropyruvate is derived hydrolytically from the pyridoxamine phosphate-bound fluoropyruvate and is then transformed to fluoroacetyl-coenzyme A (CoA) or directly to fluoro-acetate by an oxidative decarboxylation. If fluoroacetate is formed by this pathway, it could be converted to fluoroacetyl-CoA and could condense with oxaloacetate to form fluorocitrate or could be incorporated as the terminal group of long-chain ω-fluoro fatty acids. Later, Vickery *et al.* (1979) proposed a pathway for fluoroacetate synthesis by fluoro-decarboxylation of malonic acid. They suggested the following sequence, starting with malonate:

$$HO_2CCH_2CO_2H \quad \overset{F^-}{\underset{[O]}{\rightarrow}} \quad HO_2CCH_2F + CO_2$$

Using a strong oxidizing agent, such as sodium hypochlorite or sodium peroxidisulphate, a small amount of fluoroacetate was detected by gas chromatography. Of course, this synthesis was hardly conducted under biological conditions.

O'Hagan and Harper (1999) have reviewed other proposed pathways for organofluorine synthesis, including fluoride displacement of phosphate from phosphoglycolate, generation of fluorophosphate as a fluoride carrier and fluoro-decarboxylation of malonic acid. Most of the recent research has involved microorganisms and there

have been few attempts to investigate higher plants. One of the reasons for this is the great variability in concentrations between individuals. This was illustrated by Weinstein *et al.* (1972) working with *A. georginae* seedlings. Aseptic cultures were grown under controlled environmental conditions, NaF solutions were added at various time periods and the cultures were grown for different lengths of time before harvest. Of the 42 seedlings supplied with 10^{-2} M NaF, after about 50 days, 11 seedlings were classed as susceptible, 20 as intermediate and 11 as tolerant to fluoride, based upon the degree of toxic symptoms. The greatest amount of fluoroacetate synthesized was in the tolerant seedlings (about 1% of the total fluoride absorbed). In other experiments, it was noted that the oldest group of seedlings (58 days old at the beginning of the experiment) accumulated the most fluoride, were generally the least efficient in synthesizing fluoroacetate and were the most susceptible to endogenously added NaF. In younger seedlings, the greatest accumulation of fluoroacetate was 830 mg/kg dry wt, accounting for about 15% of the total absorbed fluoride. For the most part, these studies with *A. georginae* showed that biosynthesis of fluoroacetate was related to the age of the seedlings, the length of the incubation period in the presence of NaF, the relative susceptibility of the seedlings to fluoride and the fact that fluoroacetate biosynthesis is not dependent on organisms in the plant's rhizosphere. In an attempt to develop a tissue that did not have this variability, Preuss *et al.* (1970) grew a stem-callus culture from a seedling of *A. georginae*. The callus was grown in aseptic culture with and without added NaF. Gas-chromatographic analysis showed that fluoroacetate was produced by the *Acacia* stem callus and was present in the culture medium. These results contradicted the earlier suggestion by Peters *et al.* (1965b) that fluoroacetate is synthesized in the roots. Others have also produced callus cultures capable of synthesizing organofluorides (Bennett *et al.*, 1983; Grobbelaar and Meyer, 1989), and analytical methods have improved enormously, so it is hoped that progress will be made in the near future.

Why is Fluoroacetate Toxic?
'Lethal Synthesis'

The toxicity of fluoroacetate to animals presented a problem to the biochemist, for, unlike the usual biological poisons, such as arsenicals, iodoacetate and others, fluoro-acetate itself had no inhibitory effect on enzymes. The search for the mechanism led to one of the most elegant solutions in forensic biochemistry (Buffa and Peters, 1949; Liébecq and Peters, 1949; Peters, 1952, 1954, 1957). Using guinea-pig kidney extract, Liébecq and Peters (1949) found that addition of fluoroacetate had no effect on acetate metabolism. What occurred, however, was a massive accumulation of citrate. This was confirmed by Buffa and Peters (1949) in rats *in vivo*. Injection of fluoroacetate was lethal to the rats and all organs showed a remarkable accumulation of citrate. As an example, heart tissue rose from an average of 25 to 677 mg/kg and kidney from 14 to 1036 mg/kg of fresh tissue. It was considered that the mode of action of fluoroacetate could not be due to the liberation of the fluoride ion and its subsequent toxicity, because of the extreme stability of the carbon–fluorine bond, a bond so stable that, as stated earlier, fluoroacetate can be distilled from sulphuric acid. From results with rats, it was suggested that fluoroacetate was activated in intermediary metabolism in a manner comparable to acetate, it combined with oxaloacetate to form a 6-carbon acid and this led to the accumulation of citrate. This hypothesis integrated well with an earlier one of Bartlett and Barron (1947) and Kalnitsky and Barron (1947, 1948) that there was competition between acetate and fluoroacetate for an enzyme reaction. It also helped to explain the results of Saunders (1947), who had noted that the toxicity of ω-fluoro-acids was related to the length of the carbon chain and that esters with an even number of carbons were toxic and with odd numbers were not. Buckle *et al.* (1949), in a study on the chemical synthesis of ω-fluorocarboxylic acids of the type $F.(CH_2)_n.CO_2R$, concluded that, if n is odd, the compound is toxic and causes

fluoroacetate-like symptoms in animals and that if n is even, no such toxic properties are shown. They concluded that this verified the theory of β-oxidation of fatty acids in animals. In this theory, oxidation occurs at the β-carbon followed by hydrolytic chain cleavage. It is now accepted that fatty acids are degraded by the successive removal of two-carbon fragments. Thus, it seems evident that the toxicity of the naturally occurring ω-fluoro-fatty acids in plants is due to β-oxidation to fluoroacetate and then to fluorocitrate.

Buffa and Peters (1949) concluded that:

> some compound, which is not fluoroacetate but which is formed from this, blocks the action *in vivo* either of aconitase (aconitate hydratase {citrate [isocitrate] hydrolyase}) or isocitrate dehydrogenase; the most probable being the isocitrate dehydrogenase. The most likely possibility is that the compound is fluorocitric acid.

We know now that the enzyme blocked by fluorocitrate is aconitate hydratase. Of the four possible isomers of fluorocitrate, only one is active: (–)-erythrofluorocitrate. This same compound also prevents citrate transport into and out of mitochondria (Kirsten *et al.*, 1978). Thus, one can summarize the cascade of salient events in 'lethal synthesis' as follows:

1.   Fluoroacetate replaces acetate (which derives from the oxidative decarboxylation of pyruvate) and reacts with CoA to form fluoroacetyl-CoA (instead of acetyl-CoA).
2.   Fluoroacetyl-CoA condenses with oxaloacetate (catalysed by citrate synthase) to form fluorocitrate.
3.   The cycle is blocked by fluorocitrate inhibition of aconitate hydratase.
4.   Citrate accumulates and energy production ceases (see also Buffa *et al.*, 1977).

Although this scheme has been generally accepted and is usually cited as being the basis for toxicity, evidence has also been presented (Kun *et al.*, 1977; Kirsten *et al.*, 1978; Twigg and King, 1991) that suggests that the principal mode of toxic action of fluorocitrate results, not from its inhibition of aconitate hydratase, but from its inhibition of citrate transport through mito-chondrial membranes. The relative impor-tance of these two mechanisms has not been resolved. It is highly likely that both effects are important in fluoroacetate intoxication, but that their importance may vary between species and/or ages (Twigg and King, 1991). Furthermore, citrate accumulation and blockage of citrate transport are just the starting points for a further cascade of pathological effects that have not been fully described in any species. Interestingly, fluorocitrate has also an inhibitory effect on succinate dehydrogenase (Fanshier et al., 1964), but the relative importance of this in fluoroacetate intoxication has never been investigated.

Relative Toxicity of Monofluoroactetae to Different Organisms

For obvious reasons, much of the interest has centred on the effects of fluoroacetate on native animals and domestic livestock. In their comprehensive review, Twigg and King (1991) listed the tolerance of many groups of animals and concluded that birds are the least sensitive of endothermic (warm-blooded) vertebrates and that ecto-therms (fish, amphibians, reptiles) are rela-tively insensitive. However, in an Austra-lian context the tolerance is also related to other factors (see later). Chenoweth (1950) reported cold-blooded vertebrates to be generally insensitive to fluoroacetate. Some have long been known to be relatively toler-ant to fluoroacetate, such as the South Afri-can clawed toad (Xenopus laevis), which has a median lethal dose (LD$_{50}$) of 5 mM/kg (Chenoweth, 1950), and fish, such as bass and bream (King and Penfound, 1946). The LD$_{50}$ for steel head trout (Salmo gairdnerii) was estimated to be 500 μM/kg by intraperi-toneal injection (Baumeister et al., 1977).

The differences in tolerance between groups of animals could be due to variations in physiology or in their metabolism of fluoroacetate (Twigg and King, 1991).

Theoretically, the concentration of reduced glutathione might help account for toler-ance, but there is little evidence to support this idea. A much more researched area is fluoroacetate metabolism. Twigg and King (1991) considered that sensitive and tolerant animals may differ in: the rate of conversion of fluoroacetate to fluorocitrate; the sensitiv-ity of their aconitate hydratase to fluoro-citrate; effects of fluorocitrate on citrate transport; and their capacity to detoxify fluoroacetate. The most comprehensive exploration of these possibilities is provided by Twigg et al. (1986) and Twigg and King (1991), who compared the sensitivity to fluoroacetate of the skink (an ectotherm lizard, Tiliqua rugosa; LD$_{50}$ ~800 mg/kg) and the Norway rat (Rattus norvegicus; LD$_{50}$ ~2 mg/kg), which have similar body weights. Administration of fluoroacetate at 100 mg/kg body weight to the skink resulted in a 3.4-fold increase in plasma citrate levels after 48 h, while 3 mg/kg body weight in the rat produced a fivefold increase in plasma citrate in less than 4 h. The difference in metabolic rate of about ten times results in slower conversion of fluoroacetate into fluorocitrate and therefore greater time for excretion and defluorination (Twigg and King, 1991). Aconitate hydratase from skink liver was less inhibited by (−)-erythro-fluorocitrate than that from rat liver, and defluorination of fluoroacetate was much more rapid in erythrocytes and liver prepa-rations from the skink than in those from the rat. However, comparisons of populations of skinks that differed in tolerance by a large factor showed that they had similar sensitiv-ity of aconitate hydratase and similar abili-ties to convert fluoroacetate to fluorocitrate and to defluorinate fluoroacetate. It was argued that the natural selection for such high tolerance in this skink may have been acting through the effects of fluoroacetate on fertility rather than direct effects of mortality per se (Twigg and King, 1991). The authors concluded that 'the biochemical mecha-nisms responsible for the development of tolerance to fluoroacetate are yet to be fully understood'.

Invertebrates

Fluoroacetate is lethal to many insect species, including the common housefly (*Musca domestica*). The LD_{50} for a fluoroacetate-sensitive strain is about 0.25–0.30 µg per fly (Zahavi *et al.*, 1968). In studies with sarcosomes from fluoroacetate-sensitive and -resistant house flies, Zahavi *et al.* (1964) found that, following treatment with fluoroacetate, the fluoroacetate-sensitive organelles accumulated citrate while the resistant ones did not. Selection of an insecticide-susceptible strain by exposing it to fluoroacetate produced moderate resistance to fluoroacetate after 65 generations. By the 81st generation, the fluoroacetate resistance level had attained 500 µg per fly. As a parallel effect, after 57 generations, there was a 200-fold increase in resistance to dichlorodiphenyltrichloroethane (DDT) (Tahori, 1966) and increased resistence to methoxychlor and dieldrin (Tahori, 1965, 1966).

Microorganisms

Most studies involving microorganisms have been concerned with metabolism and defluorination, rather than effects on growth and the environment, so there is relatively little information available on the subject. However, there are several reports of inhibition of bacteria (e.g. Liébecq and Osterrieth, 1963, 1965) and yeasts (e.g. Aldous, 1963; Stewart and Brunt, 1968; Ketring *et al.*, 1968) grown in culture. Some cyanobacteria are also sensitive, as reported for *Azotobacter vinelandii* (Broomier, 1962) and *Gloeocapsa* sp. $_{LB}$795 (Gallon *et al.*, 1978). In contrast, unicellular algae, such as *Chlorella* sp. and *Schenedesmus* sp. have been shown to tolerate fluoroacetate concentrations up to 20 mM (Bong *et al.*, 1979), but the environmental significance of effects on microorganisms is not known. In Australia, over 20 species of microorganisms are known to be able to detoxify fluoroacetate (see Chapter 9). The most important fungi are species of *Fusarium*, and species of *Pseudomonas* and *Bacillus* appear to be the most important bacteria. The distribution of these microorganisms is independent of that of the fluoroacetate-bearing vegetation that occurs in Australia (Twigg and Socha, 2001).

Higher plants

There are two important questions in relation to plants, one being why fluoro-organic acids are not toxic to the plants that produce them and the other being whether the compounds are toxic to other plants that grow in the vicinity or that are exposed to manufactured compounds. Treble *et al.* (1962) pointed out that 'the mystery remains why *D. cymosum* and *Acacia georginae* which make fluoroacetate are not poisoned by their own products'. The obvious possibility is that the potentially toxic compounds are compartmented in the vacuoles or cell walls, out of the way of sensitive enzymes, but the cellular location of these compounds is unknown. Using modern methods for extracting cell-wall fluid and isolating organelles, it would be feasible to investigate this now that analytical methods are more sensitive and specific.

Kuhn *et al.* (1981) suggested that, in species that synthesize organofluorides, the acetyl-CoA synthetase is not located in the mitochondria but in the chloroplasts. Eloff and von Sydow (1971) found that application of fluoroacetate to leaf discs of *D. cymosum* did not lead to inhibition of oxygen uptake, nor was there an accumulation of citrate, while in the non-fluoroacetate-producing plant *P. capense* oxygen uptake was inhibited and there was an accumulation of citrate. However, when fluorocitrate was applied to either species, oxygen uptake was inhibited and citrate accumulated. The authors concluded that lethal synthesis does not take place in *D. cymosum* because citrate synthetase has a much lower affinity for fluoroacetyl-CoA than for acetyl-CoA. When intact leaves of

D. cymosum and *P. capense* were provided with fluoroacetate-2-¹⁴C and acetate-2-¹⁴C, Eloff and Grobbelaar (1972) found that fluoroacetate was relatively inactive metabolically with respect of the rate at which it was metabolized and the range of labelled compounds that arose. On the other hand, acetate was found in nearly all important metabolic pools in both species. The authors concluded that although having a normal citric acid cycle (Eloff, 1972), *D. cymosum* and *P. capense* were insensitive to fluoroacetate because they did not readily convert fluoroacetate to fluorocitrate. More recently, Meyer *et al.* (1992), using a crude mitochondrial extract from *D. cymosum* as an enzyme source, found what appeared to be a fluoroacetyl-CoA hydrolase, which could explain the plant's tolerance to fluoroacetate. If some fluoroacetate entered the mitochondria of the plant and was converted to fluoroacetyl-CoA, it would be immediately broken down by the fluoroacetyl-CoA hydrolase-like enzyme.

Most plants never encounter the toxic fluoro-organic acids, but it does happen where fluorocompounds are accidentally or deliberately released into the environment. Reports of the toxicity of fluoroacetate are mixed, some reporting that it is innocuous and others that it is highly toxic. For example, Ramagopal *et al.* (1969) found that, although potassium fluoroacetate was inhibitory to young wheat roots, potassium fluoride was about ten times more toxic. In contrast, Bong *et al.* (1980) reported that duckweeds are far more sensitive to fluoroacetate than to sodium fluoride. Fluoroacetate at 0.5 mM completely inhibited frond multiplication in *Spirodela polyrrhiza* and at 1.0 mM for *Spirodela oligorrhiza* and *Lemna minor*. Sodium fluoride at the same concentration, on the other hand, had a barely discernible effect on the growth of any species. In the case of fluoroacetate, frond multiplication followed the logarithmic relationship:

$$K_{fa} = a_{fa} + b_{fa}.\log_{10}[NaFac]$$

while, for sodium fluoride, it followed the linear relationship:

$$K_F = a_F + b_F[F^-]$$

Similarly, Cooke (1972, 1976) found that fluoroacetate was far more toxic than sodium fluoride to the pasture plants yarrow (*Achillea millefolium*) and ryegrass (*Lolium perenne*) (Fig. 8.2). Although growth was greatly reduced by as little as 20 mg F/l as sodium monofluoroacetic acid (NaMFA), even 250 mg F/l as NaF had comparatively little effect.

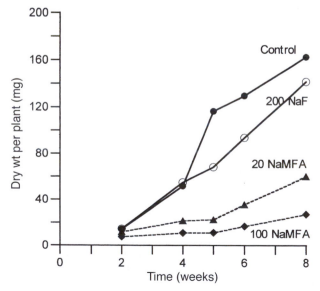

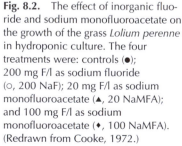

Fig. 8.2. The effect of inorganic fluoride and sodium monofluoroacetate on the growth of the grass *Lolium perenne* in hydroponic culture. The four treatments were: controls (●); 200 mg F/l as sodium fluoride (○, 200 NaF); 20 mg F/l as sodium monofluoroacetate (▲, 20 NaMFA); and 100 mg F/l as sodium monofluoroacetate (♦, 100 NaMFA). (Redrawn from Cooke, 1972.)

Biological Cleavage of Carbon–Fluorine Bonds; Defluorination

In view of the toxicity of the natural organo-fluorides and the strength of carbon–fluorine bonds, an important question is the fate of these bonds in the environment. What happens to fluoroacetate when it is eaten by a herbivore or when leaves fall to the ground? Do carbon–fluorine bonds persist in soil and water? The answer is that many organisms, from microorganisms to humans, possess a remarkable capacity to defluorinate fluoroacetate.

Microorganisms

Most of the studies of defluorination have involved microorganisms, so it is appropriate to start with them. In most cases, the microorganisms were those that can utilize fluoroacetate or fluoroacetamide as their sole carbon source (Goldman, 1965; Kelly, 1965; Tonomura *et al.*, 1965). Goldman (1965, 1972), utilizing enrichment culture, was able to isolate an organism that had the capacity to grow rapidly with fluoroacetate as its sole carbon source. He stated (Goldman, 1972) that:

> An aerobic organism which utilizes a potential tricarboxylic acid cycle inhibitor as a source of energy and carbon can understandably be regarded with scepticism or amusement. Nevertheless, it is reasonable from teleological considerations that fluoroacetate, a natural product [Marais 1944], should not accumulate indefinitely and that bacteria might be found in soil which are capable of degrading it.

Logic might have dictated that an organism capable of degrading fluoroacetate would have been most likely to be found in the rhizosphere of a fluoroacetate-synthesizing plant (see Meyer *et al.*, 1990), but Goldman (1965) isolated his pseudomonad from soil in a creek in Maryland, USA, and others have isolated organisms from Japan (Horiuchi, 1960; Tonomura *et al.*, 1965) and Great Britain (Kelly, 1965), where, to

the best of our knowledge, fluoroacetate-synthesizing plants do not grow.

The organisms isolated in the laboratories in the USA, Japan and Great Britain were found to have an inducible enzyme capable of cleaving fluoroacetate. The enzyme found in Goldman's laboratory (1965, 1966, 1969, 1972), a haloacetate halidohydro-lase, cleaved fluoroacetate according to the following equation:

$$FCH_2COO^- + HO^- \Rightarrow HOCH_2COO^- + F^-$$
Fluoroacetate   Glycolate

The same enzyme was found to catalyse the analogous release of the halide from chloroacetate and iodoacetate, although fluoroacetate was the preferred substrate based upon the kinetic constants (Table 8.5). The same enzyme that cleaved fluoroacetate to glycolate and fluoride ion, however, was not capable of catalysing the reverse reaction, i.e. yielding fluoroacetate (Goldman, 1966).

Walker and Lien (1981) isolated and partially purified halidohydrolases from a soil pseudomonad and from the common soil fungus *Fusarium solani*. Both enzymes readily released fluoride from fluoroacetate and fluoroacetamide but were without effect on a wide range of other organofluorine compounds. The enzymes also cleaved the halogen–carbon bonds from monochloroacetate and monobromoacetate.

An unexpected example of defluorination was found by Vickery and Vickery (1975). They reported that an organism

Table 8.5. Kinetic constants for substrates of haloacetate halidohydrolase. (From Goldman, P. (1972) The use of microorganisms in the study of fluorinated compounds. In: *Carbon–Fluorine Compounds: Chemistry, Biochemistry, Biological Activity Symposium* 1971, pp. 335–356.)

Substrate	V_{max}[a]	K_m (mM)
Fluoroacetate	100	2.4
Chloroacetate	15	20
Iodoacetate	0.8[b]	

[a]Relative to fluoroacetate taken as 100.
[b]Rate measured with substrate at 10^{-1} M, based upon the kinetic constants.

present in the air was responsible for defluorination of organofluorides present in the leaves of *D. cymosum* and *Dichapetalum heudelotii*. They also reported that organofluorides in *D. toxicarium* were defluorinated even under herbarium conditions. More recently, Meyer *et al.* (1990) isolated a bacterium, identified as *Pseudomonas cepacia*, from the seeds and stem of *D. cymosum* that was capable of growing in a 50 mM solution of fluoroacetate with essentially no effect on its growth. The organism not only cleaved the carbon–fluorine bond, producing after 52 h 2.69 mg F/10^9 cells/h, but all of the ^{14}C evolved was found as $^{14}CO_2$. Of interest is the fact that *P. cepacia* is a soil organism responsible for root rot of onions and may be an opportunistic pathogen of humans. Meyer *et al.* (1990) suggested that the organism is capable of living within the tissues of *D. cymosum* because it can metabolize fluoroacetate, deriving energy from the 2-carbon defluorinated fragment.

In an environmental context, the most important question is whether there is sufficient defluorination in soil or water to keep concentrations below toxic levels. Several publications from Australia and New Zealand support the idea that fluoroacetate is rapidly defluorinated in soil, though the rate depends on the soil and climate (see papers in Seawright and Eason, 1994; Bowman, 1999; Eason *et al.*, 1999; Twigg and Socha, 2001). For example, Twigg and Socha (2001) isolated 13 species of bacteria and 11 fungi from soils collected in arid, central Australia that were capable of defluorinating fluoroacetate when it was the sole carbon source. The abundance of the organisms varied with time, and they appeared to increase in abundance after rain, so the rate of defluorination varied from month to month (Fig. 8.3). Microorganisms with a similar capacity for detoxifying fluoroacetate have also been found in temperate Australia (see Wong *et al.*, 1992). The profile of fluoroacetate in the environment where fluoroacetate-bearing vegetation occurs in temperate Australia has also been examined, including the determination of fluoroacetate in soil samples from beneath *Gastrolobium* (Twigg *et al.*, 1996a). Twigg *et al.* (1996a) stated that:

> Despite the high concentrations of fluoroacetate in many [of the toxic *Gastrolobium*] species, only one of nine soil samples collected from beneath these plants contained fluoroacetate. None of the 16 water samples collected from

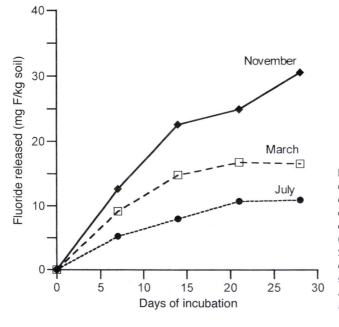

Fig. 8.3. The rate of defluorination of sodium monofluoroacetate containing 380 µg F added to 5 g of soil collected at different times of the year in central Australia. (Redrawn from Twigg, L.E. and Socha, L.V. (2001) Defluorination of sodium monofluoroacetate by soil microorganisms from central Australia. *Soil Biology and Biochemistry* 33, 227–234.)

nearby streams and catchment dams contained fluoroacetate. This suggests that fluoroacetate does not persist in this environment.

Similar conclusions have been reached by other workers (see next chapter for further comment).

Plants

As indicated previously, most plants never encounter organofluorides in their environment but, because of interest in the metabolism of fluoroacetate, there have been several studies of the defluorination capacity of plants. Two methods have been used: ^{14}C labelling and detection of inorganic fluoride ions.

Preuss *et al.* (1968) carried out a series of experiments to determine if germinated seeds of *A. georginae*, groundnut (*Arachis hypogaea*), castor bean (*Ricinus communis*) and pinto bean (*Phaseolus vulgaris*) had the capacity to cleave the carbon–fluorine bond. Surface-sterilized seeds and sterile conditions were used during germination and the introduction of 2-^{14}C-fluoroacetate over a 21 or 22 h incubation period. $^{14}CO_2$ was evolved by each species, indicating that there was cleavage of the carbon–fluorine bond. There were great differences between species, however. For example, of the total amount of ^{14}C absorbed, 0.2% was liberated as $^{14}CO_2$ by pinto bean, 3.3% by *A. georginae*, 8.8% by castor bean and 42.0% by groundnut. The time course of $^{14}CO_2$ evolution is shown in Fig. 8.4. The rest of the ^{14}C was incorporated into a variety of metabolites, notably lipids, so Preuss and Weinstein (1969) continued their studies to determine if incorporation of ^{14}C in lipids indicated the formation of fluoro-fatty acids or if incorporation occurred after cleavage of the carbon–fluoride bond. Germinated seeds of groundnut were set up in aseptic culture. The culture vessels were connected to an alkali trap to collect any volatile fluoride released. No fluoride was detected in any of the traps, eliminating the possibility that there was an evolution of volatile inorganic fluoride, but not eliminating the possibility of an emission of fluoroacetone (Peters and Shorthouse, 1971). About 15% of the fluoride provided as fluoroacetate was found in the inorganic form in the seedlings or in the incubation medium. Where boiled seeds or no seeds were used, between 2 and 5% of the fluoride was found. In germinated seeds, about 29% of the total fluoride present was in the inorganic form, while, in boiled seeds, all the fluoride remained in the organic form. In *A. georginae*, there were no qualitative differences in fatty acids in seeds with or without added fluoroacetate. Gas

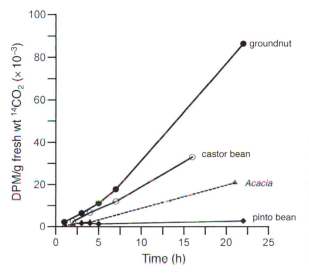

Fig. 8.4. The time course of $^{14}CO_2$ evolution (DPM, disintegrations per minute) by seedlings of four species of plants supplied with 2-^{14}C-fluoroacetate for 21 or 22 h. (Redrawn from Preuss, P.W., Lemmens, A.G. and Weinstein, L.H. (1968) Studies on fluoro-organic compounds in plants. I. Metabolism of 2-^{14}C-fluoroacetate. *Contributions from the Boyce Thompson Institute* 24, 25–32.)

chromatography and mass spectrometry showed that there was no evidence for the presence of fluoro-fatty acids. The study supported the view that groundnut seedlings have a haloacetate halidohydrolase that can cleave the carbon–fluorine bond and that subsequent metabolism of the 2-carbon fragment would account for the relatively large amount of $^{14}CO_2$ evolved from the seedlings in the presence of 2-^{14}C-fluoroacetate (Preuss *et al.*, 1968).

Cooke (1972, 1976) detected defluorination by analysing plants for inorganic fluoride ions (Fig. 8.5). He grew sunflower (*Helianthus annuus*) from seed for 4 weeks in nutrient culture. The plants were given NaMFA at a concentration of 100 mg F/l. The total fluoride content quickly reached 1200 mg/kg in the leaves but it continued to increase in the cotyledons until it exceeded 1600 mg/kg (Fig. 8.5a). Concentrations were much lower in the stem and root. Figure 8.5b demonstrates that a high percentage of the NaMFA was defluorinated, because over 90% of fluoride was in the inorganic form in the cotyledons. It was lower in the leaves and other organs. This suggested that defluorination was mostly in the photosynthetic tissues, though defluorination in the roots, coupled with transport to the shoot, could not be ruled out in this experiment. In later experiments with aseptically grown

roots and cotyledons, Cooke (1972) confirmed that roots had a very low capacity and cotyledons a high capacity for defluorination (Table 8.6). Defluorination was greater when cotyledons were in the light. Interestingly, in the same study, sunflower leaves developed necrosis of the tips and margins, symptoms that closely resembled those caused by inorganic fluoride (Weinstein *et al.*, 1998). It appears, therefore, that at least some of the effect of fluoroacetate on the growth of sunflower may have been caused by defluorination and the accumulation of inorganic fluoride in the leaves.

Overall, studies with plants support the idea that they, like microorganisms, contain a haloacetate halidohydrolase and that the

Table 8.6. The concentration (μg/10 ml bathing solution) of inorganic fluoride detected after 100 h in tissues incubated aseptically in a culture medium containing 200 mg F/l as sodium monofluoroacetate (data from Cooke, 1972).

Material	μg F in tissue	μg F in bathing solution	Total μg F
Cotyledons in light	1.1	35.9	37.0
Cotyledons in dark	0.95	15.1	16.0
Roots	0.0	7.2	7.2
Bathing solution only	–	0.0	0.0

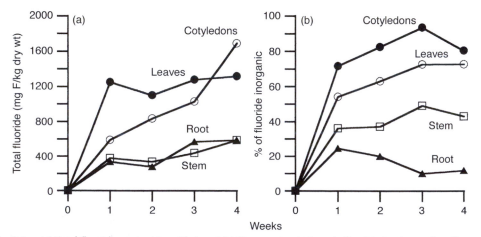

Fig. 8.5. (a) Total fluoride content (mg F/kg) and (b) % present as inorganic fluoride in plants of sunflower (*Helianthus annuus*) grown in hydroponic culture containing sodium monofluoroacetate at a concentration of 100 mg F/l (redrawn from Cooke, 1972, 1976). (Courtesy of *Fluoride* 9, 204–212, 1976.)

dehalogenated end-product is metabolized and evolved as CO_2. It is highly unlikely that incorporation into normal plant metabolites occurs before defluorination, since many enzymes of the tricarboxylic acid cycle and other pathways would be required to utilize fluorinated metabolites. Furthermore, no more than trace amounts of such metabolites have been found (with the exception of the reports of Cheng et al., 1968; Lovelace et al., 1968; Yu and Miller, 1970; Yu et al., 1971) in plants. Thus, it is reasonable to conclude that the fluorine atom is removed from fluoroacetate prior to its further metabolism, giving rise to a 2-carbon compound, such as glycolate, much as suggested by Goldman (1972). This compound could then traverse the tricarboxylic acid cycle by conversion to acetate or glyoxylate and be incorporated into fatty acids and lipids with great facility. Except for fluoroacetate-producing species, it is unlikely that fluoroacetate persists for very long in vegetation in those rare circumstances where plants might be exposed to it.

Animals

Although there is less evidence than for other organisms, it is clear that defluorination does occur in some animal tissues. In an experiment with rats, Gal et al. (1961) found that, during a 4-day period following injection of $2\text{-}^{14}C$-fluoroacetate, 3% of the label appeared in respiratory CO_2 and 32% was excreted in urine. The urine contained at least seven radioactive substances, indicating extensive metabolism of the fluoroacetate. Fluorocitrate in the urine was found not to exceed 3% of the total radioactivity administered. ^{14}C accumulated in heart, kidneys, liver and brain. In mitochondrial preparations, there was four times as much ^{14}C in kidney than in liver. Likewise, kidney had 15 times as much fluorocitrate as did liver. There was ^{14}C incorporation into fatty acids and small amounts in cholesterol. Clearly, the rat metabolic systems were able to cleave the carbon–fluorine bond with some facility.

Smith et al. (1977), Kostyniak et al. (1978) and Kostyniak (1979) have also reported defluorination of fluoroacetate in the rat. Kostyniak (1979) suggested that the defluorinating system in rats played a role in the detoxification of fluoroacetate by reducing the amount available for biotransformation to fluorocitrate. Thus, the tolerance of some mammals to fluoroacetate may be the result of enzymatic defluorination, though the evidence so far is mixed. For example, Twigg and Mead (1990) found that liver preparations from populations of the skink, T. rugosa, that differed in tolerance to MFA were about equally able to defluorinate fluoroacetate and to convert fluoroacetate to fluorocitrate. In addition, (–)-erythro-fluorocitrate inhibited the liver aconitate hydratase from both conspecifics about equally. It was also suggested, however, that the observed differences in sensitivity could arise from differing effects of fluorocitrate on citrate transport mechanisms, but this is yet to be tested (Twigg and King, 1991).

Mead et al. (1979) compared the relative tolerance of Western and South Australian populations of brush-tailed possum and found that the Western Australian species was nearly 150-fold more tolerant than the South Australian species. None the less, preparations from the livers of animals from both populations were equally able to convert fluoroacetate to fluorocitrate. Aconitate hydratase from both populations was competitively inhibited by fluorocitrate. Both populations were capable of defluorinating fluoroacetate at similar rates by a glutathione-dependent enzymatic reaction resulting in free fluoride ion and S-carboxymethylcysteine. The defluorination appeared to result from a nucleophilic attack on the β-carbon of fluoroacetate, with the consequent release of fluoride ion and S-carboxymethylglutathione. The latter compound would then be metabolized into its component amino acids by normal metabolism. Activation of fluoroacetate to fluoroacetyl-CoA was found not to be a necessary prerequisite for defluorination.

Finally, in 1987 Sykes et al. used ^{18}F-labelled fluoroacetate to study the disposition, metabolism and excretion of fluoroacetate in mice. They reported that there was no tissue-specific accumulation of the compound itself, but bone accumulation of ^{18}F activity was significant and that their data supported the concept of metabolic defluorination.

Evolution of Organofluorine Compounds in Plants

Although the production of organofluorine compounds is restricted to a relatively small number of species, it occurs in six[2] different families that occur in widely separated regions of the world. Therefore, the capacity to produce or accumulate appreciable quantities of fluoroacetate appears to have arisen independently several times in geographically distinct regions. The presence of toxins is usually interpreted as being an animal defence mechanism that has evolved in response to herbivory (Mead et al., 1985a; Harborne, 1977). Twigg and King (1991) have suggested this in the case of fluoroacetate, commenting that its presence evolved in response to herbivory from vertebrate and/or invertebrate herbivores or from some combination of both of these selection pressures. Plants often use a combination of chemical and physical defences against herbivory and, because there are limited resources in most natural environments, there may be a trade-off between the two mechanisms. It is interesting, therefore, that Twigg and Socha (1996) found that there was a strong suggestion of a negative correlation between these two mechanisms in the toxic Gastrolobium species: that is, the species with the greatest investment in physical defences (spines, etc.) had the lowest concentrations of fluoroacetate, and vice versa.

Having a defence strategy based upon a highly toxic chemical may appear to be the perfect solution for a plant, but the evolutionary story does not end there, because herbivores may evolve mechanisms to counter the defence, leading an evolutionary 'arms race' (Harborne, 1977). During the early colonization of Western Australia, it was observed that some of the native animals, such as the bronze-wing pigeon, fed on the poisonous Gastrolobium with apparent impunity (Marchant, 1992). This was the first illustration of the great difference between species in their ability to cope with organofluorides and it gave a hint of what was later found by experimentation; which is that many native Australian herbivores have evolved a tolerance to this potent toxin. Many of the native animals that coexist with fluoroacetate-containing plants are unusually tolerant to the poison. In contrast, animals of the same species living in other areas of Australia outside the range of the toxic plants are highly sensitive to it (Oliver et al., 1977; King et al., 1978). The tolerance to fluoroacetate is found at all trophic levels and in different taxonomic groups, including insects, reptiles, birds and mammals (Twigg and King, 1991). A prime example (cited by Mead et al., 1979) is the comparison of the brush-tailed possum (Trichosurus vulpecula) from south-western and eastern Australia: the LD_{50} for the possum in eastern Australia is 0.68 mg/kg and, in contrast, the LD_{50} for the south-western Australian conspecifics from Western Australia is 100 mg/kg, a difference of nearly 150-fold. Mead et al. (1985b) also showed this remarkable adaptive tolerance in geographically separated populations of the macropodid marsupial Setonix brachyurus (= the quokka). The authors compared populations of quokka from mainland Western Australia, where toxic species of Gastrolobium are widespread; from Bald Island, off the coast of Western Australia, where G. bilobum (heart-leaf poison, the most toxic Gastrolobium) is common; and from Rottnest Island off Western Australia, where fluoroacetate-containing species do not occur. The latter population has been isolated from fluoroacetate-containing species for 5000–7000 years. The mainland and Bald Island populations have a similar tolerance to fluoroacetate (LD_{50} values of about 40–60 mg/kg). The Rottnest population, although separated from fluoroacetate-containing plants for many millennia, had an LD_{50} value of about 10 mg/kg. This was much lower than that of the mainland population but was higher than anticipated. The authors suggested that the Rottnest Island population may have been derived from a previous coastal population that had received a genetic contribution from a population that had been exposed to fluoroacetate-containing vegetation in the adjacent hills.

In their excellent review of the subject, Twigg and King (1991) concluded that the degree to which fluoroacetate tolerance has developed in native animals is: herbivores > omnivores > carnivores. However, the tolerance is also dependent on the length of time the population has been exposed to fluoroacetate, the specificity of their diet (some will eat only toxic species, while in others they may be a minor component of the diet), the size of the home range (which affects the amount of toxic material they encounter) and the degree of mobility. In some instances, such as where animals have exceptional tolerance to fluoroacetate (e.g. *T. rugosa*, *T. vulpecula*), Twigg and King (1991) argued that the detrimental effects of fluoroacetate on animal fertility may have been an important component of the selection pressure for fluoroacetate tolerance because reproductive tissues have high energy demands. Recent work by Twigg *et al.* (2002) on tolerance in rabbit populations is covered in the next chapter.

Management of Naturally Occurring Fluoroacetate Toxicity

Since the toxic plants were identified in South Africa and Australia, efforts have been made to decrease the economic and social effects of these plants. In South Africa, areas with *Dichapetalum* were fenced off during the season when the plants were most toxic, while in Australia, once the toxic species were known, a programme of eradication was started along main roads and major stock routes (Marchant, 1992). Where it was impossible to eradicate plants, the known stands were

avoided. However, because of past management, many of these plants now only occur in conservation reserves. Restricting the access of livestock to areas containing *A. georginae* is more problematic. This is mainly because the greater variability in toxicity makes predicting the time when the plants are toxic more difficult. Because of the importance of animal poisoning, efforts have been made to find an antidote, but without much success. The administration of glycerol monoacetate was shown to reduce mortality a half-century ago (Chenoweth *et al.*, 1951), but this and other related compounds which exert their antidotal effects through the conversion to acetate have not proved to be very effective against fluoroacetate poisoning. In cases of poisoning of humans (which is usually due to the manufactured compound 1080), treatment consists of alleviating the symptoms (e.g. giving oxygen, anticonvulsants), emptying the stomach, administering fluids to support excretion and intravenous application of calcium gluconate. On the other hand, there may not be a pressing need for an antidote to protect non-target species. Recent studies suggest that consumption of 1080 by non-target species may present only a low potential risk to native animals (L.E. Twigg, personal communication).

Recently, a more sophisticated approach to prevention of toxicity in cattle and sheep was produced by Gregg and his colleagues, using genetic manipulation of rumen bacteria (reviewed by Gregg, 1995). This was triggered by the high losses of cattle and an initiative by the cattle producers. A gene encoding fluoroacetate dehalogenase isolated from the soil bacterium *Moraxella* sp. (Kawasaki *et al.*, 1981), together with a gene conferring resistance to the antibiotic erythromycin (as a marker), were introduced into strain OB156 of *Butyrivibrio fibrisolvens*, a bacterium that is normally present in cattle and sheep. A plasmid (a self-replicating piece of DNA) was used to introduce the genes and it included sequences from the bacteria *Escherichia coli* and *Enterococcus faecalis*. The modified bacterium was able to degrade fluoroacetate and it persisted in the rumen of sheep for at least 5 months (Gregg,

1995; Gregg *et al.*, 1993, 1998). Five inoculated sheep exhibited markedly reduced toxicological symptoms (with respect to behavioural, physiological and histological effects) after fluoroacetate poisoning. Although this technique shows promise, the release of the genetically modified organisms in the field is a highly contentious issue and there have been some important criticisms.

In 1994, Gregg proposed to the Genetic Manipulation Advisory Committee (GMAC) a field trial to determine the field effectiveness of the modified organism in protecting cattle against poisoning. The experiment was to involve inoculating the modified organism, using conventional drenching procedures, into approximately 10% of animals within herds on a series of grazing properties in Queensland and the Northern Territory. All research on genetically modified organisms must be approved by GMAC, but at that stage the proposal was submitted in order to determine the concerns and questions and to define what further work was needed. Not unexpectedly, GMAC did not allow the trial to proceed, on the grounds of biosafety (GMAC, 1994). The grounds were principally that the modified organism might be transferred to feral pests, with adverse environmental consequences. It was argued that, if feral pests acquired the ability to detoxify fluoroacetate, the possible consequences would include the potential for the pests to become tolerant to the pesticide 1080 and increased pressure on fluoroacetate-containing plants by grazing animals. Furthermore, the transfer to feral ruminants, such as goats and camels, could increase their resistance to fluoroacetate and therefore cause increases in the feral populations due to decreased toxicity of native plants. At the time, it was considered by some scientists that it was unlikely that the organism would spread to feral non-ruminants, such as rabbits, but GMAC considered that experimental evidence was needed to confirm this expectation. A second, modified proposal was reviewed in 1999 (GMAC, 1999). This proposal was to use several modified strains of *B. fibrisolvens* and *Bacteriodes* sp. to compare the

protection against fluoroacetate provided for cattle, sheep and goats and to confirm that there were no adverse effects on the animals. The plasmids were modified to prevent transfer to other bacteria and there were to be stringent controls to prevent release of the organisms outside the fenced experimental area. The decision of GMAC was that, although much more information had been supplied and containment was proposed, there were still concerns about the possible risks to the environment if the modified organisms were able to spread from the test site. They asked for more information about transmission between animals, the browsing behaviour of the animals and the potential for the use of compound 1080 to be compromised. There was also concern about the potential effects of the antibiotic marker genes on the use of the antibiotics in veterinary and clinical practice. Other concerns have been raised. If the modified organisms give only partial protection it would allow animals to consume more toxic plants before they sickened, which might increase the overall level of suffering. On the other hand, if the treatments work, there could be increased grazing pressure on what is already marginal land, possibly leading to ecosystem collapse.

This case is an excellent example of the dilemma and problems raised by the use of genetically modified organisms. On the one hand, there is very serious loss of livestock by a poison that causes pain to the affected animals and the modified bacteria show signs of being able to alleviate the problem. On the other hand, there is a real basis for concern about the effects of releasing the organisms into the environment, so the proponents must provide robust evidence that the technique will work and that the organisms will not have an adverse effect on the environment.

Notes

[1]   Monofluoroacetic acid probably occurs in nature as sodium or potassium monofluoroacetate. In the literature it is mostly referred to

as fluoroacetate, so we use that term or MFA for convenience.

2   The number depends on interpretation of the status of *Gastrolobium* and *Acacia*. Although both are legumes, they are in widely separated groups, which are often, but not always, classified as being in different families, the *Fabaceae* and *Mimosaceae*.

9

Manufactured Organofluorine Compounds

Introduction

Organofluorine compounds are used so extensively in industry, commerce and the home that we cannot avoid them; they touch every part of our lives. Common fluorinated compounds include pharmaceuticals, pesticides, herbicides, surfactants, fire-extinguishing agents, aerosol propellants, refrigerants, adhesives, surface coatings for paper and fabrics, fibres, membranes, insulating materials and lubricants (Table 9.1). This chapter is concerned with the effects of the organofluorides that are released into the environment and that may affect animal or plant health or the functioning of ecosystems. It does not deal with occupational exposure or the effects of pharmaceuticals on humans or agrochemicals on target organisms.

Release, Dispersion and Effects of Organofluorine Compounds in the Environment

When a compound is released into the environment, its effects depend on its dispersion and dilution in air, water and soil and on its subsequent metabolic or chemical alteration. Some organofluorine compounds are readily altered chemically but others are inert under ambient conditions, and the range of chemical and physical properties is

such that some find their way into every part of the biosphere. The potential for widespread dissemination and the environmental effects will be illustrated by reference to five groups of compounds: fluoropolymers; surfactants; anaesthetics; agrochemicals; and halocarbons.

Fluoropolymers

Probably the most familiar fluoropolymer is Teflon (a Du Pont version of polytetrafluoroethylene or PTFE), but there are many others with different properties such as fluorinated ethylene propylene (FEP), a copolymer of tetrafluoroethylene and hexafluoropropylene. They are unreactive solids with properties that suit them admirably for uses such as manufacture of weatherproof fabrics, non-stick coatings for cooking utensils, chemical-resistant tubing, inert medical materials and cable insulation. Fiering (1994) estimated global production of fluoropolymers to be 40,000 t in 1988, and Jordan and Frank (1999) cited contemporary annual consumption in Europe as 16,000 t. Ellis *et al.* (2001b) suggested that sales indicated a rise of 220% from 1988 to 1997, with a projected annual increase of 7%. Some is recycled but most of the waste, such as worn-out fabrics and utensils, ends up in a landfill or is incinerated, but there do not seem to be any quantitative estimates

Table 9.1. Examples of organofluorine compounds that are in common use.

Category	Compound	Uses
Pharmaceuticals	Isoflurane $CF_3CHClOCHF_2$	Inhalation anaesthetic
	Efavirenz $C_{14}H_9ClF_3NO_2$	Treatment of HIV and AIDS. An HIV-1-specific, non-nucleoside, reverse transcriptase inhibitor
	Ofloxacin $C_{18}H_{20}FN_3O_4$	Antibacterial activity due to inhibition of DNA gyrase, which assists in DNA replication, repair, deactivation and transcription
	Prozac – fluoxetine hydrochloride $C_{17}H_{18}F_3NO.HCl$	Antidepressive, antiobsessive, anticompulsive and antibulimic action. Mechanism unknown
Agrochemicals	Trifluralin $C_{13}H_{16}F_3N_3O_4$	Pre-emergence selective aromatic herbicide. Disrupts cell division
	Compound 1080, sodium monofluoroacetate $CH_2FCOONa$	Fluorinated aliphatic pesticide, control of mammalian pests. Inhibition of aconitate hydratase and citrate transport
Miscellaneous compounds with industrial, commercial and domestic uses	Perfluoro-octane sulphonyl fluoride $C_8F_{17}SO_2F$	Feed chemical for perfluoroalkylsulphonates, used as surfactants and surface protectants (carpets, leather, paper, etc.)
	Perfluoro-octane sulphonate $C_8F_{17}SO_3^-$	Surfactant
	Chlorofluorocarbons (CFCs)	Aerosol propellants, foam-blowing agents, refrigerants
	Poyltetrafluoroethylene (PTFE)	Weatherproof fabrics, non-stick cookware, insulation, inert medical tubing/equipment

HIV, human immunodeficiency virus; AIDS, acquired immune deficiency syndrome.

of the global rates of disposal by each route. However, Jordan and Frank (1999) estimated that 2500 t of fluoropolymer waste is incinerated in Europe each year.

These polymers are so resistant to breakdown that in landfill they can be considered to last indefinitely, with no significant environmental effects. On the other hand, excessive heating or incineration leads to the release of a spectrum of fluorine compounds, some of which may react further in the atmosphere and disperse over large distances. The earliest research on this subject was in the 1950s in relation to the potential toxicity of fumes from PTFE. Since then, there have been many publications reporting combustion tests and studies of the toxicity of the products. A recent study by Sultan et al. (2000) will illustrate the nature of the combustion products of one polymer, FEP. It is used as cable and plenum insulation and it starts to decompose above

about 400°C in air. The composition of the fire gases in non-ventilated conditions at 575°C is shown in Table 9.2. The predominant fluorine compound in this test was HF but there was a sizeable fraction of organo-fluorides, especially trifluoroacetic acid (TFA). Release of fluorinated compounds by fires makes a very small contribution to the atmospheric burden of organofluorine compounds but it does raise the question of the fate of fluorinated polymers that are routinely exposed to high temperatures in applications such as ovens and engines or that are incinerated at the end of their life. This was investigated recently by Ellis et al. (2001b), who exposed fluoropolymers to conditions that were similar to those found in burning domestic waste (isothermal 360°C and heated to 500°C in an air stream). The main gases that the authors determined were tetrafluoroethene, hexafluoropropene (HFP, 10.8%) and cyclo-octafluorobutane,

Table 9.2. The composition of the fire gases emitted from fluorinated ethylene propylene copolymer (FEP) in non-ventilated conditions at 575°C. (Data from Sultan *et al.* (2000), reprinted with permission of the 49th International Cable and Wire Symposium.)

Product	g product/g polymer
CO_2	0.49
CO	0.034
HF	0.49
Perfluoroaliphatics	Estimated 0.23
Fluoro-oxygenated compounds	Estimated 0.14
Airborne particles	0.009
Residue	0
Total	1.393

together with several other compounds, including TFA and traces of monofluoro-acetic acid (MFA) and difluoroacetic acid. Some products, such as COF_2, were probably not captured by their condensation system so they were not estimated. The authors then produced a model for the reactions of these compounds in the atmosphere, suggesting that thermal decomposition of fluoropolymers may lead to significant production of haloacids, notably TFA and chlorodifluoroacetic acid, contributing to the concentrations that have been observed in urban areas. They also suggested that chlorofluoropolymers may lead to production of chlorofluorocarbons (CFCs), so these non-toxic, inert fluoropolymers may be leading to the widespread dispersal of a large number of fluorinated compounds in the atmosphere. While it is clear that thermolysis does lead to the release of a range of organofluorides, the model has not been independently verified and the environmental significance of most of the compounds is not clear.

Surfactants

Many common materials, such as carpets, upholstery, leather goods and paper, are treated with sulphonate-based fluorochemical protectants, which decrease soiling by repelling oil and water. The same group of chemicals is used in food packaging, specialized fire-extinguishing foams and many other products. Total annual production in 2000 was around 5000 t. The main manufacturer, 3M, uses an electrochemical process in which an organic feedstock is fluorinated, replacing the hydrogen atoms on the carbon skeleton with fluorine atoms:

$$C_8H_{17}SO_2F + 17HF \rightarrow C_8F_{17}SO_2F + 17H_2$$

1-octanesulphonyl fluoride feedstock → perfluoro-octanesulphonyl fluoride (POSF)

POSF is then used as the building block for a large and diverse family of compounds that have many different end uses. Some of these fluorochemicals are released into the environment at the point of manufacture (Hansen *et al.*, 2002) and during product treatment. Some are dispersed by user activities, such as vacuum-cleaning carpets, but most probably ends up in a landfill or is incinerated after use. In addition, they have low volatility and they are mostly used in urban/periurban areas, so widespread dispersal would seem to be unlikely. The strength of the carbon–fluorine bonds makes the products environmentally very stable but all POSF-derived fluoro-chemicals have the potential to degrade chemically or metabolically to perfluoro-octane sulphonate ($C_8F_{17}SO_3^-$, PFOS). This compound is also released directly into the environment, because it is used as a surfactant and it is present in relatively small amounts in other products. A key feature of this compound is that it does not degrade any further, so it is very persistent in the environment (Remde and Debus, 1996; Key *et al.*, 1997; Olsen *et al.*, 1999).

Just over 30 years ago, Taves (1968) suggested that human serum contained measurable concentrations of organic fluorine. This was based on the fact that there appeared to be two forms of fluorine in serum, one exchangeable with radioactive fluorine (i.e. inorganic) and the other not. The non-exchangeable form was released as fluoride anions only by combustion, so it was considered to contain carbon–fluorine bonds. Several similar reports in the 1970s and 1980s (cited in Belisle, 1981) confirmed that there was a form of fluoride in the blood of the general population that was released only by ashing, but the compounds were not characterized any further. In 1976, the use of nuclear magnetic resonance (NMR) spectroscopy led to the tentative identification of perfluoro-octanoic acid (PFOA) in serum (Guy et al., 1976). The authors speculated that the source was manufactured fluorochemicals, but Belisle (1981) considered that it was more likely to be natural, largely because there were measurable concentrations in samples from a rural Chinese population. Improvements in analytical techniques, especially in the last decade, have since allowed better characterization of the compounds, and it is now thought that the main organofluorine compound is PFOS. Olsen et al. (1999) reported concentrations of PFOS in the serum of workers in US manufacturing industry in 1995 from 0 to 12.8 mg/l (mean = 2.19, n = 178) and in 1997 they ranged from 0.10 to 9.93 mg/l (mean = 1.75, n = 149). More recently, it has been shown that PFOS, PFOA, perfluorohexanesulphonate (PFHS) and perfluorooctane sulphonylamide (PFOSA) are detectable in the general population of the USA. Using serum samples of 65 nonindustrially exposed humans, Hansen et al.

(2001) found a mean PFOS concentration of 28.4 ng/ml (standard deviation 13.6). The other compounds were present in much lower concentrations and were not detectable in some samples. Concern about sulphonyl-based fluoro-compounds grew in the late 1990s and 3M launched a major research effort, which included a survey of concentrations of PFOS and related compounds in animal tissues (Giesy and Kannan, 2001, 2002; Giesy et al., 2001; Kannan et al., 2001a,b, 2002a,b). As a result, it is clear that several perfluorinated compounds are present in animal tissues collected not only from industrialized regions but also from remote areas (Tables 9.3 and 9.4). The concentrations of PFOS range from the detection limit (1 ng/g wet tissue) up to several thousand ng/g. In general, the lowest concentrations are in the more remote regions, but are detectable even in the Arctic, and the highest are in industrial/periurban regions. Analysis of predators and prey suggests that PFOS concentrations are magnified in the food chain. The widespread occurrence of PFOS and related compounds raises important questions about their sources in the general environment, the routes of dispersal and the potential effects on the general human population and other organisms.

Table 9.3. Examples of the concentrations of perfluoro-octane sulphonate (PFOS, $C_8F_{17}SO_3^-$) in various animal tissues (ng/g wet tissue). (Reprinted with permission from Giesy J.P. and Kannan, K. (2001) *Environmental Science and Technology* 35, 1339–1342. © 2001, American Chemical Society.)

Group	Species	Location	Tissue	n	Range of concentrations
Aquatic mammals	Ringed seal	Canadian Arctic	Plasma	24	< 3–12
		Norwegian Arctic	Plasma	18	5–14
		Baltic Sea	Plasma	18	16–230
	California sea lion	California	Liver	6	< 35–49
	Mink	Mid-western USA	Liver	18	970–3680
Birds	Polar skua	Antarctica	Plasma	2	< 1–1.4
	Black-tailed gull	Hokkaido, Japan	Plasma	24	2–12
	Cormorant	Italy	Liver	12	33–470
	Bald eagle	Mid-western USA	Plasma	26	1–2570
Fish	Yellow-fin tuna	North Pacific	Liver	12	< 7
	Blue-fin tuna	Mediterranean	Liver	8	21–87
	Chinook salmon	Michigan waters, USA	Liver	6	33–170

Table 9.4. The concentrations of perfluoro-octane sulphonate (PFOS), perfluoro-octane sulphonamide (FOSA), perfluorohexane sulphonate (PFHxS) and perfluoro-octanate (PFOA) in mink liver from Illinois and Massachusetts (ng/g wet tissue). (Reprinted with permission from Kannan, K. *et al.* (2002a), *Environmental Science and Technology* 36, 2566–2571. © 2002, American Chemical Society.)

State	n		PFOS	FOSA	PFHxS	PFOA
Illinois	65	% detects	100	33.8	17	8
		Range or maximum	47–4870	< 37–590	< 7–85	24
Massachusetts	31	% detects	100	12.9	16	58
		Range or maximum	20–1100	170	< 4.5–12	< 4.5–27

The sources and routes of transport of PFOS remain obscure, but it is thought that more volatile precursors reach remote locations by atmospheric or aquatic transport processes and then they are metabolized to PFOS (Giesy and Kannan, 2002). Available information on the toxicity of PFOS and related compounds was summarized by Giesy and Kannan (2002) and DePierre (2002) reviewed literature on the effects of perfluoro-fatty acids on rodents. The evidence, based on surveillance of industrial workers and experiments with laboratory animals, suggests that concentrations of PFOS are currently less than required to cause significant effects but the data are inadequate for assessing the risk to the wide spectrum of wild animal species, and there have been no studies of multicompound toxicity or of effects on animals that are under stress. Giesy and Kannan (2002) concluded that 'knowledge of the critical mechanisms of toxic effects is needed to select appropriate endpoints and bio-markers of functional exposure and to assess complex PFC [perfluoro-chemical] mixtures and their relationship to one another and to other environmental residues'.

As a result of data supplied to the US Environmental Protection Agency (EPA) by 3M, the EPA issued a press release (16 May 2000) stating that:

Following negotiations between EPA and 3M, the company today announced that it will voluntarily phase out and find substitutes for perfluorooctanyl sulfonate (PFOS) chemistry used to produce a range of products, including some of their Scotchgard lines. 3M data supplied to EPA indicated that these chemicals are very persistent in the environment, have a strong tendency to accumulate in human and animal tissues and could potentially pose a risk to human health and the environment over the long term. EPA supports the company's plans to phase out and develop substitutes by year's end for the production of their involved products.

The EPA informed other agencies and governments of the potential dangers, and there are signs that action to decrease exposure to PFOS products is being taken. For example, the Danish EPA (Vejrup and Lindblom, 2002) undertook a survey of the occurrence of several compounds in fabric/leather-impregnating agents and wax and floor polishes. They reported that several retailers and manufacturers have substituted other compounds for PFOS products or removed them from the market. The latest development (2003) is that the EPA has ordered companies to conduct more research.

Anaesthetics

Fluorinated general anaesthetics have been in medical and veterinary use for decades and they have been subjected to much more research than most other organofluorides. The fluorinated anaesthetics methoxyflurane, halothane, isoflurane and desflurane have different degrees of fluorine substitution (Table 9.5) and this affects their metabolism and toxicity. Increasing fluorine substitution leads to a reduction in metabolism, lower incidence of toxicity and less defluorination (Park *et al.*, 2001). The

Table 9.5. Examples of common fluorinated anaesthetics, the extent to which they are metabolized, the products of metabolism and the waste gases emitted into the atmosphere. (Data from Jordan and Frank (1999) and Park *et al.* (2001). Park, B.K., Kitteringham N.R. and O'Neill, P.M. (2001) Metabolism of fluorine-containing drugs. *Annual Reviews of Pharmacology and Toxicology* 41, 443–470, with permission.)

Compound	Formula	% of dose metabolized	Main metabolic products	Products of waste gas in the atmosphere	Toxicity of the anaesthetic
Methoxyflurane	$CHCl_2$-CF_2-O-CH_3	Up to 50	Oxalic acid and inorganic fluoride	?	Nephrotoxicity
Halothane	CF_3-CHBrCl	20–50	CF_2 = CHCl Trifluoroacetyl chloride, trifluoroacetic acid	Trifluoroacetic acid	Reversible and irreversible hepatotoxicity
Isoflurane	CF_3-CHCl-O-CF_3H	< 1	Trifluoroacetic acid, inorganic fluoride	Trifluoroacetic acid	Rare hepatotoxicity
Desflurane	CF_3-CHF-O-CF_2H	< 1	Resistant to metabolism and exhaled	Trifluoroacetic acid	None reported

oldest of the flurane anaesthetics, methoxyflurane ($CHCl_2$-CF_2-O-CH_3), was widely used in the 1960s. Up to 50% of the dose is metabolized by the liver and kidneys, and the rest is exhaled to the atmosphere unchanged. The products of metabolism are oxalic acid and inorganic fluoride ions, demonstrating that the body is capable of defluorinating difluoromethyl groups, but the chemical fate of waste methoxyflurane in the atmosphere does not appear to have been investigated.

Halothane (CF_3-CHClBr), with a single trifluoromethyl group, has also largely been superseded because of liver toxicity caused by its metabolism. About 50–80% of inhaled halothane is eliminated unchanged by the lungs and is oxidized in the atmosphere to TFA (Jordan and Frank, 1999). The remainder is metabolized by two pathways mediated by the cytochrome P450 system but it is not defluorinated. The metabolic products are the hydrochlorofluorocarbon (HCFC) CF_2 = CHCl and trifluoroacetylchloride (CF_3COCl), which binds to proteins or is converted to TFA. The latter is excreted in urine. Halothane hepatitis is thought to be an immunological phenomenon initiated by trifluoroacetylchloride.

In contrast, isoflurane (CF_3-CHCl-O-CF_3H) and desflurane (CF_3-CHF-O-CF_2H)

have greater degrees of fluorine substitution, the % of the dose that is metabolized is very low and the incidence of toxicity is much lower. Isoflurane is metabolized to a small extent and the products are TFA and inorganic fluoride. In both cases, the exhaled waste gases are transformed in the atmosphere to TFA. These examples show the medical advantage of increasing fluorine substitution – lower rates of metabolism and toxicity – but also that there is greater release of exhaled gases and TFA into the environment.

Agrochemicals

Fluorinated agrochemicals include aliphatic and aromatic compounds with different degrees of fluorine substitution, although Key *et al.* (1997) stated that most are trifluoromethyl-substituted aromatic compounds. Key *et al.* estimated that 9% of all agrochemicals were fluorinated and that the proportion is probably increasing. Roberts (1999) and Roberts and Hutson (1999) give outlines of the metabolism of most of the agrochemicals that are currently in use. They possess a huge range of chemical properties but, unlike the previous compounds, they are all designed to be toxic to a suite of target organisms. We shall use a rodenticide, an insecticide and a herbicide to illustrate the environmental significance of the degree of fluorine substitution and the difference between aliphatic and aromatic compounds.

Aliphatic pesticides – sodium monofluoroacetate and fluoroacetamide

Probably the best researched fluorinated aliphatic pesticide is sodium monofluoroacetate (NaMFA or 1080). As discussed in the previous chapter, it is a natural product but it is also manufactured and used in some countries as a pesticide, principally against mammals. A related compound, fluoroacetamide (FCH_3CONH_2), has also been widely used. As previously mentioned, MFA was first synthesized by Swarts in Belgium in 1896 and it is said to have been patented as a moth-proofing agent in Germany in 1930 (US EPA, 1995d). In the same decade, Polish scientists working on chemical warfare agents discovered that methylfluoroacetate was toxic to rabbits. Simultaneously, Schrader was working in Germany on fluoroacetate in relation to the development of new pesticides (Timperley, 2000). In 1941 one of the Polish workers, Sporzynski, fled to England, where he told British intelligence about the Polish work on methylfluoroacetate (Timperley, 2000). Sporzynski's information led to a team at Cambridge working secretly, investigating the synthesis and toxicity of fluoroacetates and related compounds (McCombie and Saunders, 1946). Among the compounds they synthesized were fluoroacetamide and NaMFA. The method that they devised formed the basis for later commercial production. At the end of the war, Schrader reported on his work to British scientists (David, 1950) and both MFA and fluoroacetamide began to be used in Britain, principally as rodenticides in areas such as warehouses, that could be sealed. There is some indication that MFA was considered for use as an insecticide, because David (1950) reported experiments where NaMFA was applied to aphids by contact and by absorption via plant roots. Although he showed it was effective as a contact insecticide, he cautioned that it could not be used with the 'present state of knowledge' because its hazards were completely unknown.

Simultaneously with the work in Britain, MFA was investigated in the USA as a result of a drive to find rodenticides because of shortages brought about by the Second World War. Over 1000 compounds were tested and among them sodium fluoroacetate had the invoice number 1080 and fluoroacetamide was 1081. These codes have been used ever since. When 1080's properties were discovered, it was regarded as a very important finding for the nation. Kalmbach (1945) said that 'It is reasonably certain that the discovery of 1080 assures this nation of a highly effective economic poison which can not be denied this country through any future interruptions of world trade'. By the time Kalmbach published his paper, 1080 was already in use for killing rodents in several states, so it was put into commercial use very quickly after its discovery. Some data on the toxicity of 1080 are shown in Table 9.6. It is notable that some mammals, such as coyotes, are much more sensitive than most birds or fish. However, the median lethal dose (LD_{50}) is only one component that determines the susceptibility of animals to 1080 (McIlroy, 1994). Body weight and the ability of populations to recover after loss are both of major importance. McIlroy (1994) cites two bird species, the silver-eye and Australian magpie, as having similar sensitivity in terms of LD_{50} (9.25 and 9.5 mg/kg body weight, respectively), but the large difference in body weight means that the silver-eye would have to consume only 0.13 mg to reach the LD_{50} while the magpie would need 3.18 mg. In addition, McIlroy (1994) and others have commented that species with poor reproductive and dispersal capacities are at greater risk than those with the opposite traits, and the former are often those in need of conservation. In contrast, one of the characteristics of pests is their high reproductive rate and they can recover quickly from poison operations, resulting in the need

Table 9.6. Toxicity of compound 1080 (sodium monofluoroacetate) to North American species (data from US EPA, 1995d).

Common name	Binomial	LD_{50} (mg/kg)
Birds		
Mallard duck	*Anas platyrhynchos*	9.1
Chukar partridge	*Alectoris graeca*	3.6
Ring-necked pheasant	*Phasianus colchicus*	6.4
Widgeon	*Mareca americana*	3.0
Golden eagle	*Aquila chrysaetos*	5.0
Black vulture	*Cartharista urubu*	15.0
Black-billed magpie	*Pica pica*	1.0–2.3
Mammals		
Coyote	*Canis latrans*	0.01
Cotton rat	*Sigmodon hispidus*	0.1
Deer mouse	*Peromyces* sp.	4.0
Raccoon	*Procyon lotor*	1.1
Opossum	*Didelphis virginiana*	41.6
Skunk	*Mephitus mephitus*	1.0
Fish and aquatic invertebrates		
Rainbow trout	*Oncorhynchus mykiss*	54 (96 h LC_{50} (mg/l))
Blue-gill sunfish	*Lepomis macrochirus*	> 970 (96 h LC_{50} (mg/l))
Daphnid	*Daphnia magna*	350 (48 h LC_{50} (mg/l))

LD_{50}, dose in mg/kg body weight killing 50% of test animals; LC_{50}, concentration that kills approximately 50% of test organisms.

for continued maintenance treatment. As a result, field studies indicate that there are often differences between the potential and actual susceptibility of species to 1080 (McIlroy, 1994).

The use of 1080 and its environmental impact have changed considerably since the 1940s. Currently, approval and regulation vary greatly from country to country. There are few or no data available on amounts released into the environment for most countries, but 1080 is used in the largest quantities in New Zealand and Australia, with a very small amount being released in the USA. In most other countries, the use of 1080 is severely limited or banned. However, even where 1080 is banned, there is still some illegal use and this has resulted in human fatalities (Chi *et al.*, 1996, 1999; C. Chi, personal communication).

A brief history of the use of 1080 and its regulation in the USA is given in Fagerstone *et al.* (1994) and US EPA (1995d). In the last

two decades, there have been numerous changes to US state and federal registration and currently its use is allowed only in livestock protection collars (LPCs), which greatly restrict its release into the environment. These are rubber neck collars for sheep and goats that contain enough 1080 to kill a coyote when ruptured by the teeth. The reregistration process demanded a thorough assessment of all aspects of collar use, including risks to non-target animals (US EPA, 1995d). A strong case is presented for there being minimal effects on non-target animals, including predators and scavengers, but there are specific areas of the USA where collars may not be used because of potential risks to endangered species. The US regulations state that, when predation is anticipated, up to 20 collars may be used in fenced pastures up to 100 acres (40 ha) in size, up to 50 in 101–640 acres (21–260 ha) and up to 100 in 641–10,000 acres (261–4046 ha). A drawback to this regulation is that the larger the area, the more difficult it is to maintain oversight of the collars and the fate of the 1080. However, an estimate of the typical use of LPCs given in US EPA (1995d) is about 800–900 per year, which contain in total less than 1 lb (0.45 kg) of 1080, so the amount is minute in comparison with that used in New Zealand and Australia. The US EPA (1995d) also gave examples of the degree of success of collars and their fate. Very few of the collars are punctured and successful, as two examples cited in US EPA (1995d) illustrate:

> From 1981 to 1983, the New Mexico Department of Agriculture conducted an experimental field program evaluating the efficacy and safety of the toxic collar. A total of 330 collars were used over approximately 1,000 days. Twelve collared lambs were attacked, but only 5 collars were actually punctured by predators (1.5%). A total of 18 collars were accidentally punctured while 21 collars were lost. Three predators were found dead, two coyote and one bobcat (*Lynx rufus*). The only non-target animal believed to have been poisoned during the study was a skunk. The results of this study suggest that exposure of non-target species by feeding on the coyote carcass or the neck area of the collared livestock was low and did not result in any significant adverse effects . . .
>
> During three years of monitoring in Wyoming, Montana, Texas, and New Mexico only 294 out of 2257 collars (13%) were punctured and had their contents released . . . What is even more significant is that only 108 of the 294 collars punctured, or 36.7% (4.8% of the total number of collars), were punctured by coyotes.

The situation in Australia and New Zealand is quite different because both countries have large-scale agricultural and conservation problems caused by introduced mammals. Much more 1080 is used. Settlers took European animals with them for food and game and as reminders of home. Many have become serious pests, such as the rabbit (*Oryctolagus cuniculus*). The magnitude of the rabbit problem in Australia and New Zealand is well known in other countries, but it is perhaps less appreciated that there are several species that are also important

pests, such as the red fox (*Vulpes vulpes*), red deer (*Cervus elaphus scoticus*) and goat (*Cervus hiscus*). New Zealand has a particular problem with the brush-tail possum (*Trichosurus vulpecula*). This species was introduced, not from Europe, but from Australia from 1837 onwards in order to establish a fur trade. In the absence of natural enemies, it has become a most destructive pest, consuming and threatening the native flora and bird populations (by eating eggs and chicks). It also acts as a vector for bovine tuberculosis. At its peak, the population was estimated at around 70 million, but now it is probably around 50 million. The sheer size of the areas of Australia and New Zealand affected by these pests and often their inaccessibility limit the choice of control systems, so poisoning is usually considered to be the most viable option. Several poisons have been used in both countries and usually control strategies combine poisoning with other methods, such as trapping and shooting. Recently in New Zealand there has been attempt to increase sales of possum fur marketed as 'New Zealand mink'.

Compound 1080 was first used in both countries in the 1950s for rabbit control. Since then it has been used for other species, regulations have become stricter and baiting methods have been improved to increase effectiveness and decrease deaths of non-target species (for information about New Zealand, see Livingstone, 1994; Eason *et al.*, 1999). There is no doubt about the effectiveness of 1080, as shown by the data on rabbit populations (Table 9.7). Livingstone (1994) gave the following estimate of the amounts of 1080 that were to be used in 1993/94:

> approximately 5,000 tonnes of 1080 carrot and 1,200 tonnes of cereal based baits will be sown over 350,000 ha, mostly on Crown Estate, for Tb related possum control. Overall, approximately 7.5 g of active 1080 is applied per ha. An additional 40 tonnes of 1080 paste will be used as ground bait for Tb-related control of possums over 1 million ha of largely arable land. The Department of Conservation will also apply 400 tonnes of cereal-based bait over 82,000 ha to protect conservation values

Table 9.7. Surveys of rabbit numbers and % bare ground before and after a major 1080 poisoning programme (data from Ross, and Manaaki Whenua Landcare Research, cited by Livingstone, 1994). The use of 1080 in new Zealand. In: Seawright, A.A., Eason, C.T. (eds) *Proceedings of the Science Workshop on 1080.* The Royal Society of New Zealand Miscellaneous series 28. SIR Publishing, Wellington, pp. 1–9.)

	Canterbury	Otago
Rabbit counts/km		
Autumn 1990	45	16
Spring 1993	1.5	4
Per cent bare ground		
1990/91	40	35
1992/93	16	10

from possums in primary areas. An additional 90,000 ha will be treated by a variety of methods including 1080 paste.

Obviously the size and importance of the pest control problems in New Zealand are huge and this is matched by the scale of use of 1080. Because of concerns about safety and effects on non-target species, there has been more research on the environmental impact of 1080 in New Zealand than anywhere else.

The advantages of 1080 that led to its widespread use are that it is cheap to manufacture and toxic in very low amounts to the target organisms. It is colourless and odourless, so target animals are unable to detect it, and, because there is a latent period before the animal develops symptoms, they consume larger amounts, giving a high kill rate. Two advantages over other poisons are the fact that 1080 is broken down reasonably rapidly in the environment and it does not accumulate in food chains. Both of these are important because water contamination has been one of the main public concerns in countries where 1080 is used extensively. Largely because of this concern, environmental breakdown of 1080 and the potential for contamination of water have been comprehensively studied in both Australia and New Zealand (see several papers in Seawright and Eason, 1994; Bowman, 1999;

Eason *et al.*, 1999; Twigg and Socha, 2001). The immediate fate of unconsumed 1080 depends on the type of bait and the weather. For example, it is more readily leached by rain from oat- than from carrot-based baits, but in the process it is diluted (Eason *et al.*, 1999). However, control operations are planned for periods of dry weather, which prolongs the field life of the 1080. There may be some microbial defluorination if the bait is moist, but once it is in the soil 1080 breaks down in 1–2 weeks under optimum conditions and several weeks under less amenable conditions (Livingstone, 1994). Eason *et al.* (1999) cited the results of field monitoring programmes undertaken after 40 large-scale possum and one rabbit operation using aerial dispersal of 1080 in New Zealand. Over 800 water samples were collected and no evidence was found of prolonged contamination of surface or groundwaters; 94.7% of samples contained no detectable 1080. Residues were found in 5.3% of samples but in most of these concentrations were below 1 p.p.b. (1 µg/l). The authors concluded that monitoring of waterways showed that there is a negligible risk of human contact with 1080 through this route. In a later study, Bowman (1999) followed the breakdown of 12,000 kg of bait that had been buried in a purpose-built landfill site. *In situ* sampling indicated that the 1080 concentration was less than 10% of the original after 12 months. Low concentrations were detected in shallow-bore groundwaters 5 and 13 m from the site, but none was detected after 10 months.

In Australia, 1080 is considered to have a very specific advantage over other poisons because some of the native fauna has a resistance to it, which the target mammals do not. This means that the risk to native animals is minimal. It is often said that another advantage of 1080 is that it is humane, but this is challenged by opponents to its use because it usually takes a significant time for animals to die and it is claimed that the death is painful. Many reports indicate a period of at least 2–5 h for death to occur. Table 9.8 shows the time to death of sheep after eating different doses of 1080. The average time when sheep were given the minimum lethal dose was 720–960 min. However, these concerns must be balanced against the magnitude of the economic and ecological problems posed by the pests and the effects of alternative methods.

The main disadvantages of 1080, apart from those mentioned above, are the potential risks to non-target organisms – humans, domestic animals, wildlife – and the fact that there is still not a very effective or easily administered antidote (Feldwick *et al.*, 1994; Livingstone, 1994). Where the use of 1080 is well regulated and those regulations are enforced, the evidence indicates, as mentioned above, that there is a negligible risk to humans. The greatest risk is to other animals, notably domestic dogs, cattle and

wildlife. Experience in New Zealand shows that deaths of dogs and livestock are due to aircraft overflying target areas, bait being left in paddocks, animals breaking into paddocks and animals being allowed into treated areas before the poison has detoxified (Livingstone, 1994). Although there are many press reports of non-target deaths in the USA, Australia and New Zealand, the effects on wildlife have been most thoroughly investigated in New Zealand (Spurr, 1994a,b; Eason *et al.*, 1999). Spurr's review (1994b) indicates that very few dead birds were found after most 1080 operations from the early 1950s until 1976/77, when 748 dead birds were found after four operations and birds that were tested contained residues of 1080. This unusual death rate was attributed largely to the use of undyed, raspberry-lured, unscreened carrot baits that were attractive to non-target species. Baits have been dyed to reduce their attractiveness since the early years of use and then subsequently carrot baits have been screened to remove chaff, cereal baits have been used more than carrot and cinnamon has been added as a repellent. Later studies show much lower numbers of dead animals found after 1080 operations (Table 9.9). Spurr's

Table 9.8. The time to death of sheep after eating different doses of 1080 (data from VPC, 2000).

Number of lethal doses	Approximate time to die (min)
50	97
22	120
10	40
5	360
1	720–960

Table 9.9. The number of dead birds found after 1080 poisoning operations for control of brush-tail possum (*Trichosurus vulpecula*) in New Zealand, 1978–1993, using different baits. (Reprinted from Spurr, E.B. (1994b) Review of the impacts on non-target species of sodium monofluoroacetate (1080) in baits used for brushtail possum control in New Zealand. In: Seawright, A.A. and Eason, C.T. (eds) *Proceedings of the Science Workshop on 1080*. The Royal Society of New Zealand Miscellaneous series 28. SIR Publishing, Wellington, pp. 124–133.)

	Carrot bait	Mapua cereal bait	Wanganui No. 7 bait	Waimate RS5 bait
Systematic searches				
No. of operations	3	2	3	3
No. of dead birds	52	6	6	1
Average dead birds per operation	17.3	3	2	0.3
Incidental observations				
No. of operations	24	5	28	2
No. of dead birds	17	1	0	0
Average dead birds per operation	0.7	0.2	0	0
Totals				
No. of operations	27	7	31	5
No. of dead birds	69	7	6	1
Average dead birds per operation	2.6	1	0.2	0.2

(1994b) data for birds found dead after 71 operations between 1978 and 1993 indicate the low numbers and the importance of the bait in reducing non-target deaths. More birds were found dead where carrot bait was used and slightly more after using Mapua cereal bait than with the other two cereal baits. Although the numbers of individuals killed are significant, monitoring led Spurr (1994b) to conclude that there was no evidence that 1080 had detrimental effects on populations of adequately monitored bird species. However, less common species and some other groups (bats, lizards and frogs) have not been adequately monitored. Spurr (1994a) has also reported no effect on a range of invertebrates, but concern about non-target species continues (e.g. Powlesland et al., 2000).

The use of 1080 always raises justifiable concerns because of its toxicity, but it is also very controversial and many conservation organizations are opposed to its use. Apart from the rights and wrongs of using a poison that causes an agonizing death, the case for using it clearly depends on the potential benefits and the risks. On the one hand, the benefits in the USA for coyote control seem to be minimal because of its inefficiency and the small numbers taken (in 2001 the USDA reported 27 killed by 1080 plus two feral dogs, while 4279 were shot), but, in the case of possum control in New Zealand, the agricultural and conservation problems are so great that there is a very strong case for its use until alternatives can be found. The greatest controversy arises when 1080 is used to control native species, especially when the control is not related to conservation. A good example, cited by Stewart et al. (1994), is where 1080 was proposed for control of the native swamp wallaby in part of Victoria, Australia, in order to protect plantations of non-native Pinus radiata trees. The plantations themselves were controversial and attracted public criticism, so adding 1080 and killing a native animal produced a major conflict.

Finally, there is a recent twist to the 1080 story that may affect its future use. The development of resistance to toxins by animal populations is well known, an example being warfarin anticoagulant resistance in rats. Although many Australian animals have evolved resistance to naturally occurring MFA, there has not been any indication that populations of vertebrate pests might develop resistance to 1080. However, Twigg et al. (2002) have compared the sensitivity of rabbit populations with those that were reported for the same populations over 25 years previously. They found that for three out of the four populations the LD_{50} had increased from a range of 0.34–0.46 to 0.744–1.019 mg pure 1080/kg. They concluded that genetic resistance to 1080 is developing in at least some populations of Australian rabbits. They also stated that this has worldwide implications for agricultural protection and wildlife conservation programmes that depend on a 1080-baiting strategy.

Fluoroacetamide, compound 1081, has a long history of use as a rodenticide. It was developed in parallel to 1080 and was used initially because it was thought to be safer for humans than 1080. A trial by Chapman and Phillips (1955) showed that it was effective against rats and the authors considered that 'it is probable that it is safer to handle by operators than fluoroacetate as indicated by lower mammalian toxicity'. Field trials using a 0.25% solution on bran baits gave rodent kills in warehouses and a ship of 70–85%. They did, however, point out that there was no antidote. Fluoroacetamide began to replace MFA in Britain because of the supposed greater safety and it was widely used in closed areas, such as warehouses and ships where there is minimum risk of accidental exposure. It was adopted in the UK as an insecticide after David and Gardiner (1958) showed that it was effective against Aphis fabae (black bean aphid) and that it was as effective as MFA (which had not been recommended because of its toxicity and rapidity of action). It is difficult to believe now, but in 1959 fluoroacetamide came into use for application to crops, such as broad bean (Vicia faba), brassicas, strawberries and sugar beet, provided a minimum of 4 weeks was allowed between application and harvest of edible crops. It was even approved for use with beet destined for animal feed, and it was available as a general insecticide for ornamental plants (roses and chrysanthemums). However, its use was greatly restricted in 1964, probably as a result of the so-called Smarden incident. This occurred when a dairy herd was poisoned by drinking water and eating forage contaminated by fluoroacetamide leaking from a factory in Smarden, Kent (Allcroft and Jones, 1969; Allcroft et al., 1969). After that, the Ministry of Agriculture recommended that it should not be used as an insecticide in agriculture, home gardens or food storage. Products were withdrawn from the market and its use was limited to authorized persons, operating only in ships or sewers. It remained in use in the UK until about 1992, when it was deregistered. Fluoroacetamide has a similar history in other countries and it is now banned in most, including the European Union (EU). It was first registered in the USA in 1972 and was restricted in 1978 because of its acute oral toxicity and then registration was cancelled in 1989 (US EPA, 1995d). It is still discussed as a possible rodenticide in some special cases, such as the control of plague (Poland and Dennis, 1999), but, in general, its use is declining rapidly. It will be further decreased by implementation of the Rotterdam Convention on the Prior Informed Consent Procedure (PIC) (United Nations Environment Programme/Food and Agriculture Organization press release 01/67 dated 11 October 2001, Rome). This is designed to reduce the health and environmental risks associated with certain hazardous chemicals. Fluoroacetamide is one of the chemicals covered by PIC. However, it is disturbing to find that it still appears to be recommended in some countries. For example, the Indian Defence Research and Development Organization still considers that fluoroacetamide is the best weapon for communal rodent control and it has recently filed a patent for a preparation process for fluoroacetamide and included it in the schedule of the Central Insecticide Board for commercial viability.

Aromatic compounds

Most fluorinated agrochemicals are aromatic compounds. Pesticides and herbicides are strictly regulated in most countries and the manufacturers and regulators subject them to a battery of toxicological and environmental assessments. These include defining the sources of exposure, environmental transport, metabolism in test organisms, effects on target and non-target organisms, carcinogenicity, mutagenicity and degradation in the environment. Two illustrative examples are the pesticide Diflubenzuron ($C_{14}H_9ClF_2N_2O_2$) and the herbicide Trifluralin ($C_{13}H_{16}F_3N_3O_4$).

Diflubenzuron is used to control insect pests of many crops, including cotton and tea, and it is also used to control mosquito larvae. It is a chitin-synthesis inhibitor, so it acts as an antimoulting agent. The International Programme on Chemical Safety (IPCS, 1995) states that the most sensitive toxicological effect is that it may cause methaemoglobinaemia and sulphaemoglobinaemia, with a no-observed-effect level of 2 mg/kg body weight/day in rats and dogs. It is metabolized by a range of test animals to produce two major products, 2,6-difluorobenzoic acid and 4-chlorophenylurea, and several minor ones, including (in some species) 4-chloroaniline. The latter is carcinogenic in rats and mice and it has tested positive for mutagenicity. The toxicity of Diflubenzuron to fish, earthworms, various aquatic invertebrates, birds and duckweed is said to be low, so the environmental risk appears to be minimal. In soil, microbial hydrolysis is rapid, producing the same two major products as in animal metabolism and minor amounts of parachloroaniline. The products are stated to be irreversibly bound to soil and are not metabolized any further (IPCS, 1995), so the fluorinated breakdown compounds persist and accumulate in the surface layers of soil. The regulatory authorities assume that these compounds have no further environmental impact.

Trifluralin disrupts plant cell division and acts as a pre-emergence herbicide on crops. Jordan and Frank (1999) cite the production in 1986 as being 50,000 t and Key et al. (1997) state that, in 1995, 6600 t were used on cotton, soybean, maize and wheat in the USA. It has been subjected to the same tests as other pesticides. Based on animal studies, it does not cause genetic damage, birth defects, etc., but some animal studies found it to be carcinogenic, so the US EPA consider it to be a possible human carcinogen (US Department of Agriculture undated fact-sheet). Where the environment is concerned, Trifluralin is very toxic to fish and it may bioaccumulate. The lowest concentration that kills approximately 50% of test organisms (LC_{50}) is 58 p.p.b. (µg/l) for blue-gill sunfish, but chronic exposure of minnows to 5.1 p.p.b. affected survival. Clearly there is a potential for environmental harm if the herbicide is not used properly. When released, it is subject to photo-oxidation and in soil it breaks down to produce over 30 known products, most of which bind to soil (Roberts, 1999). There is no evidence of cleavage of the aromatic rings or of defluorination of the trifluoromethyl groups, so the large array of fluorinated compounds will also accumulate in soil. It has been suggested that metabolism of trifluorinated aromatics might lead to the production of TFA (Key *et al.*, 1997), but Jordan and Frank (1999) consider it unlikely. The effects of the breakdown products on soil biota are not known.

Most other fluorinated agrochemicals, and especially aromatic compounds with trifluoromethyl groups, have an environmental fate similar to that of Diflubenzuron and Trifluralin. Although they have been through the standard tests and the immediate hazards are known, knowledge of the nature, fate and toxicity of breakdown stops at the soil. There is no evidence that aromatic rings are cleaved, and trifluoromethyl groups are not defluorinated, so hundreds of persistent compounds are being released into the environment to accumulate in various environmental compartments. Their long-term fate and the effects of these compounds on soil biota and processes are not known.

Halocarbons

CFCs have caused great environmental concern in the last 20 years, so they are the best-known example of the environmental dispersion and effects of organofluorides. Halons are fully saturated compounds containing C, F and Br, and they are mostly used as specialized fire-extinguishing agents. The chemistry for producing CFCs was invented by the Belgian chemist Swarts in the 1890s, when he showed that dry SbF_3 in the presence of $SbCl_5$ causes the substitution of fluorine for chlorine:

$$3 CCl_4 + 2 SbF_3 \xrightarrow{SbCl_5} 3 CCl_2F_2 + 2 SbCl_3$$

However, the industrial use of CFCs began in the early part of the 20th century, when refrigeration was in its infancy and domestic refrigerators were not common. One of the problems at that time was that all of the cooling fluids, such as ammonia, SO_2, methylchloride and butane, were toxic or flammable. There were numerous reports in the USA of deaths due to refrigeration leaks, particularly from methylchloride, and owners often kept their refrigerators outside for safety reasons. Stable, non-corrosive, non-toxic, non-flammable replacements were needed and in 1928 the Frigidaire Division of General Motors asked Thomas Midgley and Albert Henne to investigate the possibilities. In 1930 Midgley and Henne reported that fluorohalo-derivatives of aliphatic hydrocarbons had the appropriate properties and selected dichlorodifluoromethane (CCl_2F_2 or CFC-12). They pointed out that, contrary to what was expected, substitution of fluorine for hydrogen atoms decreased toxicity. Midgley even gave a public demonstration of the non-toxicity and non-inflammability by inhaling a lung-full of CFC vapour and then breathing it out on to a candle flame, which was extinguished. CFCs were eminently successful and over the following decades more and more uses were found for the various compounds (Table 9.10).

Although many of the uses might be classed as non-essential, CFCs played a definitive, positive role in 20th-century history, industry and society. For example, they were used as aerosol propellants for the delivery of insecticides. This use was developed during the Second World War, initially because of the massive loss of troops due to malaria. It was so successful that, by 1945, approximately 40 million dichlorodiphenyl-trichloroethane (DDT) 'bug bombs' were produced (Tedder *et al.*, 1975). Although the adverse effects of both components are now well recognized, the combination of the CFC propellant and DDT saved hundreds of thousands, perhaps millions, of lives. Similarly, air-conditioning allowed areas in hot climates to be inhabitable and become economically viable. CFCs have also had an impact on health care because a significant proportion of the estimated 300 million asthma sufferers depend on metered-dose

Table 9.10. Examples of the uses of chloro-fluorocarbons and halons.

Propellants for aerosol cans – paint sprays, oven cleaner, etc.
Blowing agents for insulating foam and packaging, foam padding in furniture
Fire-extinguishing agents (halons)
Refrigeration and air-conditioning systems
Cleaning solvents
Metered-dose inhalers used by asthma sufferers

inhalers that use CFC propellants. Halon fire extinguishers have been the first choice for protecting works of art, electronics and computer equipment that would be damaged by water. At the peak of production, in 1974, the chemical industry was producing over 800,000 t of CFC-11 and CFC-12 per year (Midgley, 1995). Figure 9.1 shows the annual production of CFC-12 from 1930 onwards.

CFCs are gases under ambient conditions, so they are released into the atmosphere on use or when there is leakage from sealed units. Many common pollutant gases, such as SO_2 and HF, are water soluble and are scavenged from the atmosphere relatively quickly. Some react in the lower atmosphere to form other compounds, such as sulphates, they impact with and are held on surfaces and they are washed out in cloud and rain. However, the stability and lack of water solubility of CFCs mean that they are not readily scavenged and that they are persistent in the atmosphere for many decades. For instance, CFC-12 molecules have an average lifetime of 102 years in the atmosphere (Midgley, 1995). This persistence allows them to enter the stratosphere, where

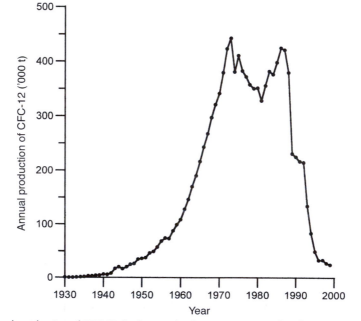

Fig. 9.1. Annual production of CFC-12 (in thousands of tonnes per year) (data from AFEAS at http://www.afeas.org).

there is sufficient short-wave radiant energy to cause breakdown and the release of atoms of chlorine and/or bromine. It is these two elements that cause the loss of ozone:

$$CF_2Cl_2 + h\upsilon \rightarrow Cl + CF_2Cl$$
$$(\lambda < 220 \text{ nm})$$

Ozone is created and destroyed naturally in the stratosphere by the dissociation of oxygen molecules to produce atomic oxygen:

$$O_2 + h\upsilon \rightarrow O + O$$
$$(\lambda < 242 \text{ nm})$$

which reacts with O_2 to form O_3:

$$O + O_2 \rightarrow O_3$$

Destruction is catalysed by nitrogen oxides and other molecules:

$$O_3 + h\upsilon \rightarrow O + O_2$$
$$O + O_3 \rightarrow 2\,O_2$$
$$(\lambda < 200\text{–}300 \text{ nm})$$

This reaction reduces the ultraviolet (UV) light that reaches the earth's surface, protecting plants and animals.

However, chlorine may also start the destruction of ozone:

$$Cl + O_3 \rightarrow ClO + O_2$$

Chlorine monoxide (ClO) starts a catalytic cycle in which it regenerates Cl, which is free to attack another ozone molecule:

$$ClO + O \rightarrow Cl + O_2$$

A single chlorine molecule will, on average, remove between 10,000 and 100,000 molecules of ozone before it randomly moves down into the troposphere or is removed by reaction with molecules such as NO_2 or methane: hence the ozone-depleting effect of CFCs.

There was little concern about the huge amounts of CFCs that were being released until Lovelock (1971) reported detecting CFC-11 in the air in south-west Ireland. He was exploring the possibility of using such gases as tracers of air movement and, ironically, he commented that 'The presence of . . . carbon fluorides in the atmosphere is not in any sense a hazard' – a reflection of the then current view. However, his data led to Du Pont organizing a seminar on the ecology of fluorocarbons for the world's CFC producers in 1972 (Anon., 1990). The invitation said:

> Fluorocarbons are intentionally or accidentally vented to the atmosphere worldwide at a rate approaching one billion pounds per year. These compounds may be either accumulating in the atmosphere or returning to the surface, land or sea in the pure form or as decomposition products. Under any of these alternatives it is prudent that we investigate any effects which the compounds may produce on plants or animals now or in the future.
>
> (Anon., 1990).

The symposium led to a major research programme, under the auspices of the Chemical Manufacturers Association Fluorocarbon Panel, between 1972 and 1990, which cost $26 million, to investigate the fate and impact of CFCs in the environment. The research was undertaken by universities, government agencies and independent laboratories in 14 countries. It played a vital part in understanding the problem of fluorocarbons and ozone depletion.

Shortly after the Fluorocarbon Panel was set up, Molina and Rowland (1974) proposed the theory that the release of chlorine from CFCs in the stratosphere may cause ozone depletion. This was contested by some organizations but several US companies, such as Johnson's Wax, and several states halted the use of CFCs. By the late 1970s, CFCs were banned for almost all propellant uses in the USA, Canada, Norway and Sweden, though use in refrigeration increased. In the EU, production was capped at the then current capacity, with a requirement to reduce aerosol propellants by 30% (Midgley and McCulloch, 1999). There was still no proof that CFCs were causing damage but, in the early 1980s, Farman et al. (1985) of the British Antarctic Survey observed extremely low ozone levels over Antarctica. At first, they thought their instruments were faulty but in 1985 they reported a dramatic loss of ozone in the lower stratosphere. This triggered international concern about the effects of ozone thinning and increased ultraviolet B (UVB) radiation, which eventually led to the signing of the Montreal

Protocol in 1987 by 24 countries (UNEP, 1987). It has now been ratified by 165 countries (Midgley and McCulloch, 1999). The Protocol had the aim of reducing the production of CFCs to 50% of the 1986 level by 1998. Continued research and debate led to the decision to phase out CFCs and halons by 2000. The steep decline in production of CFC-12 after the mid-1980s is evident in Fig. 9.1. Production and consumption of other ozone-depleting substances (e.g. carbon tetrachloride, methyl chloroform) were also covered by the original Protocol and subsequent amendments.

A second, but perhaps less publicized, reason for discontinuing CFCs is that they can act as greenhouse gases, contributing to climate forcing (Table 9.11). The concept of global-warming potential (GWP) was developed to compare the ability of each gas to trap heat in the atmosphere relative to a reference gas, carbon dioxide. As Table 9.11 shows, most CFCs have warming potentials several thousand times greater than CO_2.

The alternatives to CFCs

The planned phase-out of CFCs resulted in a major effort to find alternatives that were effective but which were environmentally safe. Midgley (1995) cited the options

as being: conservation (improved design, recovery and recycling); non-fluorocarbon compounds (e.g. hydrocarbons for foam-blowing); HCFCs; and hydrofluorocarbons (HFCs). It was estimated (cited by Midgley, 1995) that about three-quarters of future demand could be met through conservation and the use of non-fluorocarbon alternatives. The remainder required the use of compounds such as HCFCs and HFCs, which would allow phase-out of CFCs, with existing equipment to be operated for its useful economic life (Midgley, 1995). These two groups of compounds decompose in the lower atmosphere, so they have much shorter lifetimes than CFCs, but because of their ozone-depleting potential, under the Montreal Protocol, HCFCs were to be transitional, with a phase-out date of 2030 (Midgely, 1995). Figure 9.2 shows the predicted decrease in the production of CFC-12 and the production of the transitional HCFC-22 and the replacement HFC-134a over a century. After 2050, HFCs and perfluorinated compounds are predicted to account for about 1.4 and 0.6%, respectively, of the total radiative forcing, so their contribution will be relatively small (McCulloch, 2000).

In order to assess the environmental effects of the alternative compounds, a

Table 9.11. Ozone-depletion (ODP) and global-warming potentials (GWP) of selected halocarbons and their potential replacements. (Data from McCulloch, A. (2000) Halocarbon greenhouse gas emissions during the next century. In: van Ham, J., Baede, A.P.M., Meyer, L.A. and Ybema, R. (eds) *Non-CO$_2$ Greenhouse Gases: Scientific Understanding Control and Implementation*. Kluwer Academic Publishers, The Netherlands, pp. 223–230, with kind permission of Kluwer Academic Publishers.)

Compound	Formula	Principal uses	Atmospheric lifetime (years)	ODP relative to CFC-11	GWP relative to CO_2 at 100 years
To be phased out under Montreal Protocol by 2000					
CFC-11	CCl_3F	Foam blowing propellant	50	1	4000
CFC-12	CCl_2F_2	Refrigerant, propellant	102	1	8500
Halon-1301	$CBrF_3$	Fire extinguisher	65	10	5600
Transitional compounds to be phased out under Montreal Protocol					
HCFC-22	$CHClF_2$	Alternative to CFC-12	13.3	0.055	1700
HCFC-124	$CHClFCF_3$	Alternative to CFC-11	5.9	0.022	480
Potential alternatives, emissions controlled under Kyoto Protocol					
HFC-134a	CH_2FCF_3	Alternative to CFC-12	14.6	0	1300
HFC-227ea	CF_3CHFCF_3	Alternative to CFC-11 and halon-1301	36.5	0	2900

consortium of manufacturers set up the Alternative Fluorocarbons Environmental Acceptability Study (AFEAS). Early in the study it was established that the alternatives would predominantly break down to produce inorganic compounds – CO_2, water and inorganic chlorides and fluorides. Estimates of inorganic fluoride concentrations and deposition rates showed that they were not sufficient to have an environmental effect (WMO, 1989). However, it was also predicted that in the atmosphere some of the alternatives (Table 9.12) would be oxidized to produce trifluoroacetyl halides that would be taken up by clouds and quickly hydrolysed to yield TFA, which would then

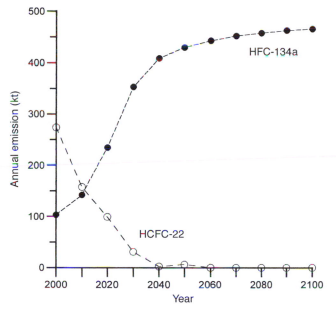

Fig. 9.2. Predicted changes in the production of two substitutes for CFC-12: the transitional HCFC-22 and the replacement HFC-134a over a century. o HCFC-22, ● HFC-134a. (Data from McCulloch, A. (2000) Halocarbon greenhouse gas emissions during the next century. In: van Ham, J., Baede, A.P.M., Meyer, L.A. and Ybema, R. (eds) *Non-CO₂ Greenhouse Gases: Scientific Understanding Control and Implementation*. Kluwer Academic Publishers, The Netherlands, pp. 223–230, with kind permission of Kluwer Academic Publishers.)

Table 9.12. Compounds that were predicted to result in the formation of trifluoroacetic acid (TFA, CF_3COOH) in the atmosphere. Reprinted with permission of CRC Press, from Boutonnet *et al.* (1999) *Human and Ecological Risk Assessment* 5; permission conveyed through Copyright Clearance Center Inc.

Compound	Formula	Approximate atmospheric lifetime (years)	Molar yield of TFA	Comments
Halothane	$CF_3CHClBr$	1.2	0.6	Current global deposition of TFA from
Isoflurane	$CF_3CHClOCHF_2$	5	0.6	anaesthetics estimated 800 t/year
HCFC-123	CF_3CHCl_2	1.4	0.6	All HCFC precursors currently at detection limits in atmosphere. HCFC-123 and 124 to be phased out
HCFC-124	CF_3CHFCl	5.9	1	Current deposition of TFA from HCFC-123 = 760 t/year HCFC-124 = 320 t/year
HFC-134a	CF_3CH_2F	14.6	0.13	Currently 1.2–2.5 parts per trillion by volume in atmosphere Deposition of TFA = 960 t/year
HFC-227ea	CF_3CHFCF_3	36.5	1	Minor usage, not detected in atmosphere Assumed zero deposition

be wet-deposited in precipitation. TFA was known to be stable to oxidation and to have relatively low mammalian toxicity, largely because of research on anaesthetics. The main biological concerns that were expressed at the 1989 AFEAS conference (WMO, 1989) were that TFA might be concentrated in some part of the environment, that it might be toxic to biota and/or that it might be metabolized to produce MFA, which, as already discussed, has very high mammalian toxicity. Partial defluorination was considered unlikely on theoretical grounds and MFA is rapidly broken down by many organisms, so accumulation of MFA seemed extremely unlikely. Nevertheless, a major study was started in 1992 to assess the environmental risk posed by TFA (Boutonnet et al., 1999).

In 1999, Boutonnet et al. presented the environmental risk assessment of TFA. This involved an estimation of the environmental sources of TFA and prediction of concentrations in rainwater. At the time there were only three recognized sources: industrial manufacture, oxidation of anaesthetics and the halocarbon alternatives. TFA is manufactured for use in derivatization, as a catalyst and as a solvent for proteins. About 1000 t of TFA are produced each year, but it is either consumed in the processes or released in very small quantities on a local scale, so it was not considered to be a significant source. It is difficult to assess the amount produced by oxidation of anaesthetics, because of uncertainties about amounts used and the fact that emissions are not measured. The best estimate is that anaesthetics may produce a maximum global deposition of 800 t TFA/year (Boutonnet et al., 1999). Jordan and Frank (1999) calculated that anaesthetics would be expected to produce an average concentration of TFA in rain of less than 5 ng/l. TFA production from fluorocarbon oxidation can be estimated more accurately and Boutonnet et al. (1999) calculated the maximum global deposition as 2800 t/year.

Clearly, there is great regional variation in the emission and oxidation of TFA precursors and in precipitation, but several attempts have been made to predict average concentrations of TFA in rain, and they are reasonably consistent. For example, Jordan and Frank (1999) calculated that HCF-134a should result in an average concentration in the northern hemisphere of 7 ng TFA/l. Similarly, Kotamarthi et al. (1998) calculated that HCFC-123 and HCFC-124 should produce a western European average of 14 ng/l in 1995. For the AFEAS risk assessment, Boutonnet et al. (1999) calculated that the global maximum concentration by the year 2020 would be 100 ng TFA/l. This was adopted as the standard for the risk assessment.

The AFEAS programme included measurements of TFA in the environment to establish a baseline. Because of the low concentrations, accurate, reliable determinations are technically demanding and great care must be taken to avoid contamination. Nevertheless, consistent data have accumulated over the last decade and there is now a reasonable picture of global concentrations (examples in Table 9.13). In the mid-1990s it became clear that TFA was detectable in the air, rain, fresh water and oceans in most parts of the world and that the concentrations were far higher than could be accounted for by CFC replacements and anaesthetics by a very large margin. However, the sources of this 'extra' TFA were unknown (Boutonnet et al., 1999). In their discussion of other sources of TFA, Jordan and Frank (1999) concluded that the contribution from anaesthetics would be to produce an average rain concentration of 5 ng/l, that pesticides are not a relevant source but that thermolysis of perfluorinated polymers might be an important source in industrial regions. The experimental thermolysis by Ellis et al. (2001b) supports the idea that perfluorinated polymers are an important source of TFA in contemporary urban precipitation.

The overall pattern of global TFA in the environment (Table 9.13) is that, in general, concentrations in the air and water are higher in industrial than in remote regions, but it is detectable even in Antarctica. Younger spring/mineral water has higher concentrations than older water, suggesting a recent origin related to industrial processes.

Table 9.13. Examples to demonstrate the range of trifluoroacetate concentrations in the environment. Reprinted with permission from Grimvall *et al.*[©] (1997) American Chemical Society,

Source and location	TFA	Source
Air (pg/m^3)		
Bayreuth, Germany, 1995–1996	10–125, mean = 44	All cited by Jordan and
Pretoria, South Africa, 1996	40	Frank (1999)
Zurich, Switzerland, 1995	65	
Precipitation (ng/l)		
Bayreuth Botanic Garden, Germany, 1995–1996	< 10–410 mean = 110	Cited by Jordan and Frank (1999)
Switzerland, 1995	50–80	
Mace Head, Ireland, 1996	2–93	Grimvall *et al.*, cited by Boutonnet *et al.* (1999)
Queen Maude Land, Antarctica (snow), 1996	2–9	Grimvall *et al.*, cited by Boutonnet *et al.* (1999)
Reno, Nevada, USA, 1994	31–37	Grimvall *et al.*, cited by Boutonnet *et al.* (1999)
Springs and mineral waters (ng/l)		
Various springs, Fichtelgeberge, Bavaria, Germany, 1995	70–320, age of water *c.* 4 years	All cited by Jordan and Frank (1999)
Bayern source, Kondrau, Germany, 1996	23, age of water *c.* 200 years	
Thuringer Wald, Thuringia, Germany, 1996	< 9, age of water *c.* 400 years	
Antonien source, Kondrau, Germany, 1996	< 10, age of water *c.* 700 years	
Rivers (ng/l)		
Rio Tocantins, Brazil, 1996	< 15	All cited by Jordan and
Tweede Tol, South Africa, 1996	< 15	Frank (1999)
Apies River, Pretoria, South Africa, 1996	500	
Jennesej, Russia, 1996	35	
Rhine, Bregenz, Austria, 1995	55	
Rhine, Duisberh-Ruhrort, Germany, 1995	630	
Elbe, Hamburg, Germany, 1995	100	
Elbe, Wittenberg, Germany, 1995	200	
Lakes (ng/l)		
Lough Ahalia, Ireland, 1995	< 10	All cited by Jordan and
Lake Constance, Germany, 1995	60	Frank (1999)
Lake Fichtel, Germany, 1995	70	
Dead Sea, Israel, 1995	6,400	
Pyramid Lake, Nevada, USA	40,000	
Oceans (ng/l)		
N. Atlantic, Mace Head, Ireland, 1995	70	Cited by Jordan and
Atlantic, Isle d'Yeu, France, 1995	250	Frank (1999)
Pacific, Australia, 1995	200	
Changes with depth, Elephant Island, Weddell Sea, South Atlantic, 1999		
Depth (m)		
10	196 ± 6	Frank *et al.* (2002)
100	200 ± 6	
1000	205 ± 6	
2000	210 ± 6	

However, the amount in the oceans is so large that it suggests that there must be a natural source of TFA, so Frank *et al.* (2002) analysed the change in concentrations with depth in remote oceans to clarify this issue. Clearly, if the TFA is recent, there should be a gradient with depth. The data (Table 9.13) showed that in fact there is no gradient and that the mean concentration is relatively high (*c.* 200 ng/l), supporting the idea that the source of this massive reservoir of TFA is ancient, predating industrial production. Frank *et al.* (2002) concluded that TFA is in two major compartments, ocean water on the one hand and precipitation, fresh water and vegetation on the other. They consider that the oceans are a final sink for a natural source whose nature is unknown and that the TFA in the other compartment is mostly of anthropogenic origin.

Movement, concentration and metabolism of TFA in the environment

The bioavailability of TFA to terrestrial organisms would be expected to depend on the extent to which it is adsorbed to soil, so this was investigated in the AFEAS study (Boutonnet *et al.*, 1999). Overall, it was found that mineral soils did not show significant retention of TFA but that there was some retention by highly organic soils. Boutonnet *et al.* (1999) concluded that 'typical soils do not exhibit significant retention of TFA, but that some partitioning to organic-rich soils is observed'. Presumably 'typical' meant mineral soils used for intensive agriculture, but many soils in cooler, temperate climates have a significant organic layer and large areas are covered by peat, so this statement may be misleading. If the observations are generally true, there should be faster movement of TFA into groundwater in agricultural areas and significant retention in less-managed, temperate ecosystems, such as boreal forest. None of the studies investigated the reversibility of soil-binding, so there are still some questions about soil–TFA interactions.

One of the important questions about TFA in view of its resistance to degradation is whether it might accumulate and concentrate in some ecosystems. A study conducted by Tromp *et al.* (1995) suggested that it might accumulate to very high levels in vernal pools (wetlands that dry out and are replenished only by rainfall). They calculated that a wetland receiving rain with 1 mg TFA/l, (a concentration at least 10,000 times greater than occurs in rain) would reach 100 mg/l after 30 years. The basis for the proposal was strongly criticized by Boutonnet *et al.* (1999) on the grounds that it was 'simply a mathematical calculation that has been conducted by adding together the consequences of extreme conditions without regard for the probability that they could occur simultaneously'. Boutonnet *et al.* (1999) concluded that, although some accumulation may occur in aquatic systems, such as vernal pools, accumulation to high concentrations such as those described by Tromp *et al.* (1995) appears to be highly improbable. There has been no other evidence produced to support Tromp *et al.*'s work and attempts to do so have given inconclusive results. For example, Cahill *et al.* (2001) showed that, although TFA concentrations rose as pools dried out, accumulation from year to year was not significant, while Ellis *et al.* (2001a) found that the TFA concentration in pond waters remained approximately constant.

Bioconcentration by plants was examined by Thompson *et al.* (cited in Davison and Pearson, 1997 and Boutonnet *et al.*, 1999), using a range of species grown in hydroponic culture (Table 9.14). The studies showed that TFA is carried by the transpiration stream to the most rapidly transpiring part of leaves, so the concentration factor is variable, depending on which part of the plant is analysed and its age. A major source of variation, and one that is not mentioned by Boutonnet *et al.* (1999), is the ratio between the surface area of the transpiring leaf and its dry weight. This varies by a factor of at least ten times between a grass leaf and a conifer needle. Therefore, while it is safe to conclude that bioconcentration does occur in plants, great caution has to be used in interpreting these factors, because they are so variable.

Table 9.14. Bioconcentration of TFA by plants grown in hydroponic culture. (Data from Thompson *et al.* and Davison and Pearson, 1997).

Species	Duration (days)	Exposure concentration (mg/l)	Tissue concentration (mg/kg dry wt)	Bioconcentration factor
Wheat shoot/leaf	43	1	43	5.4
		5	195	4.9
		10	770	9.6
Soybean leaf	33	1	191	24
	43	5	620	16

Work on the biodegradation and abiotic mineralization of TFA was reviewed in Boutonnet *et al.* (1999). Biodegradation was examined in aerobic and anaerobic bacterial cultures and environmental samples. The studies did not provide evidence of a widespread, environmentally significant biological mechanism for defluorination. One report (Visscher *et al.*, 1994) indicated rapid defluorination of TFA in sediments, but the results could not be repeated on samples from the same sites despite numerous attempts. Boutonnet *et al.* (1999) concluded that TFA will be very long-lived in the environment and that the evidence produced by Visscher *et al.* (1994) must be considered hypothetical until the original result can be confirmed. Since then, Kim *et al.* (2000) have demonstrated degradation over a 90-day period using a continuous-flow anaerobic reactor. This is an important step in understanding the movements and fate of TFA, but much more research is needed.

Two studies have provided evidence that TFA is not readily metabolized or defluorinated by plants. Thompson *et al.* (cited in Boutonnet *et al.*, 1999) exposed sunflower (*Helianthus annuus*) seedlings to 2 μg/l ^{14}C-radiolabelled TFA/l in an aqueous medium. More than 80% of the ^{14}C residues in the leaves were water-extractable and they coeluted with a TFA-spiked leaf extract. Davison and Pearson (1997) analysed the fluoride content of leaves of soybean (*Glycine max*) and wheat (*Triticum aestivum*) that had been exposed to 1.0–10.0 mg NaTFA/l in hydroponic culture for up to 43 days. Some material was used to determine the inorganic fluoride and the rest was fused in sodium carbonate for several hours to cleave C–F bonds and estimate the

TFA content. The inorganic fluoride content was at background level, indicating no significant defluorination, even in leaves containing as much as 770 mg TFA/kg dry weight. Furthermore, the toxicity symptoms produced by high concentrations of NaTFA were different from those produced by monofluoroacetate. Consequently, the AFEAS studies of higher plants do not provide any indication of defluorination of TFA.

Toxicity – effects of TFA on processes and biota

AFEAS targeted key processes and species that it was thought might be affected by TFA (Boutonnet *et al.*, 1999). The processes included carbon and nitrogen fixation, methanogenesis, biogeochemical cycling, acetate metabolism, biodegradation by activated sludge and mutagenesis, while the species included algae, crustaceans, fish, higher plants and mammals. The research allowed estimation of no-observed-effect concentrations (NOECs) for a wide range of organisms and with one exception, the NOECs ranged from about 1 to > 1000 mg of NaTFA/l. The exception was the green alga *Selenastrum capriconutum* (= *Raphidocelis subcapitata*) with a NOEC of 0.12 mg NaTFA/l. However, this is still about 1000 times higher than the highest average concentration of 100 ng/l that has been observed in rain in Europe. The AFEAS report (Boutonnet *et al.*, 1999) also showed that the toxicity of TFA is much lower than di- and mono-fluoroacetate, illustrating once again that increasing fluorination tends to reduce toxicity.

It is particularly important to understand the potential effects on higher plants because they are exposed directly to TFA in the atmosphere and in rain and indirectly via the roots. Atmospheric exposure of yew (*Taxus baccata*) to very high concentrations of TFA, which were increased from 100 to 2100 mg/l over 21 days, produced only slight necrosis of needle tips at 2100 mg/l and it was attributed to the acidity of the condensation water (Freist, 1986, cited in Boutonnet *et al.*, 1999). Foliar exposure to simulated rain was studied by Davison and Pearson (1997), using a spray with 60 µm droplets containing NaTFA. A wide range of crops (e.g. wheat, rice, maize, soybean) and a weed (*Plantago major*) were given prolonged exposure to this aerial treatment, while the roots and soil were protected from contact with the NaTFA. After 3 weeks, there was no effect on height, harvest weight, stomatal conductance or chlorophyll content of any species, even at the highest concentration of 100 mg/l. When the NaTFA was applied to the soil, but not the leaves, simulating 10 mm of rain every 3 days over 44 days, there was no effect at 1, 5 or 10 mg/l, but toxicity symptoms developed at 100 mg/l in soybean. No other species showed any effect. In contrast, when soybean was exposed in hydroponic culture, growth was reduced by prolonged exposure to 5 mg/l (Fig. 9.3) and symptoms were produced in some individual plants at the much lower concentration of 1 mg/l. This

illustrates the importance of the soil in mitigating the effects of the TFA. The difference was probably due to the fact that the soil was an organic-based compost that bound a fraction of the TFA. Binding may mean that plants growing on wet, mineral soils might be more susceptible to TFA than those growing on highly organic media, but the lowest concentration that caused toxicity in hydroponic culture was over 10,000 times higher than the TFA content of rain, so the risk seems to be insignificant.

Several experiments reported in Boutonnet *et al.* (1999) indicated that TFA is transported in the xylem to the leaves, where it concentrates towards the margins in the same way as inorganic fluoride. In a 44-day hydroponics experiment (Davison and Pearson, 1997) using soybean, the concentration of TFA reached a plateau in the first trifoliate leaves after 20 days. Each successive set of leaves reached a higher concentration and then reached a plateau, so that at the final harvest the youngest fully expanded leaves (fourth trifoliates) had about 600 mg TFA/kg dry weight (Fig. 9.3). Symptoms developed only in leaves that were still expanding when the roots were exposed to the TFA, suggesting that the effect was via cell wall expansion, but otherwise the mechanism of toxicity is not known. At 5 mg TFA/l, a significant effect

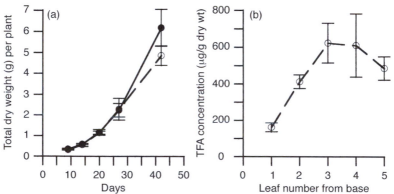

Fig. 9.3. (a) The effect of 5 mg/l NaTFA on the total dry weight of soybean cv. 'Ranson' in hydroponic culture. ○ = controls, ● = 5 mg/l NaTFA. Bars represent 1 standard error. (b) The concentration of TFA (µg/g dry weight) in successive leaves at final harvest (44 days). The fourth and fifth leaves were not fully expanded at 42 days. (Data from Davison and Pearson, 1997.)

on growth began to develop after about 40 days, and about 30–40% of the youngest leaves had distorted, crinkled tips and upper margins (Plate 29). At 1 mg/l there was no effect on growth, but about 25% of the fourth and fifth trifoliate leaves showed slight rounding of the tips and a barely discernible necrosis on the margins. This was used to define the NOEC of 1 mg/l. Analysis of soybean leaves grown in 5 mg NaTFA/l showed that, at the time when the first trifoliate leaves started to develop symptoms, the tissue concentration of TFA was approximately 150 mg NaTFA/kg and, when growth was beginning to be inhibited, it was about 190 mg/kg. The only data with which these can be compared are the analyses of TFA in spruce needles reported by Frank *et al.* (cited in Boutonnet *et al.*, 1999). They ranged from 98 to 195 ng/g, which is six orders of magnitude lower than the toxic threshold in soybean, giving a strong indication that current concentrations do not present a serious risk.

After considering all of the data, the final conclusion of the AFEAS risk assessment (Boutonnet *et al.*, 1999) concluded that TFA was of low toxicity to alga, higher plants, fish and humans, and that there is a 1000-fold difference between the no-effect concentration and the projected environmental levels of TFA from HFCs and HCFCs (0.0001 mg/l). They also drew attention to the lack of information about the origins of the present levels of TFA that are in the environment.

Conclusion

The examples involve just a few of the enormous range of organofluorine compounds that are manufactured and that find their way into the environment. Many have been or still are of vital importance to society: refrigerants and aerosols changed society and saved lives; many pharmaceuticals play a major part in health care and will probably play an even more important part in the future; and many other compounds are part of everyday life. But there is a downside because some of the very properties that are essential to the functioning of organofluorides also allow them to disperse on a global scale and, in some cases, make them recalcitrant, so that they accumulate, often with unknown effects.

References

ACGIH (1996) *Threshold Limit Values for Chemical Substances and Physical Agents; Biological Exposure Indices.* American Conference of Government Industrial Hygienists, Cincinnati, Ohio.

Adams, D.F. (1961) A quantitative study of the limed filter paper technique for fluorine air pollution studies. *International Journal of Air and Water Pollution* 4, 247–255.

Adams, D.F. and Sulzbach, C.W. (1961) Nitrogen deficiency and fluoride susceptibility of bean seedlings. *Science* 133, 1425–1426.

Adams, D., Hendrix, J. and Applegate, H. (1957) Relation among exposure periods, foliar burn, and fluoride content of plants exposed to hydrogen fluoride. *Journal of Agriculture and Food Chemistry* 5, 108–116.

AEI (1999) Alternative Energy Institute Coal Fact Sheet, January 1999. Electronic document: www.altenergy.org/2/nonrenewables/fossil-fuel/facts/coal/coal.htm

Aftab Ahamed, C.A. and Chandrakala, M.V. (1999) Effect of oral administration of sodium fluoride on food and water utilisation in silkworm, *Bombyx mori* L. *Insect Scientific Applications* 19, 193–198.

Agate, J.M., Bell, G.H., Boddie, G.F., Bowler, R.G., Buckele, M., Cheeseman, E.A., Douglas, T.H.J., Druett, H.A., Garrad, J., Hunter, D., Perry, K.M.A., Richardson, J.D. and Weir, J.B. de V. (1949) *Industrial Fluorosis: a Study of the Hazard to Man and Animals near Fort William, Scotland.* Medical Research Council Memorandum No. 22. HM Stationery Office, London.

Alcan Surveillance Committee (1979) *Environmental Effects of Emissions from the Alcan Smelter at Kitimat, B.C.* Ministry of the Environment, Vancouver, Province of British Columbia, 151 pp.

Aldous, J.G. (1963) Nature of metabolites of fluoroacetic acid in bakers yeast. *Biochemical Pharmacology* 12, 627–632.

Allcroft, R. and Jones, J. (1969) Fluoracetamide poisoning. Part I. Toxicity in dairy cattle: clinical history and preliminary investigations. *Veterinary Record* 84, 399–402.

Allcroft, R.A., Burns, K.N. and Hebert, N.C. (1965) *Fluorosis in Cattle. 2. Development and Alleviation: Experimental Studies.* Her Majesty's Stationery Office, London.

Allcroft, R., Salt, F., Peters, R. and Shorthouse, M. (1969) Fluoroacetamide poisoning. Part II. Toxicity in dairy cattle: confirmation of diagnosis. *Veterinary Record* 84, 403–409.

Allmendinger, D.F., Miller, V.L. and Johnson, F. (1950) Control of fluorine scorch of gladiolus with foliar dusts and sprays. *Proceedings. American Society for Horticultural Science* 56, 427–432.

Alonso, C. and Godinho, R. (1992) A evolucão do ar Cubatão. *Quimica Nova* 15, 126–136.

Alstad, D., Edmunds, G.F., Jr and Weinstein, L.H. (1982) Effects of air pollutants on insect populations. *Annual Review of Entomology* 27, 369–384.

AMS (1994) *The Norwegian Aluminium Industry and the Local Environment.* Aluminium industries Miljøsekretariat, Oslo.

Amthor, J.S. (2000) The McCree–de Wit–Penning de Vries–Thornley respiration paradigms: 30 years later. *Annals of Botany* 86, 1–20.

Amundson, R.G. and Weinstein, L.H. (1980) Effects of airborne F on forest ecosystems. In: *Proceedings. Symposium on Effects of Air Pollutants on Mediterranean and Temperate*

Forest Ecosystems. Pacific Southwest Range and Forest Experiment Station, Berkeley, California, pp. 63–78.

An, D., He, Y. and Xu, Q. (1997) Poisoning by coal smoke containing arsenic and fluoride. *Fluoride* 30, 29–32.

Ando, M., Tadano, M., Asanuma, S., Tamura, K., Matsushima, S., Watanabe, T., Kondo, T., Sakurai, S., Ronji, J., Liang, C. and Cao, S. (1998) Health effects of indoor fluoride pollution from coal burning in China. *Environmental Health Perspectives* 106, 239–244.

Ando, M., Tadano, M., Yamamoto, S., Tamura, K., Asanuma, S., Watanabe, T., Kondo, T., Sakurai, S., Ji, R., Liang, C., Chen, X., Hong, Z. and Cao, S. (2001) Health effects of fluoride pollution caused by coal burning. *Science of the Total Environment* 271, 107–116.

Andrews, S.M., Cooke, J.A. and Johnson, M.S. (1989) Distribution of trace element pollutants in a contaminated ecosystem established on metalliferous fluorspar tailings. 3. Fluoride. *Environmental Pollution* 60, 165–179.

Anon. (1990) *Searching the Stratosphere*. Report by the Fluorocarbon Panel of the Chemical Manufacturers Association, 18 pp [no place of publication stated]

Aplin, T.E.H. (1967) Poison plants of Western Australia. The toxic species of the genera *Gastrolobium* and *Oxylobium*. 1. Characteristics of the group. *Journal of Agriculture of Western Australia* 8, 45–52.

Aplin, T.E.H. (1968a) Poison plants in the West Midlands. *Journal of Agriculture of Western Australia* 9, 24–26.

Aplin, T.E.H. (1968b) Poison plants of Western Australia. The toxic species of the genera *Gastrolobium* and *Oxylobium*. Crinkle-leaf poison (*Gastrolobium villosum* Benth.), runner poison (*Gastrolobium ovalifolium* Henfr.), horned poison and Hill River poison (*Gastrolobium polystachyum* Meissn.), wooly poison (*Gastrolobium tomentosum* C.A. Gardn.). *Journal of Agriculture of Western Australia* 9, 69–74.

Aplin, T.E.H. (1968c) Poison plants of Western Australia. The toxic species of the genera Gastrolobium and Oxylobium. Heart-leaf poison (*Gastrolobium bilobum* R. Br.), river poison (*Gastrolobium forrestii* A.J. Ewart), Stirling Range poison (*Gastrolobium velutinum* Lindl.). *Journal of Agriculture of Western Australia* 9, 356–362.

Aplin, T.E.H. (1969a) The toxic species of the genera *Gastrolobium* and *Oxylobium*. Champion Bay poison (*Gastrolobium oxylobioides* Benth.), sandplain poison (*Gastrolobium microcarpum* Meissn.), cluster poison (*Gastrolobium bennettsianum* C.A. Gardn.), Hutt River poison (*Gastrolobium propinquum* C.A. Gardn.), Gilbernine poison (*Gastrolobium rotundifolium* Meissn.). *Journal of Agriculture of Western Australia* 10, 248–257.

Aplin, T.E.H. (1969b) Poison plants of Western Australia. The toxic species of the genera *Gastrolobium* and *Oxylobium*. Berry poison (*Gastrolobium parvifolium* Benth.), spike poison (*Gastrolobium glaucum* C.A. Gardn.}, hook-point poison (*Gastrolobium hamulosum* Meissn.), scale-leaf poison (*Gastrolobium appressum* C.A. Gardn.). *Journal of Agriculture of Western Australia* 10, 517–522.

Aplin, T.E.H. (1971a) Poison plants of Western Australia. The toxic species of the genera *Gastrolobium* and *Oxylobium*. Thickleaf poison (*Gastrolobium crassifolium* Benth.), narrow-leaf poison (*Gastrolobium stenophyllum* Turcz.), mallet poison (*Gastrolobium densifolium* C.A. Gardn.), wall-flower poison (*Gastrolobium grandiflorum* F. Muell.). *Journal of Agriculture of Western Australia* 12, 12–18.

Aplin, T.E.H. (1971b) Poison plants of Western Australia. The toxic species of the genera *Gastrolobium* and *Oxylobium*. Wodjil poison (*Gastrolobium floribundum* S. Moore), breelya or kite-leaf poison (*Gastrolobium laytonii* J. White), Roe's poison (*Oxylobium spectabile* Endl.), granite poison (*Oxylobium graniticum* S. Moore). *Journal of Agriculture of Western Australia* 12, 154–159.

Aplin, T.E.H. (1971c) *Poison Plants of Western Australia*. Gastrolobium *and* Oxylobium. Western Australian Department of Agriculture, Bulletin No. 3772. Perth, Western Australia.

Applegate, H.B. and Adams, D.F. (1960a) Nutritional and water effect on fluoride uptake and respiration in bean seedlings. *Phyton* 14, 111–120.

Applegate, H.B. and Adams, D.F. (1960b) Effect of atmospheric fluoride on respiration of bush beans. *Botanical Gazette* 121, 223–227.

Applegate, H.B., Adams, D.F. and Carriker, R.C. (1960) Effects of aqueous fluoride solutions on respiration of intact bush bean seedlings. *American Journal of Botany* 47, 339–345.

Araya, O., Wittwer, F., Villa, A. and Duncom, C. (1990) Bovine fluorosis following volcanic activity in the southern Andes. *Veterinary Record* 126, 641–642.

Ares, J.O. (1978) Fluoride cycling near a coastal emission source. *Journal of the Air Pollution Control Association* 28, 344–349.

Ares, J.O., Villa, A. and Mondadori, G. (1980) Air pollutant uptake by xerophytic vegetation: fluoride. *Environmental and Experimental Botany* 20, 259–269.

Arndt, U., Nobel, W. and Schweizer, B. (1987) *Bioindikatoren.* Eugen Ulmer GmbH, Stuttgart.

Arndt, U., Flores, F. and Weinstein, L. (1995) *Fluoride Effects on Plants. Diagnosis of Injury in the Vegetation of Brazil.* Editora de Universidade/UFRGS, Porto Alegre, Rio Grand do Sul, Brazil.

Arndt, W. (1970) Concentration changes of free amino acids in plants under the effect of hydrogen fluoride and sulfur dioxide. *Staub-Reinhalt der Luft* 30, 28–32.

Arnesen, A.K.M. (1998) Availability of fluoride to plants grown in contaminated soils. *Plant and Soil* 191, 13–25.

Ashenden, T.W. and Mansfield, T.A. (1977) Influence of wind speed on the sensitivity of rye grass to SO_2. *Journal of Experimental Botany* 28, 729–735.

AWWA (1985) *Fluoride Removal by Means of Activated Alumina.* American Water Works Association Research Foundation, Denver, Colorado, USA.

Ball, M., Howe, D. and Bauer, J. (1999) Measurement and speciation of fluoride emissions at a phosphate fertilizer manufacturing plant using open-path and FTIR and TDL. Report 99–660 for SF Phosphates Company, Wyoming. Electronic document: www.boreal-laser.com/docs/HF%20monitoring%20in%20Phosphates%20manufacture%20-%20Ball%20et%20al.pdf

Ballantyne, D.J. (1972) Fluoride inhibition of the Hill reaction in bean chloroplasts. *Atmospheric Environment* 6, 267–273.

Banks, D., Frengstad, B., Midtgård, A.K., Krog, J.R. and Strand, T. (1998) The chemistry of Norwegian groundwaters: 1. The distribution of radon, major and minor elements in 1604 crystalline bedrock groundwaters. *Science of the Total Environment* 222, 71–91.

Barnard, W.R. and Nordstrom, D.K. (1982) Fluoride in precipitation: 2. Implications for geochemical cycling of fluoride. *Atmospheric Environment* 16, 105–111.

Barnes, J.E. (1958) Georgina poisoning of cattle in the Northern Territory. *Australian Veterinary Journal* 34, 281–290.

Bartlett, B.R. (1951) The action of certain 'inert' dust materials on parasitic Hymenoptera.

Journal of Economic Entomology 44, 891–896.

Bartlett, G.R. and Barron, E.S.G. (1947) Effect of fluoroacetate on enzymes and on tissue metabolism: its use for the study of the oxidative pathway of pyruvate metabolism. *Journal of Biological Chemistry* 170, 67–84.

Baud, C.-A., Lagier, R., Boivin, G. and Boillat, M.-A. (1978) Value of the bone biopsy in the diagnosis of industrial fluorosis. *Virchows Archiv. A Pathological Anatomy and Histology* 80, 283–297.

Baumeister, A., Thompson, C.J. and Nimmi, I.A. (1977) The susceptibility of rainbow trout to fluoroacetate. *Biochemical Society Transactions* 5, 304–306.

Baxter, P., Baubron, J. and Coutinho, R. (1999) Health hazards and disaster potential of ground gas emissions at Furnas volcano. *Journal of Volcanology and Geothermal Research* 92, 95–106.

Béjaoui, M. and Pilet, P.E. (1975) Effets du fluor sur l'absorption de l'oxygène de tissus de Ronce cultivés *in vitro. Comptes Rendus Hebdomadaires des Séances de l'Académie des Sciences, Series D* 280, 1457–1460.

Belisle, J. (1981) Organic fluorine in human serum: natural versus industrial sources. *Science* 212, 1509–1510.

Bell, A.T., Newton, L.G., Everist, S.L. and Legg, J. (1955) *Acacia georginae* poisoning of cattle and sheep. *Australian Veterinary Journal* 31, 249–257.

Bell, M., Largent, E., Ludwig, T., Muhler, J. and Stookey, G. (1970) *Fluorides and Human Health.* Monograph Series No. 59, World Health Organization, Geneva.

Benedict, H.M. and Breen, W.H. (1955) The use of weeds as a means of evaluating vegetation damage caused by air pollution. In: *Proceedings, Third National Air Pollution Symposium.* Pasadena, California, pp. 177–190.

Benedict, H.M., Ross, J.M. and Wade, R.W. (1964) The disposition of atmospheric fluorides by vegetation. *International Journal of Air and Water Pollution* 8, 279–289.

Benedict, H.M., Ross, J.M. and Wade, R.W. (1965) Some responses of vegetation to atmospheric fluorides. *Journal of the Air Pollution Control Association* 15, 253–255.

Bennett, J.H. and Hill, A.C. (1973) Inhibition of apparent photosynthesis by air pollution. *Journal of Environmental Quality* 2, 526–553.

Benson, N.R. (1959) Fluoride injury or soft suture and splitting of peaches. *Proceedings,*

American Society for Horticultural Science 74, 184–198.

Beyer, W.N., Miller, G.W. and Fleming, W.J. (1987) Populations of trap-nesting wasps near a major source of fluoride emissions in western Tennessee. *Proceedings of the Entomological Society of Washington* 89, 478–482.

Bi, S., Liu, F., Chen, F. and Gan, N. (2000) Speciation of aluminum equilibria with kaolinite in acidic natural water. *Journal of Environmental Science and Health* A35, 1849–1857.

Bian, Y. and Wang, J. (1987) Characteristics of fluoride accumulation in mulberry leaves and pollution forecasting. *Journal of Environmental Sciences China* 8, 71–74.

Blakemore, J. (1978) Fluoride deposition to grass swards under field conditions. PhD thesis, University of Newcastle, Newcastle upon Tyne, UK.

Bligny, R., Bisch, A.M., Garrec, J.-P. and Fourcy, A. (1973a) Observations morphologiques et structurales des effets du fluor sur les cires épicuticulaires et sur les chloroplastes des aiguilles de sapin (*Abies alba* Mill.). *Journal de Microscopie (Paris)* 17, 207–214.

Bligny, R., Garrec, J-P. and Fourcy, A. (1973b) Effect of calcium on migration and accumulation of fluorine in *Zea mays*. *Comptes Rendus Hebdomadaires des Séances de l'Académie des Sciences, Série D* 276, 961–964.

Bohne, H. (1964) Fluor-Emission und Tunnelofen. *Staub* 24, 261–265.

Bohne, H. (1970) Fluorides and sulfur oxides as causes of plant damage. *Fluoride Quarterly Reports* 3, 137–142.

Bohne, H. (1972) Klärung eines Rauschadenfalles bei Kiefernbestanden im Ruhrgebiet. *Mitteilungen Forstliche Bundes-Versuchsanstalt* 97, 141–150.

Boillat, M., Boud, C., Lagier, R., Garcia, J., Rey, P., Bang, S., Boivin, C., Demeunisse, D., Goessi, M., Tochon-Danguy, H., Very, J., Burkhardt, P., Voiner, B., Donath, A. and Courvoisier, B. (1979) Fluorose industrielle. Etude multidisciplinaire de 43 ouvriers de l'industrie de l'aluminium. *Schweizeriche. Medizinische Wochenschrift* 109 (Suppl. 8), 5–28.

Bolay, A. and Bovay, E. (1965) Observations sur la sensibilité aux gases fluorés de quelques espèces végétales du Valais. *Phytopathologie Zeitschrift* 53, 289–298.

Bolay, A., Bovay, E., Quinche, J.P. and Zuber, R. (1971a) Teneurs en fluor et en bore des feuilles et des fruits d'arbres fruitiers et des vignes fumés avec certain engrais composés boriqués fluorés. *Revue Suisse de Viticulture et Arboriculture* III, 54–61.

Bolay, A., Bovay, E., Neury, G., Quinche, J.P. and Zuber, R. (1971b) Dégâts causés aux abricots et à d'autres fruits par les composés fluorés. *Revue Suisse de Viticulture et Arboriculture* III, 82–89.

Bong, C.L., Cole, A.L.J., Walker, J.R.L. and Peters, A.J. (1979) Effect of sodium fluoroacetate ('Compound 1080') on the soil microflora. *Soil Biology and Biochemistry* 11, 13–18.

Bong, C.L., Walker, J.R.L. and Peters, J.A. (1980) The effect of fluoroacetate ('Compound 1080') on fluoride upon duckweeds. *New Zealand Journal of Science* 23, 179–183.

Bonte, J. (1982) Effects of air pollutants on flowering and fruiting. In: Unsworth, M.H. and Ormrod, D.P. (eds) *Effects of Gaseous Pollutants on Agriculture and Horticulture.* Butterworths, London, pp. 207–223.

Bonte, J. and Garrec, J.P. (1980) Pollution de l'atmosphère par les composés fluorés et fructification chez *Fragaria* L.: mise en évidence par analyse directe à la microsonde électronique d'une forte accumulation de fluor à la surface des stigmates. *Comptes Rendus Hebdomadaires des Séances de l'Académie des Sciences, Series D: Natural Sciences* 290, 815–818.

Bonte, J., Bonte, C. and de Cormis, L. (1980) Les composés fluorés atmosphériques et la fructification. Approche des méchanismes d'action sur le développement des akènes et du réceptacle de la fraise. In: *Comptes Rendus Académie Agriculture de France*, Séance de 16 Janvier. Paris, pp. 80–89.

Borsdorf, W. (1960) Beiträge zur Fluorschädendiagnistik. 1. Fluorschäden-Weiserpflanzen in der Wildflora. *Phytopathologische Zeitschrift* 38, 309–315.

Bossavy, J. (1965) Échelles de sensibilité au fluor. *Revue Forestière Français* 17, 205–211.

Boulton, I.C., Cooke, J.A. and Johnson, M.S. (1994a) Experimental fluoride accumulation and toxicity in the short-tailed field vole (*Microtus agrestis*). *Journal of Zoology* 234, 409–421.

Boulton, I.C., Cooke, J.A. and Johnson, M.S. (1994b) Age-accumulation of fluoride in an experimental population of short-tailed field voles (*Microtus agrestis* L.). *Science of the Total Environment* 154, 29–37.

Boulton, I.C., Cooke, J.A. and Johnson, M.S. (1994c) Fluoride accumulation and toxicity in wild small mammals. *Environmental Pollution* 85, 161–167.

Boulton, I.C., Cooke, J.A. and Johnson, M.S. (1995) Fluoride accumulation and toxicity in laboratory populations of wild small mammals and white mice. *Journal of Applied Toxicology* 15, 423–431.

Boulton, I.C., Cooke J.A. and Johnson, M.S. (1997) Fluoride-induced lesions in the teeth of the short-tailed field vole (*Microtus agrestis*): a description of the dental pathology. *Journal of Morphology* 232, 155–167.

Boulton, I.C., Cooke, J.A. and Johnson, M.S. (1999) Lesion scoring in field vole teeth: application to the biological monitoring of environmental fluoride contamination. *Environmental Monitoring and Assessment* 55, 409–422.

Bourbon, P. (1967) Analytical problems posed by pollution by fluorine compounds. *Journal of the Air Pollution Control Association* 17, 661–663.

Boutonnet, J.C., Bingham, P., Calamari, D., de Rooij, C., Franklin, J., Kawano, T., Libre, J.-M., McCullough, A., Malinverno, G., Odom, J.M., Rusch, G.M., Smythe, K., Sobolev, I., Thompson, R. and Tiedje, J. (1999) Environmental risk assessment of trifluoroacetic acid. *Human and Ecological Risk Assessment* 5, 59–124.

Bowman, R. (1999) Fate of sodium monofluoacetate (1080) following disposal of pest bait to landfill. *New Zealand Journal of Ecology* 23, 193–197.

Braen, S.N. and Weinstein, L.H. (1985) Uptake of fluoride and aluminum by plants grown in contaminated soils. *Water, Air and Soil Pollution* 24, 215–223.

Brandt, C.J. (1981) Wirkungen von Fluorwasserstoff auf *Lolium multiflorum*. *Landesanstalt für Immissions- und Bodennutzungsschutz der Landes Nordrhein-Westfalen Berichte* 14, 140 pp.

Brandt, J. and Heck, W.W. (1968) Effects of air pollutants on vegetation. In: Stern, A.C. (ed.) *Air Pollution*, Vol. 1: *Air Pollution and Its Effects*, 2nd edn. Academic Press, New York, pp. 401–443.

Brennan, E.G., Leone, I.A. and Daines, R.H. (1950) Fluorine toxicity in tomato as modified by alterations in nitrogen, calcium, and phosphorus nutrition of the plant. *Plant Physiology* 25, 736–747.

Brewer, R.F. (1965) Fluorine. *Agronomy Journal* 9, 1135–1148.

Brewer, R.F., Creveling, R.K., Guillemet, F.B. and Sutherland, F.H. (1960) The effects of hydrogen fluoride on seven citrus varieties. *Proceedings of the American Society for Horticultural Science* 75, 236–243.

Brewer, R.F., Garber, M.J., Guillemet, F.B. and Sutherland, F.H. (1967) The effects of accumulated fluoride on yields and fruit quality of 'Washington' navel oranges. *Proceedings American Society for Horticultural Science* 91, 150–156.

Brewer, R.F., Sutherland, F.H. and Guillemet, F.B. (1969) Application of calcium sprays for the protection of citrus from atmospheric fluorides. *Proceedings American Society for Horticultural Science* 94, 302–304.

Bromenshenk, J.J. (1978) Entomological studies in the vicinity of Colstrip, Montana. In: Preston, E.M. and Lewis, R.A. (eds) *The Bioenvironmental Impact of a Coal-fired Power Plant. Third Interim Report, Colstrip, Montana*, Vol. 5. Report 600/3-78-021, US EPA, Corvallis, Oregon, pp. 140–212.

Bromenshenk, J.J. (1988) Regional monitoring of pollutants with honey bees. In: *Progress in Environmental Specimen Banking*, Vol. 16. NBS Special Publication 740, National Bureau of Standards, Washington, DC, pp. 156–170.

Bromenshenk, J.J. (1992) Site-specific and regional monitoring with honeybees: case study comparisons. In: *Proceedings of an International Symposium on Ecological Indicators, Fort Lauderdale, Florida, 16–19 October 1990*, Vol. 39. Elsevier, London, pp. 689–702.

Bromenshenk, J.J. (1994) Assessing ecological risks in terrestrial ecosystems with honey bees. In: Butterworth, F.M., Corkum, L.D. and Guzman-Rincon, J. (eds) *Biomonitors and Biomarkers as Indicators of Environmental Change. A Handbook*. Plenum Press, New York, pp. 9–30.

Bromenshenk, J., Carlson, S., Simpson, J. and Thomas, J. (1985) Pollution monitoring of Puget Sound with honey bees. *Science* 227, 632–634.

Bromenshenk, J., Cronin, R. and Nugent, J. (1996) Atmospheric pollutants and trace gases. *Journal of Environmental Quality* 25, 868–877.

Broomier, J.H. (1962) Inhibition of nitrogen fixation by fluoroacetate. *Biochemical and Biophysical Research Communications* 7, 53–57.

Buckenham, A.H., Parry, M.A.J. and Wittingham, C.P. (1982) Effects of aerial pollutants on the growth and yield of spring barley. *Annals of Applied Biology* 100, 179–187.

Buckle, F.J., Pattison, F.L.M. and Saunders, B.C. (1949) Toxic fluorine compounds containing the C–F link. Part VI. ω-Fluorocarboxylic acids and derivatives. *Journal of the Chemical Society* 1949, 1471–1479.

Buffa, P. and Peters, R.A. (1949) Formation of citrate *in vivo* induced by fluoroacetate poisoning. *Nature* 163, 194.

Buffa, P., Pasquali-Romchetti, I., Barasa, A. and Godina, G. (1977) Biochemical lesions of respiratory enzymes. II. Early ultrastructural modifications correlated to the biochemical lesion induced by fluoroacetate. *Cell Tissue Research* 183, 1–23.

Bulcraig, W.R. (1977) Fluorides in the vicinity of a glassworks. *Glass Technology* 18, 43–50.

Bunce, H.W.F. (1979) Fluoride emissions and forest growth. *Journal of the Air Pollution Control Association* 29, 642–643.

Bunce, H.W.F. (1983) Fluoride in air, grass and cattle. *Journal of Dairy Science* 68, 1706–1711.

Bunce, H.W.F. (1984) Fluoride emissions and forest survival, growth and regeneration. *Environmental Pollution Series A* 35, 642–643.

Bunce, H.W.F. (1985) Apparent stimulation of tree growth by low ambient levels of fluoride in the atmosphere. *Journal of Air Pollution Control Association* 35, 46–48.

Bunce, H.W.F. (1989) The continuing effects of aluminium smelter emissions on coniferous forest growth. *Forestry* 62, 223–231.

Burkinshaw, S., Cooke, J. and Johnson, M. (2001) Uptake of fluoride, zinc and lead in components of a wildflower meadow mix. Poster presented at the 6th International Conference on the Biogeochemistry of Trace Elements, Guelph, p. 397.

Burns, K.N. and Allcroft, R. (1964) *Fluorosis in Cattle in England and Wales. 1. Occurrence and Effects in Industrial Areas of England and Wales 1954–57*. Animal Disease Surveys Report No. 2, Part 1, Ministry of Agriculture, Fisheries and Food, Her Majesty's Stationery Office, London.

Burton, M.A.S. (1986) *Biological Monitoring of Environmental Contaminants*. Monitoring and Assessment Research Centre, London.

Buse, A. (1986) Fluoride accumulation in invertebrates, near an aluminium reduction plant in Wales. *Environmental Pollution Series A* 41, 199–217.

Cahill, T.M., Thomas, C.M., Schwarzenbach, S.E. and Seiber, J.N. (2001) Accumulation of trifluoroacetate in seasonal wetlands in California. *Environmental Science and Technology* 35, 820–825.

Calabrese, E.J. and Baldwin, L.A. (2002) Defining hormesis. *Belle Newsletter* 10, 25–31.

Camargo, J. (1996) Estimating safe concentrations of fluoride for three species of nearctic freshwater invertebrates: multifactor probit analysis. *Bulletin of Environmental Contamination and Toxicology* 56, 199–217.

Camargo, J.A. (2003) Fluoride toxicity to aquatic organisms: a review. *Chemosphere* 50, 251–264.

Camargo, J.A., de Jalon, D.G., Muñoz, M.J. and Tarazona, J.V. (1992) Sublethal effects of sodium fluoride (NaF) on net-spinning caddisflies (Trichoptera). *Aquatic Insects* 14, 23–30.

Cameron, J.M.R. (1977) Poison plants in western Australia and colonizer problem solving. *Journal of the Royal Society of Western Australia* 59, 71–77.

Cameron, R.S., Ritchie, G.S.P. and Robson, A.D. (1986) Relative toxicities of inorganic aluminium complexes to barley. *Soil Science Society of America* 50, 1231–1236.

Cao, J., Bai, X., Zhao, Y., Liu, J., Zhon, D., Fang, S., Jia, M. and Wu, J. (1996) Fluorosis induced by drinking brick tea. *Fluoride* 29, 139–142.

Cao, J., Zhao, Y. and Liu, J.-W. (1997) Brick tea consumption as the cause of dental fluorosis among children from Mongol, Kazak and Yugu populations in China. *Food and Chemical Toxicology* 35, 827–833.

Cao, J., Zhao, Y. and Liu, J.-W. (1998) Safety evaluation and fluorine concentration of Pu'er brick tea and Bianxao brick tea. *Food and Chemical Toxicology* 36, 1061–1063.

Cao, J., Zhao, Y. and Liu, J.W. (2000) Fluoride in the environment and brick-tea-type fluorosis in Tibet. *Journal of Fluorine Chemistry* 106, 93–97.

Carlson, C.E. and Dewey, J.E. (1971) *Environmental pollution by fluorides in Flathead National Forest and Glacier National Park*. Forest Service, US Department of Agriculture, Missoula, Montana.

Carlson, C., Cousfield, W. and McGregor, M. (1977) The relationship of an insect infestation on lodgepole pine to fluorides emitted from a nearby aluminum plant in Montana. *Fluoride* 10, 14–21.

Carpenter, R. (1969) Factors controlling the marine geochemistry of fluoride. *Geochimica et Cosmochimica Acta* 33, 1153–1167.

Carrière, D., Bird, D. and Stamm, J. (1987) Influence of a diet of fluoride-fed cockerels on reproductive performance of captive American kestrels. *Environmental Pollution* 46, 151–159.

Case Western Reserve University (2000) www.cwru,edu/pubaft/univcomm/water.htm

Cauley, J., Murphy, P., Riley, T. and Buhari, A. (1995) Effects of fluorinated drinking water on bone mass and fractures: the study of osteoporotic fractures. *Journal of Bone Mineral Research* 10, 1076–1086.

CDC (1999) Achievements in public health, 1900–1999: fluoridation of drinking water to prevent dental caries. *Morbidity and Mortality Weekly Report* 48(41).

Chamberlain, A.C. (1966) Transport of *Lycopodium* spores and other small particles to rough surfaces. *Proceedings of the Royal Society Series A* 296, 46–70.

Chamberlain, A.C. (1970) Interception and retention of radioactive aerosols by vegetation. *Atmospheric Environment* 4, 57–78.

Chamblee, J., Arey, F.K. and Heckel, E. (1984) Free fluoride of the Pamlico river in North Carolina – a new method to localize the discharge of polluted water into an estuary. *Water Research* 18, 1225–1233.

Chamel, A. and Garrec, J.-P. (1977) Penetration of fluorine through isolated pear leaf cuticles. *Environmental Pollution* 12, 307–310.

Chan, J.T. and Koh, S.H. (1996) Fluoride content in caffeinated, decaffeinated and herbal teas. *Caries Research* 30, 88–92.

Chandler, G.T., Bayer, R.J. and Crisp, M.D. (2001) A molecular phylogeny of the endemic Australian genus *Gastrolobium* (Fabaceae: Mirbelieae) and allied genera using chloroplast and nuclear markers. *American Journal of Botany* 88, 1675–1687.

Chandler, G.T., Crisp, M.D., Cayzer, L.W. and Bayer, R.J. (2002) Monograph of *Gastrolobium* (Fabaceae: Mirbelieae). *Australian Systematic Botany* 15, 619–739.

Chang, C.W. (1970) Effect of fluoride on ribosomes and ribonuclease from corn roots. *Canadian Journal of Biochemistry* 48, 450–454.

Chang, C.W. (1975) Fluorides. In: Mudd, J.B and Kozlowski, T.T. (eds) *Responses of Plants to Air Pollution*. Academic Press, New York, pp. 57–95.

Chang, C.W. and Thompson, C.R. (1966) Subcellular distribution of fluoride in navel orange leaves. *Journal of Air and Water Pollution* 9, 685–691.

Chan-Yeung, M., Wong, R., MacLean, L., Tan, F., Schulzer, M., Enarson, D., Martin, A., Dennis, R. and Grzybowski, S. (1983) Epidemiological health study of workers at an aluminum smelter at Kitimat, B.C. II. Effects on musculoskeletal and other systems. *Archives of Environmental Health* 38, 34–40.

Chapman, C. and Phillips, M.A. (1955) Fluoroacetamide as a rodenticide. *Journal of the Science of Food and Agriculture* 6, 231–232.

Chavassieux, P., Boivin, G., Sere, C. and Meunier, P. (1993) Fluoride increases rat osteoblast function and population after *in vivo* administration but not after *in vitro* exposure. *Bone* 14, 721–725.

Chen, Y., Lin, M., He, Z., Liu, Y., Xiao, Y., Zhao, J., Fan, Y., Xiao, X. and Xu, F. (1993) Air pollution-type fluorosis in the region of Pingxiang, Jiangxi, People's Republic of China. *Archives of Environmental Health* 48, 246–249.

Chen, Z.-W. (1987) Some physiological and biochemical changes of fluorosis in silkworm, *Bombyx mori* L. *Chinese Journal of Sericulture* 13, 164–168.

Chenery, E.M. (1948a) Aluminium in plants and its relation to plant pigments. *Annals of Botany, New Series* 12, 121–136.

Chenery, E.M. (1948b) Aluminium in the plant world. Part I. General survey in the Dicotyledons. *Kew Bulletin* 3, 173–186.

Chenery, E.M. (1955) A preliminary study of aluminium and the tea bush. *Plant and Soil* 6, 174–200.

Cheng, J.Y., Yu, M.-H., Miller, G.W. and Welkie, G.W. (1968) Fluoroorganic acids in soybean leaves exposed to fluoride. *Environmental Science and Technology* 2, 367–370.

Chenoweth, M.B. (1950) *Monofluoroacetic Acid and Related Compounds*. Chemical–Biological Coordination Center, National Research Council, Washington, DC.

Chenoweth, M.B., Kandel, A., Johnson, L.B. and Bennett, D.R. (1951) Factors influencing fluoroacetate: practical treatment with glycerol monoacetate. *Journal of Pharmacology and Experimental Therapies* 102, 31–49.

Chernet, T., Travi, Y. and Valles, E. (2001) Mechanism of degradation of the quality of natural water in the lakes region of the Ethiopian Rift Valley. *Water Research* 35, 2819–2832.

Chhabra, R., Singh, A. and Abrol, I.P. (1980) Fluorine in sodic soils. *Journal of the Soil Science Society of America* 44, 33–36.

Chi, C.H., Chen, K.W., Chan, S.H., Wu, M.H. and Huang, J.J. (1996) Clinical presentation and prognostic factors in sodium monofluoroacetate intoxication. *Clinical Toxicology* 34, 707–712.

Chi, C., Lin, T. and Chen, K. (1999) Hemodynamic abnormalities in sodium monofluoracetate intoxication. *Human and Experimental Toxicology* 18, 351–353.

Cholak, J. (1959) Occurrence of fluoride in air, food and water. *Journal of Occupational Medicine* 1, 501–511.

Choubisa, S.L. (1999) Some observations on endemic fluorosis in domestic animals in southern Rajastan (India). *Veterinary Research Communications* 23, 457–465.

Chung, C.W. and Nickerson, W.J. (1954) Polysaccharide syntheses in growing yeast. *Biochemistry Journal* 208, 395–407.

Clark, B. (1982) Determination of atmospheric fluoride dispersion by use of passive plate monitors. In: *75th Annual Meeting of the Air Pollution Control Association*, New Orleans, USA, paper no. 84.54M.2.

Clarkson, D.T. and Hanson, J.B. (1980) The mineral nutrition of higher plants. *Annual Review of Plant Physiology* 31, 239–298.

Collet, G.F. (1969) Biological effect of fluoride on plants. *Fluoride Quarterly Reports* 2, 229–235.

Colombini, M., Mauri, C., Olivo, R. and Vivoli, G. (1969) Observations on fluorine pollution due to emissions from aluminum plant in Trentino. *Fluoride* 2, 40–48.

COMA (1994) *Fluoride. Dietary Reference Values for Food, Energy and Nutrients for the United Kingdom*. Her Majesty's Stationery Office, London.

Comeau, G. and LeBlanc, F. (1972) Influence du fluor sur la *Funaria hygrometrica* et l'*Hypogymnia physoides. Canadian Journal of Botany* 50, 847–856.

Conover, C. and Poole, R. (1982) Fluoride induced chlorosis and necrosis of *Dracaena fragrans* 'Massangeana'. *Journal of the American Society for Horticultural Science* 107, 136–139.

Cooke, J.A. (1972) Fluoride compounds in plants: their occurrence, distribution and effects. PhD thesis. University of Newcastle, Newcastle upon Tyne, UK.

Cooke, J.A. (1976) The uptake of sodium fluoroacetate by plants and its physiological effects. *Fluoride* 9, 204–212.

Cooke, J., Johnson, M., Davison, A. and Bradshaw, A. (1976) Fluoride in plants colonising fluorspar mine waste in the Peak District and Weardale. *Environmental Pollution* 11, 9–23.

Cooke, J.A., Johnson, M.S. and Davison, A.W. (1978) Uptake and translocation of fluoride in *Helianthus annuus* L. grown in sand culture. *Fluoride* 11, 76–88.

Cooke, J.A., Andrews, S.M. and Johnson, M.S. (1990) Lead, zinc, cadmium and fluoride in small mammals from contaminated grassland established on fluorspar tailings. *Water, Air and Soil Pollution* 51, 43–54.

Cooke J.A., Boulton, I.C. and Johnson, M.S. (1996) Fluoride in small mammals. In: Beyer, W.N.,

Heinz, G.H. and Redmon-Norwood, A.W. (eds) *Environmental Contaminants in Wildlife*. CRC Press, Boca Raton, Florida, pp. 473–482.

Cowling, D.W. and Koziol, M.J. (1982) Mineral nutrition and plant responses to air pollutants. In: Unsworth, M.H. and Ormrod, D.P. (eds) *Effects of Gaseous Pollutants on Agriculture and Horticulture*. Butterworths, London, pp. 349–376

Craggs, C. and Davison, A.W. (1985) The effect of simulated rainfall on grass fluoride concentrations. *Environmental Pollution Series B* 9, 309–318.

Craggs, C. and Davison, A.W. (1987a) Autocorrelation and univariate time series modelling for grass fluoride and airborne fluoride concentrations. *Environmental Pollution* 43, 115–128.

Craggs, C. and Davison, A.W. (1987b) Multivariate stochastic modelling of grass fluoride and airborne fluorides. *Environmental Pollution* 44, 279–296.

Craggs, C., Blakemore, J. and Davison, A.W. (1985) Seasonality in the fluoride concentrations of pasture grass subject to ambient airborne fluorides. *Environmental Pollution Series B* 9, 163–177.

Crane, G.B., Goodwin, D.R. and Rook, J.H. (1970) *Atmospheric Emissions from Wet-process Phosphoric Acid Manufacture*. NAPCA Publication No. AP-57, US Public Health Service, Raleigh, North Carolina.

Crichton-Browne, J. (1892) An address on tooth culture. *Lancet*, 2 July, 6–10.

Crisofolini, M., Piscioli, F. and Urbani, F. (1980) 'Macchie do Chizzola' e patologia cutanea da fluoro: revisione critica. In: *Proceedings, First National Congress of Environmental Aspects in the Aluminium Industry, Venice, Italy*, pp. 99–103.

Crisp, M.D. and Weston, P.H. (1995) Mirbelieae. In: Crisp, M.D. and Doyle, J.H. (eds) *Advances in Legume Systematics*, Part 7, *Phylogeny*. Royal Botanic Gardens, Kew, pp. 245–282.

Cronin, S.J., Manoharan, V., Hedley, M.J. and Loganathan, P. (2000) Fluoride: a review of its fate, bioavailability and risks of fluorosis in grazed-pasture systems in New Zealand. *New Zealand Journal of Agricultural Research* 43, 295–321.

Cronin, S. and Sharp, D. (2002) Environmental impacts on health from continuous volcanic activity at Yasur (Tanna) and Anbrym, Vanuatu. *International Journal of Environmental Health Research* 12, 109–123.

Cross, F. and Ross, R. (1969) New developments in fluoride emissions from phosphate plants. *Journal of the Air Pollution Control Association* 19, 15–17.

Crossley, H.E. (1944) Fluorine in coal. III. The manner of occurrence of fluorine in coals. *Journal of the Society of Chemistry and Industry (London)* 63, 289–292.

Daines, R.H., Leone, I. and Brennan, E. (1952) The effect of fluorine on plants as determined by soil nutrition and fumigation studies. In: McCabe, L.C. (ed.) *Air Pollution*. McGraw-Hill, New York, pp. 97–105.

Darrall, N.M. (1989) The effect of air pollutants on physiological processes in plants. *Plant, Cell and Environment* 12, 1–30.

Das, B.C. and Prasad, D.N. (1973) Evaluation of some tetraploid and triploid mulberry varieties through chemical analysis and feeding experiments. *Indian Journal of Sericulture* 13, 17–22.

Dässler, H.G. and Grumbach, H. (1967) Schaden an Obst durch Industrieabgase. *Ostbau (Berlin)* 7, 27–29.

Dässler, H.G., Raft, H. and Rein, H.K. (1972) Zur Widerstandsfähikeit von Geholzen gegenüber Fluorbidungen und Schwefeldioxid. *Flora (Jena)* 161, 289–302.

Dave, G. (1984) Effects of fluoride on growth, reproduction and survival in *Daphnia magna*. *Comparative Biochemistry and Physiology* C78, 425–431.

David, W.A.L. (1950) Sodium fluoroacetate as a systemic and contact insecticide. *Nature* 165, 493–494.

David, W.A.L. and Gardiner, B.O.C. (1958) Investigations on the systemic insecticidal action of sodium fluoroacetate and of three phosphorus compounds on *Aphis fabae* Scop. *Annals of Applied Biology* 38, 91–107.

Davies, F.B.M. (1982) Accumulation of fluoride by *Xanthoria parietina* growing in the vicinity of the Bedfordshire brickfields. *Environmental Pollution* 29, 189–196.

Davies, M.T. (1989) The accumulation of fluoride by invertebrates and its effects on some aspects of their biology. PhD thesis, University of Newcastle, Newcastle upon Tyne, UK.

Davies, M.T., Davison, A.W. and Port, G.R. (1992) Fluoride loading of pine sawfly from a polluted site. *Journal of Applied Ecology* 29, 63–69.

Davies, M.T., Port, G.R. and Davison, A.W. (1998) Effects of dietary and gaseous fluoride on the aphid *Aphis fabae*. *Environmental Pollution* 99, 405–409.

Davieson, G., Murray, F. and Wilson, S. (1990) Effects of sulphur dioxide, hydrogen, fluoride, singly and in combination, on growth and yield of wheat in open-top chambers. *Agricultural Ecosystems and the Environment* 30, 317–325.

Davison, A.W. (1982) The effects of fluorides on plant growth and forage quality. In: Unsworth, M.H. and Ormrod, D.P. (eds) *Effects of Gaseous Air Pollution in Agriculture and Horticulture*. Butterworths, London, pp. 267–292.

Davison, A.W. (1983) Uptake, translocation and accumulation of soil and airborne fluorides by vegetation. In: Shupe, J.L., Peterson, H.B. and Leone, N.C. (eds) *Fluorides: Effects on Vegetation, Animals and Humans*. Paragon Press, Salt Lake City, Utah, pp. 62–82.

Davison, A.W. (1987) Pathways of fluoride transfer in terrestrial ecosystems. In: Coughtrey, P.J., Martin, M.H. and Unsworth, M.H. (eds) *Pollutant Transport and Fate in Ecosystems*. Special Publications Series of British Ecological Society, No. 6, Blackwell Scientific, Oxford, pp. 193–210.

Davison, A.W. and Blakemore, J. (1976) Factors determining fluoride accumulation in forage. In: Mansfield, T.A. (ed.) *Effects of Air Pollution on Plants*. Cambridge University Press, Cambridge, pp. 17–30.

Davison, A.W. and Blakemore, J. (1980) Rate of deposition and resistance to deposition of fluorides on alkali impregnated papers. *Environmental Pollution Series B* 1, 305–319.

Davison, A.W. and Pearson, S. (1997) *Toxicity of TFA to Plants*. Project SP91-18.23/BP96-31, Final Report to AFEAS by University of Newcastle, Newcastle upon Tyne.

Davison, A.W., Rand, A.W. and Belts, W.E. (1973) Measurement of atmospheric fluoride concentrations in urban areas. *Environmental Pollution* 5, 23–33.

Davison, A.W., Blakemore, J. and Wright, D.A. (1976) A re-examination of published data on the fluoride content of pastures. *Environmental Pollution* 10, 209–216.

Davison, A.W., Blakemore, J. and Craggs, C. (1979) The fluoride content of forage as an environmental quality standard for the protection of livestock. *Environmental Pollution* 20, 279–296.

Davison, A.W., Takmaz-Nisancioglu, S. and Bailey, I.F. (1985) The dynamics of fluoride accumulation by vegetation. In: Susheela, A.K. (ed.) *Fluoride Toxicity*. ISFR, New Delhi, pp. 30–46.

Dean, F.P., Newcomer, E.J. and Westlake, W.E. (1946) Cryolite for codling moth control in the Pacific Northwest. *Journal of Economic Entomology* 39, 523–526.

Dean, G. and Treshow, M. (1966) Effects of fluoride on the virulence of tobacco mosaic virus *in vitro. Proceeding of the Utah Arts and Sciences Proceedings* 42, 236–239.

Dean, H.T. (1938) Endemic fluorosis and its relation to dental caries. *Public Health Reports* 53, 1443–1452.

Dean, H.T., Arnold, F.A. and Elvove, E. (1942) Domestic waters and dental caries. V. Additional studies of the relation of fluoride in domestic waters to dental caries in 4425 white children, age 12–14 years, of 13 cities in 4 states. *Public Health Report* 57, 1155–1179.

Dell, B., Malajczuk, N. and Grove, T.S. (1995) *Nutrient Disorders in Plantation Eucalyptus.* ACINAR, Canberra.

de Oliveira, M.M. (1963) Chromatographic isolation of monofluoroacetic acid from *Palicourea marcgravii* St. Hil. *Experientia* 19, 586–587.

DePierre, J.W. (2002) Effects on rodents of perfluorofatty acids. In: Neilson, A.H. (ed.) *The Handbook of Environmental Chemistry,* Vol. 3, *Anthropogenic Compounds, Part N. Organofluorines.* Springer, Berlin, pp. 203–248.

Derryberry, O.M., Bartholomew, M.D. and Fleming, R.B. (1963) Fluoride exposure and worker health. *Archives of Environmental Health* 6, 503–510.

Desai, V.K., Solanki, D.M., Kanharia, S.L. and Bhavsar, B.S. (1993) Monitoring of neighbourhood fluorosis through a dental fluorosis survey in schools. *Fluoride* 26, 181–186.

de Temmerman, L.O., Istas, J.R., Rakelboom, E.L. and Baten, H. (1978) Fluor in grassoorten en bodem van een agrarisch gebied en een industriezone. *Landbouwtijdschrift* 31, 515–524.

Dewey, J.E. (1973) Accumulation of fluorides by insects near an emission source in western Montana. *Environmental Entomology* 2, 179–182.

Dinman, B.D., Elder, M.J., Bnnoy, T.B., Bovard, P.G. and Colwell, M.O. (1976) The prevention of bony fluorosis in aluminum workers: a 15 year retrospective study of fluoride excretion and bony radiopacity among aluminum workers. Part 4. *Journal of Occupational Medicine* 18, 21–23.

Dissanayake, C.B. and Chandrajith, R. (1999) Medical geochemistry of tropical environments. *Earth-Science Reviews* 47, 219–258.

Dobrosky, I.D. (1937) Orchard experiments with natural cryolite for codling moth control. *Journal of Economic Entomology* 30, 656–658.

Dobrosky, I.D. (1943) Orchard dusting experiments with natural cryolite for codling moth. *Journal of Economic Entomology* 36, 350–351.

Doley, D. (1986) *Plant–Fluoride Relationships; An Analysis with Particular Reference to Australian Vegetation.* Inkata Press, Melbourne.

Doley, D. (1988) Fluoride-induced enhancement and inhibition of photosynthesis in four taxa of pines. *New Phytologist* 110, 21–32.

Domingos, M., Klumpp, A. and Klumpp, G. (1998) Air pollution impact on the Atlantic Forest in the Cubatão region, SP, Brazil. *Ciência e Cultura, Journal of the Brazilian Association for the Advancement of Science* 50, 230–236.

Domingos, M., Lopes, M.I.M.S. and Struffaldi-De Vuono, Y. (2000) Nutrient cycling disturbance in Atlantic Forest sites affected by air pollution coming from the industrial complex of Cubatão, Southeast Brazil. *Revista Brasileira Botanica* 23, 77–85.

Domingos, M., Klumpp, A. and Klumpp, G. (2002) Disturbances to the Atlantic rain forest in southeast Brazil. In: Ashmore, M., Emberson, L. and Murray, F. (eds) *Air Pollution Impacts on Crops and Forests: a Global Assessment.* Air Pollution Reviews, Vol. 3, Imperial College Press.

Dorsey, M.J. and McMunn, R.L. (1944) *Tree Conditioning the Peach Crop.* Bulletin 507, Illinois Agricultural Experiment Station, Urbana, Illinois.

Dreher, K. (1965) Poisoning of bees by fluorine. *Apicultural Abstracts* 18, 30.

Drowley, W.B., Rayner, A.C. and Jephcott, G. (1963) Atmospheric fluoride levels in some Ontario peach orchards. *Canadian Journal of Plant Science* 43, 547–553.

Dunphy, J.T. (1906) Report of experiments carried out to observe effects of certain poisonous plants on sheep and goats. *Transvaal Agricultural Journal* 315.

Eason C.T., Wickstrom, M., Turck P. and Wright, G.R.G. (1999) A review of recent regulatory and environmental toxicology studies on 1080: results and implications. *New Zealand Journal of Ecology* 23, 129–137.

Edmunds, G.F. (1983) Effects of fluoride on plant–insect interactions. In: Shupe, J.L., Peterson, H.B. and Leone, N.C. (eds) *Fluorides. Effects on Vegetation, Animals and Humans.* Paragon Press, Salt Lake City, Utah, pp. 151–155.

Edmunds, G.F. and Allen, R.K. (1956) Comparison of black pine leaf scale population density on normal Ponderosa pine and those weakened by other agents. *Proceedings of the 10th International Congress of Entomology. Montreal* 4, 391–392.

Eichert, T., Goldbach, H.E. and Burkhardt, J. (1998) Evidence for the uptake of large anions through stomatal pores. *Botanica Acta* 111, 461–466.

Eleftheriou, E.P. and Tsekos, I. (1991) Fluoride effects on leaf cell ultrastructure of olive trees growing in the vicinity of aluminium factory of Greece. *Trees* 5, 83–89.

Ellis, D.A., Hanson, M.L., Sibley, P.K., Shahid, T., Fineberg, N.A., Solomon, K.R., Muir, D.C.G. and Mabury, S.A. (2001a) The fate and persistence of trifluoroacetic and chloroacetic acids in pond waters. *Chemosphere* 42, 309–318.

Ellis, D.A., Mabury, S.A., Martin, J.W. and Muir, D.C.G. (2001b) Thermolysis of fluoropolymers as a potential source of halogenated organic acids in the environment. *Nature* 412, 321–324.

Ellis, D.A., Moody, C.A. and Mabury, S.A. (2002) Trifluoroacetic acid and longer chain perfluoro acids – sources and analysis. *The Handbook of Environmental Chemistry*, Vol. 3. *Anthropogenic Compounds*, Part N, *Organofluorines*. Springer, Berlin, pp. 104–120.

Eloff, J.N. (1972) Evidence for the occurrence of a Krebs cycle in the fluoroacetate containing plant *Dichepetalum cymosum* (Gifblaar). *Zeitschrift für Pflanzenphysiologie* 67, 207–211.

Eloff, J.N. and Grobbelaar, N. (1972) A preliminary evaluation of the metabolic fate of fluoroacetate and acetate in *Dichapetalum cymosum* and *Parinari capensis*. *Journal of the South African Chemical Institute* 25, 109–114.

Eloff, J.N. and von Sydow, B. (1971) Experiments on the fluoroacetate metabolism of *Dichapetalum cymosum* (Gifblaar). *Phytochemistry* 10, 1409–1415.

Emden G. and Lehnartz, E. (1924) Über dei Bedeutung von Ionen für die Muskelfunktion. I. Die Wirkung vershiedener Amionen auf den Lacktacidogenwechesl im Froschmuskelbrei. *Hoppe-Seyler's Zeitschrift für Physiologische Chemie* 134, 243–275.

Englebrecht, A.H.P. and Louw, C.W. (1973) Hydrogen fluoride on the ultrastructure of mesophyll cells of sugarcane. In: *3rd International Clean Air Congress, Dusseldorf, Germany*. pp. A157–A159.

Environment Canada (2001) *Canadian Water Quality Guidelines for the Protection of Aquatic Life: Inorganic Fluorides. Scientific Supporting Document.* Environmental Quality Branch, National Guidelines and Standards Office, Environment Canada, Ottawa.

EPS (1978) *Sources of Metals and Metal Levels in Municipal Wastewaters.* Research Report No. 80, Environment Canada, Ottawa, Canada.

EuroBionet (1999) http://www.uni-hohenheim.de/eurobionet/

Everist, S. (1981) The history of poisonous plants in Australia. In: Carr, D.J. and Carr, S.G.M. (eds) *Plants and Man in Australia.* Academic Press, Sydney, pp. 223–255.

Facteau, T.J. and Rowe, R.E. (1977) Effect of hydrogen fluoride and hydrogen chloride on pollen tube growth and sodium fluoride on pollen germination in 'Tilton' apricot. *Journal of the American Society for Horticultural Science* 102, 95–96.

Facteau, T.J., Wang, S.Y. and Rowe, R.E. (1973) The effect of hydrogen fluoride on pollen germination and pollen tube growth in *Prunus avium* L. cv. 'Royal Ann'. *Journal of the American Society for Horticultural Science* 98, 234–236.

Fagerstone, K.A., Savarie, P.J., Elias, D.J. and Schafer, E.W. (1994) Recent regulatory requirements for pesticide registration and the status of Compound 1080 studies conducted to meet EPA requirements. In: Seawright, A.A. and Eason, C.T. (eds) *Proceedings of the Science Workshop on 1080.* Miscellaneous Series 28, Royal Society of New Zealand, SIR Publishing, Wellington, pp. 33–38.

Fanshier, D.W., Gottwald, L.K. and Kun, E. (1964) Studies on specific enzyme inhibiitors VI. Characterization and mechanism of action of the enzyme-inhibitor isomer of monofluorocitrate. *Journal of Biological Chemistry* 239, 425–434.

Farman, J.C., Gardiner, B.G. and Shanklin, J.D. (1985) Large losses of total ozone in Antarctica reveal seasonable CO_x/NO_x interaction. *Nature* 315, 207–210.

Featherstone, J.D.B. (2000) The science and practice of caries prevention. *Journal of the American Dental Association* 131, 887–899.

Feder, W. (1978) Plants as bioassay systems for monitoring atmospheric pollutants. *Environmental Health Perspectives* 27, 139–148.

Feder, W.A. and Manning, W.J. (1979) Living plants as indicators and monitors. In: Heck, W.W., Krupa, S.V. and Linzon, S.N. (eds)

Handbook of Methodology for the Assessment of Air Pollution Effects on Vegetation. Air Pollution Control Association, Pittsburgh, Pennsylvania, pp. 9-1–9-14.

Feiser, A.H., Sykora, J.l., Kostalos, M.S., Wu, Y.C. and Weyel, D.W. (1986) Effect of fluoride on survival and reproduction of *Daphnia magna. Water Pollution Control Federation* 58, 82–86.

Feldwick, M.G., Mead, R.J. and Kostyniak, P.K. (1994) Biochemical effects of fluoroacetate and related pesticides: the potential of 4-methylpyrazole as an antidote. In: Seawright, A.A. and Eason, C.T. (eds) *Proceedings of the Science Workshop on 1080.* Miscellaneous Series 28, Royal Society of New Zealand, SIR Publishing, Wellington, pp. 74–81.

Fiering, A.E. (1994) Fluoroelastomers. In: Banks, R.E., Smart, B.E. and Tatlow, J.C. (eds) *Organofluorine Chemistry: Principles and Commercial Applications.* Plenum Press, New York, pp. 495–520.

Fink, S. (1988) Histological and cytological changes caused by air pollutants and other abiotic factors. In: Schulte-Hostede, N., Darrall, M., Blank, L.W. and Wellburn, A.R. (eds) *Air Pollution and Plant Metabolism.* Elsevier, pp. 36–54.

Finkelman, R., Orem, W., Castranova, V., Tatu, C., Belkin, H., Zheng, B., Lerch, H., Maharaj, S. and Bates, A. (2002) Health impacts of coal and coal use: possible solutions. *International Journal of Coal Geology* 50, 425–443.

Flagler, R.B. (1998) *Recognition of Air Pollution Injury to Vegetation: A Pictorial Atlas,* 2nd edn. Air and Waste Management Association, Pittsburgh, Pennsylvania.

Fleischer, M. (1953) *Recent Estimates of the Relative Abundance of the Elements in the Earth's Crust.* Circular No. 285, US Geological Survey, Washington, DC.

Fleischer, M. and Robinson, W.D. (1963) Some problems of the geochemistry of fluorine. *Royal Society of Canada Special Paper* 58–75.

Fleming, W.J., Grue, C.E., Shuler, C.A. and Bunck, C.M. (1987) Effects of oral doses of fluoride on nestling European starlings. *Archives of Environmental Contamination and Toxicology* 16, 483–490.

Fowle, J.R. and Sexton, K. (1992) EPA priorities for biologic markers research in environmental health. *Environmental Health Perspectives* 98, 235–241.

Fowler, D. and Unsworth, M.H. (1974) Dry deposition of sulphur dioxide on wheat. *Nature* 249, 389–390.

Francis, W. (1954) *Coal. Its Formation and Composition.* Edward Arnold, London.

Frank, H., Christoph, E.H., Holm-Hansen, O. and Bullister, J. (2002) Trifluoroacetate in ocean waters. *Environmental Science and Technology* 36, 12–15.

Fratzl, P., Roschger, P., Eschberger, J., Abendroth, B. and Klaushofer, K. (1994) Abnormal bone mineralization after fluoride treatment in osteoporosis: a small angle X-ray scattering study. *Journal of Bone Mineral Research* 9, 1541–1549.

Fridriksson, S. (1983) Fluoride problems following volcanic eruptions. In: Shupe, J.L., Peterson, H.B. and Leone, N.C. (eds) *Fluorides. Effects on Vegetation, Animals and Humans.* Paragon Press, Salt Lake City, Utah, pp. 339–344.

FSA (2001) Expert Group on Vitamins and Minerals: UK Food Standards Agency. http://www.food.gov.uk and http://archive.food.gov.uk/committees/evm/evm_01_03p.pdf

Fujii, M. and Hayashi, H. (1972) Fluorides contained in mulberry leaves and silkworms in the area around a tile factory. *Journal of the Sericultural Society of Japan* 41, 150–153.

Fujii, M. and Honda, S. (1972) The relative oral toxicity of some fluorine compounds for silkworm larvae. *Journal of the Sericultural Society of Japan* 41, 104–110.

Gal, E.M., Drewes, P.A. and Taylor, N.F. (1961) Metabolism of fluoroacetic acid -2-C^{14} in the intact rat. *Archives of Biochemistry and Biophysics* 93, 1–14.

Gallon, J.R., Ul-Haque, M.I. and Chaplin, A.E. (1978) Fluoroacetate metabolism in *Gloeocapsa* sp. LB795 and its relationship to acetylene reduction (nitrogen fixation). *Journal of General Microbiology* 106, 329–336.

Garber, K. (1970) Fluoride in rainwater and vegetation. *Fluoride* 3, 22–26.

Garrec, J.-P. (1983) Modification de la composition minérale dans la zone de 'suture molle' (suture red spot) de pêches soumises à une pollution fluorée. Étude par microsonde électronique. *Environmental Pollution* 30, 189–200.

Garrec, J.-P. and Chopin, S. (1982) Calcium accumulation in relation with fluoride pollution in plants. *Fluoride* 15, 144–149.

Garrec, J.-P. and Letourneur, L. (1981) Fluoride absorption by the root and foliar tissues of the horse bean (calcicole) and lupin (calcifuge). *Fluoride* 14, 186–197.

Garrec, J.-P. and Plebin, R. (1984) Accumulation of fluorine in earthworms living in contaminated soils. *Environmental Pollution* 7, 97–106.

Garrec, J.-P. and Vavasseur, A. (1978) Distribution of fluoride in polluted poplars – detection of fluoride accumulation in roots. *European Journal of Forest Pathology* 8, 37–43.

Garrec, J.-P., Ligeon, E., Bontemps, A., Bligny, R. and Fourcy, A. (1972) Localisation exacte de fluor le long d'aiguilles polluées d'*Abies alba* par microanalyse au moyen de protons. *Comptes Rendus Hebdomadaires des Séances de l'Académie des Sciences (Paris) Série D* 274, 3468–3471.

Garrec, J.-P., Oberlin, J.C., Ligeon, C., Bisch, A.-M. and Fourcy, A. (1974) Fluoride–calcium interaction in polluted fir needles. *Fluoride* 7, 78–84.

Garrec, J.-P., Lounowski, A. and Plebin, R. (1977a) Study of the influence of volcanic fluoride emissions on the surrounding vegetation. *Fluoride* 10, 152–156.

Garrec, J.-P., Plebin, R. and Lhoste, A.-M. (1977b) Influence of fluorine on mineral composition of polluted fir needles, *Abies alba* Mill. *Environmental Pollution* 13, 159–167.

Garrec, J.-P., Abdulaziz, P., Lavielle, E., Vandevelde, L. and Plebin, R. (1978) Fluoride, calcium and aging in healthy and polluted fir trees, *Abies alba*. *Fluoride* 11, 186–197.

Geeson, N.A., Abrahams, P.W., Murphy, M.P. and Thornton, I. (1998) Fluorine and metal enrichment of soils and pasture herbage in the old mining areas of Derbyshire, UK. *Agriculture, Ecosystems and Environment* 68, 217–231.

Gemmell, G.D. (1946) Fluorine in New Zealand soils. *New Zealand Journal of Science and Technology* 27, 302–306.

Georgsson, G. and Petursson, G. (1972) Fluorosis of sheep caused by the Hekla eruption in 1970. *Fluoride* 5, 58–66.

Gerdes, R.A. (1968) Influence of atmospheric hydrogen fluoride on the frequency of sex-linked recessive lethals and sterility in *Drosophila melanogaster*. *Journal of the International Society for Fluoride* 4, 25–29.

Gerdes, R.A., Smith, J.S. and Applegate, H.G. (1971a) The effects of atmospheric hydrogen fluoride upon *Drosophila melanogaster* – 1. Differential genotypic response. *Atmospheric Environment* 5, 113–116.

Gerdes, R.A., Smith, J.S. and Applegate, H.G. (1971b) The effects of atmospheric hydrogen fluoride upon *Drosophila melanogaster* – 2. Fecundity, hatchability and fertility. *Atmospheric Environment* 5, 117–122.

Gianessi, L.P. and Marcelli, M. (2000) *Pesticide Use in US Crop Production: 1997*. National Center for Food and Agricultural Policy, Washington, DC.

Giannini, J., Miller, G.W. and Pushnik, J.C. (1985) Effects of NaF on biochemical processes of isolated soybean chloroplasts. *Fluoride* 18, 72–79.

Gibbs, M. and Beevers, H. (1955) Glucose dissimilation in higher plants: effects of age of tissue. *Plant Physiology* 30, 343–346.

Giesy J.P. and Kannan, K. (2001) Global distribution of perfluorooctane sulfonate in wildlife. *Environmental Science and Technology* 35, 1339–1342.

Giesy, J.P. and Kannan, K. (2002) Perfluorochemical surfactants in the environment. *Environmental Science and Technology* 36, 147A–152A.

Giesy, J.P., Kannan, K. and Jones, P.D. (2001) Global monitoring of perfluorinated organics. *Scientific World* 1, 627–629.

Gilbert, O.L. (1968) Bryophytes as indicators of air pollution in the Tyne valley. *New Phytologist* 67, 15–30.

Gilbert, O.L. (1971) The effect of airborne fluorides on lichens. *Lichenologist* 5, 26–32.

Gilbert, O.L. (1973) The effect of airborne fluorides. In: Ferry, B.W., Baddeley, M.S. and Hawksworth, D.L. (eds) *Air Pollution and Lichens*. Athlone Press, University of London, London, pp. 176–191.

Gilpin, L. and Johnson, A.H. (1980) Fluorine in agricultural soils of southeastern Pennsylvania. *Journal of the Soil Science Society of America* 44, 255–258.

Giovanazzi, A., Tomasi, A., Nardelli, G. and D'Andrea, F. (1980) Il fenomeno 'Macchie Blu do Chizzola' un'ulteriore puntualizzazione epidemiologica. In: *Proceedings of the First National Congress of Environmental Aspects in the Aluminium Industry, Venice*, pp. 89–98.

Gisiger, L. (1968) The solubility of various fluorine compounds in soil. *Fluoride Quarterly Reports* 1, 21–26.

Givan, C.V. and Torrey, J.G. (1968) Fluoride inhibition of respiration and fermentation in cultured cells of *Acer pseudoplatanus*. *Physiologia Plantarum* 21, 1010–1019.

Gluskoter, H. (1977) *Trace Elements in Coal. Occurrence and Distribution*. Circular 499, Illinois State Geological Survey, Urbana, Illinois.

GMAC (1994) *PR-45: Genetic Manipulation of Rumen Bacteria for Detoxification of the Plant Poison Fluoroacetate*. Genetic

Manipulation Advisory Committee, Queensland, Australia. www.health.gov.au/ogtr/pdf/volsys/pr45.pdf

GMAC (1999) *PR-130: The Use of Genetically Modified Rumen Bacteria to Protect Livestock against Fluoroacetate Poisoning.* Genetic Manipulation Advisory Committee, Queensland, Australia. www.health.gov.au/ogtr/pdf/volsys/pr130.pdf

Godbeer, W. and Swaine, D. (1987) Fluorine in Australian coals. *Fuel* 66, 794–798.

Godbeer, W., Swaine, D. and Doodarzi, F. (1994) Fluorine in Canadian coals. *Fuel* 73, 1291–1293.

Goldman, P. (1965) The enzymic cleavage of the carbon–fluorine bond in fluoroacetate. *Journal of Biological Chemistry* 240, 3434–3438.

Goldman, P. (1966) Carbon–fluorine bond cleavage. II. Studies on the mechanism of the defluorination of fluoroacetate. *Journal of Biological Chemistry* 241, 5557–5559.

Goldman, P. (1969) The carbon–fluorine bond in compounds of biological interest. *Science* 164, 1123–1130.

Goldman, P. (1972) The use of microorganisms in the study of fluorinated compounds. In: *Carbon–Fluorine Compounds: Chemistry, Biochemistry, Biological Activity, Symposium 1971,* pp. 335–356.

Goldman, P., Milne, G.W.A. and Keister, D.B. (1968) Carbon–halogen bond cleavage. III. Studies on bacterial halidohydrolases. *Journal of Biological Chemistry* 243, 248–254.

Gordon, W.C. and Ordin, L. (1972) Phosphorylated and nucleotide sugar metabolism in relation to cell wall production in *Avena coleoptiles* treated with fluoride and peroxyacetyl nitrate. *Plant Physiology* 49, 542–545.

Gosselin, R.E., Hodge, H.C., Smith, R.P. and Gleason, M.N. (1976) *Clinical Toxicology of Commercial Products,* 4th edn. Williams and Wilkins, Baltimore, Maryland.

Grandjean, P. and Thomsen, G. (1983) Reversibility of skeletal fluorosis. *British Journal of Industrial Medicine* 40, 456–461.

Greendyke, R.M. and Hodge, H.C. (1964) Accidental death due to hydrofluoric acid. *Journal of Forensic Sciences* 9, 383–389.

Gregg, K. (1995) Engineering gut flora of ruminant livestock to reduce forage toxicity: progress and problems. *Trends in Biotechnology* 13, 418–421.

Gregg, K., Cooper, C.L., Schafer, D.J., Sharpe, H., Beard, C.E., Allen, G. and Xu, J. (1993) Detoxification of the plant toxin fluoroacetate by a genetically modified rumen bacterium. *Biotechnology* 12, 1361–1365.

Gregg, K., Harndorf, B., Henderson, K., Kopecny, K. and Wong, C. (1998) Genetically modified ruminal bacteria protect sheep from fluoroacetate poisining. *Applied Environmental Microbiology* 64, 3496–3498.

Gregson, R.P., Baldo, B.A., Thomas, P.G., Quinn, R.J., Berquist, P.R., Stephens, J.F. and Horne, A.R. (1979) Fluorine is a major constituent of the marine sponge *Halichondria moorei. Science* 206, 1108–1109.

Gribble, G.W. (2002) Naturally occurring organofluorines. In: Neilson, A.H. (ed.) *The Handbook of Environmental Chemistry. Anthropogenic Compounds,* Part N, *Organofluorines.* Springer, Berlin, pp. 121–136.

Griffin, S.W. and Bayles, B.B. (1952) Some effects of fluorine fumes on vegetation. In: McCabe, L.C. (ed.) *Air Pollution. Proceedings US Technological Conference on Air Pollution.* McGraw-Hill, New York, pp. 106–115.

Grimvall, A., Boren, H., von Sydow, L. and Lanieski, K. (1997) Analysis of TFA in samples of rain, snow and ice collected at remote sites. Paper presented to the Alternative Fluorocarbons Environmental Acceptability Workshop, Washington, DC.

Grobbelaar, N. and Meyer, J.J.M. (1989) Fluoroacetate production by *Dichapetalum cymosum. Journal of Plant Physiology* 135, 550–553.

Grunder, H.D., Mahlhop, R. and Pothmann, M. (1980) Absorption and retention of fluoride by dairy cows from brickworks dust containing calcium fluoride during one year of feeding. *Deutsche Tierarztliche Wochenschrift* 87, 329–333.

Grynpas, M.D. (1990) Fluoride effects on bone crystals. *Journal of Bone Mineral Research* 5, S169–S175.

Guderian, R. (1977) *Air Pollution – Phytotoxicity of Acid Gases and its Significance in Air Pollution Control.* Springer-Verlag, Berlin.

Guderian, R., van Haut, H. and Stratmann, H. (1969) Experimentelle Untersuchungen über Pflanzenschädigende Fluorwasserstoff-Konzentrationen. *Forschritte der Landesanstalt Nordrhein-Westfalen* 2017, 54.

Gupta, M.K., Singh, V. and Dass, S. (1994) Ground water quality of Block Bichpuri, Agra (India) with special reference to fluoride. *Fluoride* 27, 89–92

Gurnon, G.J. and Smart, R.L. (1990) Reduction of fluoride emissions and effluents from Alcan's Kitimat smelter. In: Bouchard, M.

and Tremblay, P. (eds) *Production, Refining, Fabrication and Recycling of Light Metals. 29th Annual Conference of Metallurgists of CIM, 26–30 August 1990.* Hamilton, Ontario, pp. 147–159.

Gutknecht, J. and Walter, A. (1981) Hydrofluoric and nitric acid transport through lipid bilayer membranes. *Biochimica et Biophysica Acta* 644, 153–156.

Guy, W.S., Taves, D.R. and Brey, W.S. (1976) Organic fluorocompounds in human plasma: prevalence and characterization. In: Filler, R. (ed.) *Biochemistry Involving Carbon–Fluorine Bonds.* Symposium Series No. 28, American Chemical Society, pp. 117–134.

Hagvar, S., Abrahamsen, G. and Blake, A. (1976) *Attack by the Pine Bud Moth (*Exoteleia dodecella *L.) in Southernmost Norway: Possible Effect of Acid Precipitation.* Norwegian Institute Skogforskning, Ås, Norway.

Hall, R.J. (1974) The metabolism of ammonium fluoride and sodium fluoroacetate by experimental *Acacia georginae. Environmental Pollution* 6, 267–280.

Halmer, M.M., Schminke, H.-U. and Graf, H.-F. (2002) The annual volcanic gas input into the atmosphere, in particular into the stratosphere: a global data set for the past 100 years. *Journal of Volcanology and Geothermal Research* 115, 511–528.

Hamilton, J.T.G. and Harper, D.B. (1997) Fluoro fatty acids in seed oil of *Dichapetalum toxicarium. Phytochemistry* 44, 1129–1132.

Häni, H. (1978) Interactions by fluoride by a mineral soil containing illite and alterations of maize plants grown in the soil. *Fluoride* 11, 18–24.

Hansen, K.J., Clemen, L.A., Ellefson, M.E. and Johnson, H.O. (2001) Compound-specific, quantitative characterisation of organic fluorochemicals in biological matrices. *Environmental Science and Technology* 35, 766–770.

Hansen, K.J., Johnson, H.O., Eldridge, J.S., Butenhoff, J.L. and Dick, L.A. (2002) Quantitative characterisation of trace levels of PFOS and PFOA in the Tennessee River. *Environmental Science and Technology* 36, 1681–1685.

Harborne, J.B. (1977) *Introduction to Ecological Biochemistry.* Academic Press, London.

Harnisch, J. and Eisenhauer, A. (1998) Natural CF_4 and SF_6 on earth. *Geophysical Research Letters* 25, 2401–2404.

Harnisch, J., Borchers, R., Fabian, P., Gaggeler, H.W. and Schotterer, U. (1996) Effect of natural tetrafluoromethane. *Nature* 384, 32.

Harnisch, J., Frische, M., Borchers, R., Eisenhauer, A. and Jordan, A. (2000) Natural fluorinated organics in fluorite and rocks. *Geophysical Research Letters* 27, 1883–1886.

Harper, D.B., Hamilton, J.R.G. and O'Hagan, D. (1990) Identification of threo-18-fluoro-9,10-dihydrostearic acid: a novel ω-fluorinated fatty acid from *Dichapetalum toxicarium* seeds. *Tetrahedron Letters* 31, 7661–7662.

Harvey, J.M. (1952) Chronic endemic fluorosis of merino sheep in Queensland. *Queensland Journal of Agricultural Sciences* 9, 47–141.

Hausen, H.W. (2000) Fluoridation, fractures and teeth – fluoride doesn't cause fractures but its benefits vary. *British Medical Journal* 321, 844–845.

Havas, M. and Jaworski, J.F. (1986) *Aluminium in the Canadian Environment.* National Research Council of Canada Publication No. 24759. NRCC Ottawa, Canada, 331 pp.

Heagle, A.S. (1973) Interactions between air pollutants and plant parasites. *Annual Review of Phytopathology* 11, 365–388.

Heagle, A.S., Body, D.E. and Heck, W.W. (1973) An open top chamber to assess the impact of air pollution on plants. *Journal of Environmental Quality* 2, 365–368.

Healy, W.B. (1973) Nutritional aspects of soil ingestion by grazing animals. In: Butler, G.W. and Bailey, R.W. (eds) *Chemistry and Biochemistry of Herbage*, Vol. I. Academic Press, London, pp. 567–588.

Hewitt, E.J. and Nicholas, D.J.D. (1963) Cations and anions: inhibitions and interactions in metabolism and in enzyme activity. In: Hochster, R.M. and Quastel, J.H. (eds) *Metabolic Inhibitors*, Vol. 2. Academic Press, New York, pp. 311–436.

Higgins, J., Warnken, J., Sherman, P.P. and Teasdale, P.R. (2002) Survey of users and providers of recycled water: quality concerns and directions for applied research. *Water Research* 36, 5045–5056.

Hildebrand, E.M. (1943) Peach-suture spot. *Phytopathology* 33, 167–168.

Hill, A.C. (1969) Air quality standards for fluoride vegetation effects. *Journal of the Air Pollution Control Association* 19, 331–336.

Hill, A.C. (1971) Vegetation: a sink for atmospheric pollutants. *Journal of the Air Pollution Control Association* 21, 341–346.

Hill, A.C. and Pack, M.R. (1983) Effect of atmospheric fluoride on plant growth. In: Shupe, J.L., Peterson, H.B. and Leone, N.C. (eds) *Fluorides. Effects on Vegetation, Animals and Humans.* Paragon Press, Salt Lake City, Utah, pp. 105–120.

Hill, A.C., Transtrum, L.G., Pack, M.R. and Winters, W.S. (1958) Air pollution with relation to agronomic crops: VI. An investigation of the 'hidden injury' theory of fluoride damage to plants. *Agronomy Journal* 50, 562–565.

Hill, A.C., Pack, M.R., Transtrum, L.G. and Winters, W.S. (1959) Effects of atmospheric fluorides and various types of injury on the respiration of leaf tissue. *Plant Physiology* 34, 11–16.

Hindawi, I.J. (1970) *Air Pollution Injury to Vegetation.* Publication No. AP-71, National Air Pollution Control Administration, US Department of Health Education and Welfare, Raleigh, North Carolina.

Hitchcock, A.E., Zimmerman, P.W. and Coe, R.R. (1962) Results of ten years' work (1951–1960) on the effect of fluorides on gladiolus. *Contributions from the Boyce Thompson Institute* 21, 303–344.

Hitchcock, A.E., Weinstein, L.H., McCune, D.C. and Jacobson, J.S. (1964) Effects of fluorine compounds on vegetation, with special reference to sweet corn. *Journal of the Air Pollution Control Association* 14, 503–508.

Hitchcock, A.E., McCune, D.C., Weinstein, L.H., MacLean, D.C., Jacobson, J.S. and Mandl, R.H. (1971) Effects of hydrogen fluoride fumigation on alfalfa and orchard grass: a summary of experiments from 1952 through 1965. *Contributions from the Boyce Thompson Institute* 24, 363–386.

Hocking, M.B., Hocking, D. and Smythe, T.A. (1981) Fluoride distribution and dispersion processes about an industrial point source in a forested coastal zone. *Water, Air, and Soil Pollution* 14, 133–157.

Hodge, H.C. (1986) Evaluation of some objections to water fluoridation. In: Newbrun, E. (ed.) *Fluorides and Dental Caries,* 3rd edn. Charles C. Thomas, Springfield, Illinois, pp. 221–255.

Hodge, H.C. and Smith, F.A. (1965) Biological properties of inorganic fluorides. In: Simons, J.H. (ed.) *Fluorine Chemistry,* Vol. IV. Academic Press, New York, pp. 1–186.

Hodge, H.C. and Smith, F.A. (1970) Air quality criteria for the effects of fluorides on man. *Journal of the Air Pollution Control Association* 20, 226–232.

Hodge, H.C. and Smith, F.C. (1977) Occupational fluoride exposure. *Journal of Occupational Health* 19, 12–39.

Hodge, H.C., Smith, F.A. and Chen, P.S. (1963) Biological effects of organic fluorides. In: Simons, J.H. (ed.) *Fluorine Chemistry,* Vol. III. Academic Press, New York, pp. 1–240.

Hoffman, D.J., Pattee, O.H. and Wiemeyer, S.N. (1985) Effects of fluoride on screech owl reproduction: teratological evaluation, growth, and blood chemistry in hatchlings. *Toxicological Letters* 26, 19–24.

Høgdahl, B., Karstensen, R. and Virik, E. (1977) Solution of an indemnity case for damage to vegetation around an aluminium smelter. In: *Meeting of the Metallurgical Society of the American Institute of Mining, Metallurgical and Petroleum Engineers, Atlanta, Georgia, USA,* paper A 77–84.

Holopainen, J.K., Bergman, T., Hautala, E.-L. and Oksanen, J. (1995) The ground beetle fauna (Coleoptera: Carabidae) in relation to soil properties and foliar fluoride content in spring cereals. *Pedobiologia* 39, 193–206.

Holopainen, T.H. (1984) Types and distribution of ultrastructural symptoms in epiphytic lichens in several urban and industrial locations in Finland. *Annales Botanici Fennici* 21, 213–229.

Horiuchi, N. (1960) Microdetermination of fluorine in living organisms VI. Stability of the C–F link in soil. *Takamine Kankyusho Nempo* 12, 310–331.

Horntvedt, R. (1997) Accumulation of airborne fluorides in forest trees and vegetation. *European Journal of Forest Research* 27, 73–82.

Horsman, D.C. and Wellburn, A.R. (1976) Guide to the metabolic and biochemical effects of air pollutants on higher plants. In: Mansfield, T.A. (ed.) *Effects of Air Pollutants on Plants.* Cambridge University Press, Cambridge, pp. 185–199.

Horvath, I., Klasova, A. and Navara, J. (1978) Some physiological and ultrastructural changes of *Vicia faba* after fumigation with hydrogen fluoride. *Fluoride* 11, 89–99.

Hou, P.S. (1997) The control of coal burning fluorosis in China. *Fluoride* 30, 229–232.

Huggett, R.J., Kimerle, R.A., Mehrle, P.M. and Berman, H.L. (1992) *Biomarkers: Biochemical, Physiological, and Histological Markers of Anthropogenic Stress.* SETAC Special Publications Series, Lewis Publishers, Chelsea, Maryland.

Hughes, P.R. (1988) Insect populations on host plants subjected to air pollution. In: Heinrichs, E.A. (ed.) *Plant Stress–Insect Interactions.* John Wiley and Sons, New York, pp. 249–319.

Hughes, P.R. and Laurence, J.A. (1984) Relationship of biochemical effects of air pollutants on plants to environmental problems: insect and microbial interactions. In: Koziol, M.J. and Whatley, F.R. (eds) *Gaseous Air Pollutants and Plant Metabolism*. Butterworths, London, pp. 361–377.

Hughes, P.R., Weinstein, L.H., Johnson, L.M. and Braun, A.R. (1985) Fluoride transfer in the environment: accumulation and effects on cabbage looper *Trichoplusia ni* of fluoride from water soluble salts and hydrogen fluoride-fumigated leaves. *Environmental Pollution Series A*, 37, 175–192.

Huttunen, S. and Soikkeli, S. (1984) Effects of various gaseous pollutants on plant ultra-structure. In: Koziol, M.J. and Whatley, F.R. (eds) *Gaseous Air Pollutants and Plant Metabolism*. Butterworths, London, pp. 117–127.

Imai, S. and Sato, S. (1974) On the black spots observed in the integument of silkworms poisoned by fluorine compounds. *Air Pollution Abstracts* 7, 86.

IPCS (1995) International Programme on Chemical Safety Health and Safety Guide Number 99. Diflubenzuron. http://www.inchem.org

Isidorov, V.A., Povarov, V.G. and Prilepsky E.B. (1993a) Geological sources of volatile organic components in regions of seismic and volcanic activity. *Journal of Ecological Chemistry* 1, 19–25.

Isidorov, V.A., Prilepsky, E.B. and Povarov, V.G. (1993b) Photochemically and optically active components of minerals and gas emissions of mining plants. *Journal of Ecological Chemistry* 1, 201–207.

Israel, G.W. (1974a) Evaluation and comparison of three atmospheric fluoride monitors under field conditions. *Atmospheric Environment* 8, 159–166.

Israel, G.W. (1974b) A field study of the correlation of static lime paper sampler with forage and cattle urine. *Atmospheric Environment* 8, 167–181.

Israel, G.W. (1974c) Deposition velocity of gaseous fuorides on alfalfa. *Atmospheric Environment* 8, 1329–1330.

Jackson, P.J., Harvey, P.W. and Young, W.F. (2002) *Chemistry and Bioavailability Aspects of Fluoride in Drinking Water*. Report No. 5037, Wrc-NSF Ltd, Marlow, UK.

Jacobson, J.S. and Heller, L.I. (1978) Collaborative study of three methods for the determination of fluoride in vegetation. *Journal of the Association of Official Analytical Chemists* 61, 150–153.

Jacobson, J.S. and McCune, D.C. (1969) Inter-laboratory study of analytical techniques for fluorine in vegetation. *Journal of the Association of Official Analytical Chemists* 52, 894–899.

Jacobson, J.S. and McCune, D.C. (1972) Collaborative study of analytical methods for fluoride in vegetation: effects of individual techniques on results. *Journal of the Association of Official Analytical Chemists* 55, 991–1003.

Jacobson, J.S., Weinstein, L.H., McCune, D.C. and Hitchcock, A.E. (1966) The accumulation of fluorine by plants. *Journal of the Air Pollution Control Association* 16, 412–417.

Jacobson, J.S., Troiano, J., Cosentini, C.C. and Evans, J. (1982) Evaluation of agreement among routine methods of determination of fluoride in vegetation – interlaboratory collaborative study. *Journal of the Association of Official Analytical Chemists* 65, 1150–1154.

Jamil, K. (2001) *Bioindicators and Biomarkers of Environmental Pollution and Risk Assessment*. Science Publishers, Enfield, New Hampshire.

Jenkins, G.N. (1991) Fluoride intake and its safety among heavy tea drinkers in a British fluoridated city. *Proceedings of the Finnish Dental Society* 87, 571–580.

Ji, R.D. (1993) Research on fluoride level of indoor air in burning coal fluorosis areas. *Journal of Hygiene Research* 22, 10–13.

Johansson, T.S.K. and Johansson, M.P. (1972) The effect of conditioned flour on the toxicity of sodium fluoride to *Tribolium confusum* Duv. *Journal of Economic Entomology* 53, 653–655.

Johnsson, K. (1997) Chemical dating of bones based on diagenetic changes in bone apatite. *Journal of Archaeological Science* 24, 431–437.

Jordan, A. and Frank, H. (1999) Trifluoroacetate in the environment: evidence for sources other than HFC/HCFs. *Environmental Science and Technology* 33, 522–527.

Jordan, J., Harnisch, J., Borchers, R., LeGuerm, F. and Shinhara, H. (2000) Volcanogenic halocarbons. *Environmental Science and Technology* 34, 1122–1124.

Kalmbach, E.R. (1945) 'Ten Eighty': a war-produced rodenticide. *Science* 102, 232–233.

Kalnitsky, G. and Barron, E.S.G. (1947) The effect of fluoroacetate on the metabolism of yeast and bacteria. *Journal of Biological Chemistry* 170, 83–95.

Kalnitsky, G. and Barron, E.S.G. (1948) The inhibition by fluoroacetate and fluoro-butyrate of fatty acid and glucose oxidation produced by kidney homogenates. *Archives of Biochemistry* 19, 75–87.

Kannan, K., Koistinen, J., Beckmen, K., Evans, T., Gorzelany, J.F., Hansen, K.J., Jones, P.D., Helle, E., Nyman, M. and Giesy, J.P. (2001a) Accumulation of perfluorooctane sulfonate in marine mammals. *Environmental Science and Technology* 35, 1593–1598.

Kannan, K., Franson, J.C., Bowerman, W.W., Hansen, K.J., Jones, P.D. and Geisy, P.J. (2001b) Perfluorooctane sulfonate in fish-eating water birds including bald eagles and albatrosses. *Environmental Science and Technology* 35, 3065–3070.

Kannan, K., Newsted, J.L., Hallbrook, R.S. and Geisy, J.P. (2002a) Perfluorooctane sulfonate and related fluorinated hydrocarbons in mink and river otters from the United States. *Environmental Science and Technology* 36, 2566–2571.

Kannan, K., Corsolini, S., Falandysz, J., Oehme, G., Focardi, S. and Geisy, J.P. (2002b) Perfluorooctane sulfonate and related fluorinated hydrocarbons in marine mammals, fishes, and birds from coasts of the Baltic and Mediterranean Seas. *Environmental Science and Technology* 36, 3210–3216.

Karstad, L. (1967) Fluorosis in deer (*Odocoileus virginianus*). *Bulletin of the Wildlife Disease Association* 3, 42–46.

Kastle, J.H. and Loevenhart, A.S. (1900) Concerning lipase, the fat splitting enzyme, and the reversibility of its action. *American Chemical Journal* 24, 491–525.

Kawasaki, H., Tone, N. and Tonomura, K. (1981) Plasmid-determined dehalogenation of haloacetates in *Moraxella* species. *Agricultural and Biological Chemistry* 45, 29–34.

Kay, C.E., Tourangeau, P.C. and Gordon, C.C. (1975) Industrial fluorosis in wild mule and whitetail deer from western Montana. *Fluoride* 8, 182–191.

Keller, T. (1974) Translocation of fluoride in woody plants. *Fluoride* 7, 31–35.

Keller, T. (1977) Der Einfluss von Fluor-immissionen auf die Nettoassimilation von Waldbaumarten. *Eidgenossische Anstalt für das Forstliche Versuchswesen* 53, 161–198.

Kelly, M. (1965) Isolation of bacteria able to metabolize fluoroacetate or fluoroacetamide. *Nature* 208, 809–810.

Ketring, D.L., Young, R.E. and Biale, J.B. (1968) Effects of monofluoroacetate on *Penicillium digitatum* metabolism and on ethylene biosynthesis. *Plant and Cell Physiology* 9, 617–631.

Key, B.D., Howell, R.D. and Criddle, C.S. (1997) Fluorinated organics in the biosphere. *Environmental Science and Technology* 31, 2445–2454.

Kierdorf, H. and Kierdorf, U. (1999) Reduction of fluoride deposition in the vicinity of a brown coal-fired power plant as indicated by bone fluoride concentrations of roe deer (*Capreolus capreolus* L.). *Bulletin of Environmental Contamination and Toxicology* 63, 473–477.

Kierdorf, H., Kierdorf, U., Sedlacek, F. and Erdelen, M. (1996) Mandibular bone fluoride levels and occurrence of fluoride induced dental lesions in populations of wild deer (*Cervus elaphus*) from central Europe. *Environmental Pollution* 93, 75–81.

Kierdorf, H., Kierdorf, U., Richards, A. and Sedlacek, F. (2000) Disturbed enamel formation in wild boars (*Sus scrofa*) from fluoride polluted areas in Central Europe. *Anatomical Record* 259, 12–24.

Kierdorf, U. (1988) A study on chronic fluoride intoxication in roe deer (*Capreolus capreolus* L.) caused by emissions. *Zeitschrift Jagdwisschaft* 34, 192–204.

Kierdorf, U., Kierdorf, H. and Fejerskov, O. (1993) Fluoride-induced developmental changes in enamel and dentine of European roe deer (*Capreolus capreolus* L.) as a result of environmental pollution. *Archives of Oral Biology* 38, 1071–1081.

Kierdorf, U., Kierdorf, H., Sedlacek, F. and Fejerskov, O. (1996) Structural changes in fluorosed dental enamel of red deer (*Cervus elaphus* L.) from a region with severe environmental pollution by fluorides. *Journal of Anatomy* 188, 183–195.

Kim, B.R., Suidan, M.T., Wallington, T.J. and Du, X. (2000) Is trifluoroacetic acid biologically degradable? *Environmental Engineering Science*, 17, 337–342.

King, D.R., Oliver, A.J. and Mead, R.J. (1978) The adaptation of some Western Australian mammals to food plants containing fluoroacetate. *Australian Journal of Zoology* 26, 699–712.

King, J.E. and Penfound, W.T. (1946) Effects of new herbicides on fish. *Science* 103, 487.

Kirsten, E., Sharma, M.L. and Kun, E. (1978) Molecular toxicology of (–)-erythrofluorocitrate: selective inhibition of citrate transport in mitochondria and the binding of fluorocitrate to mitochondrial proteins. *Molecular Pharmacology* 14, 172–184.

Kitchen, D. and Skinner, W.J. (1969) Retarded hydration calcium sulfite hemihydrate containing AIF_5^{2-} ions. *Nature* 224, 1297–1299.

Kitchen, D. and Skinner, W.J. (1971) Chemistry of byproduct gypsum. 2. Chemistry of co-crystalline AIF_5^{2-}, during gypsum calcination and subsequent mixing of hemihydrate with water. *Journal of Applied Chemistry and Biotechnology* 21, 65–67.

Klockow, D. and Targa, H. (1993) Air pollution and vegetation damage in the tropics: the German/Brazilian Interdisciplinary Project 'Serra do Mar' in the industrial area of Cubatão/São Paulo, Brazil. In: Junk, W.J. and Bianchi, H.K. (eds) *Studies on Human Impacts on Forests and Floodplains in the Tropics.* GKSS, Geesthacht, Germany, pp. 13–18.

Klockow, D. and Targa, H. (1998) Performance and results of a six-year German/Brazilian research project in the industrial area of Cubatão/S.P., Brazil. *Pure and Applied Chemistry* 70, 2287–2293.

Klumpp, A., Klumpp, G. and Domingos, M. (1994) Plants as bioindicators of air pollution at the Serra do Mar near the industrial complex at Cubatão, Brazil. *Environmental Pollution* 85, 109–116.

Klumpp, A., Domingos, M. and Klumpp, G. (1996a) Assessment of the vegetation risk by fluoride emissions from fertiliser industries at Cubatão, Brazil. *Science of the Total Environment* 192, 219–228.

Klumpp, A., Klumpp, G. and Domingos, M. (1996b) Bioindication of air pollution in the tropics: the active monitoring program near Cubatão (Brazil). *Reinhaltung der Luft* 56, 27–31.

Klumpp, A., Klumpp, G., Domingos, M. and Silva, M.D. (1996c) Fluoride impact on native tree species of the Atlantic Forest near Cubatão, Brazil. *Water, Air and Soil Pollution* 87, 57–71.

Klumpp, A., Modesto, I.F., Domingos, M. and Klumpp, G. (1997) Susceptibility of various *Gladiolus* cultivars to fluoride pollution and their sustainability for bioindication. *Pesquisa Agropecuária Brasileira* 32, 239–247.

Klumpp, A., Domingos, M., de Moraes, R.M. and Klumpp, G. (1998) Effects of complex air pollution on tree species of the Atlantic Rain Forest near Cubatão, Brazil. *Chemosphere* 36, 989–994.

Klumpp, G., Klumpp, A., Domingos, M. and Guderian, R. (1995) *Hemerocallis* as bio-indicator of fluoride pollution in tropical countries. *Environmental Monitoring and Assessment* 35, 27–42

Knabe, W. (1970) Natural loss of fluorine from leaves of Norway spruce (*Picea abies* Karst.). *Staub-Reinhaltung der Luft* 30, 29–32.

Kokot, Z. and Drzewiecki, D. (2000) Fluoride levels in hair of exposed and unexposed populations in Poland. *Fluoride* 33, 196–204.

Koritnig, S. (1951) Ein Beitrag zur Geochemie des Fluor. *Geochimica et Cosmochimica Acta* 1, 89–116.

Kostyniak, P.J. (1979) Defluorination: a possible mechanism of detoxification in rats exposed to fluoroacetate. *Toxicology Letters* 3, 225–228.

Kostyniak, P.J., Bosmann, H.B. and Smith, F.A. (1978) Defluorination of fluoroacetate *in vitro* by rat liver subcellular fractions. *Toxicology and Applied Pharmacology* 44, 89–97.

Kotamarthi, V.R., Rodriguez J.M., Ko M.K.W., Tromp, T.K., Sze, N.D. and Prather, M.J. (1998) Production of trifluoroacetic acid from HCFCs and HFCs: a three-dimensional modelling study. *Journal of Geophysical Research* 103, 5745–5758.

Krebs, H.C., Kemmerling, W. and Habermehl, G. (1994) Qualitative and quantitative determination of fluoroacetic acid in *Arrabidea bilabiata* and *Palicourea marcgravii* by ^{19}F NMR spectroscopy. *Toxicon* 32, 909–913.

Kronberger, W. (1988a) Kinetics of nonionic diffusion of hydrogen fluoride in plants: model estimates on uptake, distribution, and translocation of F in higher plants. *Phyton* 11, 27–49.

Kronberger, W. (1988b) Kinetics of nonionic diffusion of hydrogen fluoride in plants: experimental and theoretical treatment of weak acid permeability. *Phyton* 11, 241–265.

Kronberger, W. and Halbwachs, G. (1978) Distribution of fluoride in *Zea mays* grown near an aluminum plant. *Fluoride* 12, 129–135.

Kronberger W., Halbwachs, G. and Richter, H. (1978) Fluoride translocation in *Picea abies* Karsten. *Angewante Botanik* 52, 149–154.

Kuhn, D.N., Knauf, M. and Stumpf, P.K. (1981) Subcellular localization of acetyl CoA synthetase in leaf protoplasts of *Spinacia oleracea*. *Archives of Biochemistry and Biophysics* 209, 441–450.

Kumpulainen, J. and Koivistoinen, P. (1977) Fluorine in foods. In: Gunther, F.A. and Gunther, J.D. (eds) *Residue Reviews: Residues of Pesticides and Other Contaminants in the Total Environment.* Vol. 68. Springer-Verlag, Berlin, pp. 37–57.

Kun, E., Kirsten, E. and Sharma, M.L. (1977) Enzymatic formation of glutathione-cytryl thioester by a mitochondrial system and its inhibition by (−)erythrofluorocitrate. *Proceedings of the National Academy of Sciences USA* 74, 4942–4946.

Kunze, G. (1929) Geschmachs und Giftverkungen der Fluornatriums auf die Honigbiene. *Nachrichtenblatt für der Deutsche Pflanzenschutzdienst* 9, 13–14.

Kuribayashi, S. (1971) Environmental pollution effects on sericulture and its countermeasures. *Sericultural Science and Technology* 10, 48–49.

Kuribayashi, S. (1972) Influence of air pollution with fluoride on sericulture. *Journal of the Sericultural Society of Japan* 41, 316–322.

Kuribayashi, S. (1977) Effect of atmospheric pollution by hydrogen fluoride on mulberry trees and silkworms. *Journal of the Sericultural Society of Japan* 46, 536–544.

Kuribayashi, S., Yatomi, K. and Kadota, M. (1976) Effects of hydrogen fluoride and sulfur dioxide on mulberry trees and silkworms. *Journal of the Japanese Society of Air Pollution* 6, 155.

LaCosse, J., Herget, W., Spellicy, R. and Beitler, C. (1999) *Measurement and Modeling of HF Emissions from Phosphoric Acid Production Facilities*. Project No. 197-002, Fertilizer Institute, Washington, DC.

Lalonde, J. (1976) Fluorine – an indicator of mineral deposits. *Canadian Mining and Metallurgical Bulletin* May, 110–122.

Largent, J.J. (1961) *Fluorosis: the Health Aspects of Fluorine Compounds*. Ohio State University Press, Columbus, Ohio.

Larsen, S. and Widdowson, A. (1971) Soil fluorine. *Journal of Soil Science* 22, 210–221.

Laurence, J.A. (1981) Effects of air pollutants on plant–pathogen interactions. *Zeitschrift für Pflanzenkrankheiten und Pflanzenschutz* 88, 156–173.

Laurence, J.A. (1983) Effects of fluoride on plant–pathogen interactions. In: Unsworth, M.H. and Ormrod, D.P. (eds) *Effects of Gaseous Air Pollution in Agriculture and Horticulture*. Butterworths, London, pp. 145–149.

Laurence, J.A. and Reynolds, K.L. (1984) Growth and lesion development of *Xanthomonas campestris* pv. *phaseoli* on leaves of red kidney beans exposed to hydrogen fluoride. *Phytopathology* 74, 578–580.

Laurence, J.A. and Reynolds, K.L. (1986) The joint action of hydrogen fluoride and sulfur dioxide on the development of common blight of red kidney bean. *Phytopathology* 76, 514–517.

Lavado, R.S. and Reinaudi, N. (1979) Fluoride in salt affected soils of La Pampa (Republic of Argentina). *Fluoride* 12, 28–32.

LeBlanc, C., Comeau, G. and Rao, D.N. (1971) Fluoride injury symptoms in epiphytic lichens and mosses. *Canadian Journal of Botany* 49, 1691–1698.

LeBlanc, F. and DeSloover, J. (1970) Relation between industrialization and distribution and growth of epiphytic lichens and mosses in Montreal. *Canadian Journal of Botany* 48, 1485–1496.

LeBlanc, F., Rao, D.N. and Comeau, G. (1972) Indices of atmospheric purity and fluoride pollution pattern in Arvida, Quebec. *Canadian Journal of Botany* 50, 991–998.

Ledbetter, M.D., Mavrodineanu, R. and Weiss, A.J. (1960) Distribution studies of radioactive fluorine-18 and stable fluorine-19 in tomato plants. *Contributions from the Boyce Thompson Institute* 20, 331–348.

Lee, C.J., Miller, G.W. and Welkie, G.W. (1966) The effect of hydrogen fluoride and wounding on respiratory enzymes in soybean leaves. *International Journal of Air and Water Pollution* 10, 169–181.

Lee, G., Ellersieck, M.R., Mayer, F.L. and Krause, G.F. (1995) Predicting chronic lethality of chemicals to fishes from acute toxicity test data: multifactor probit analysis. *Environmental Toxicology and Chemistry* 14, 345–349.

Leone, N.C., Martin, A.E., Minoguchi, G., Schlesinger, E.R. and Siddiqui, A.H. (1970) Fluorides and general health. In: *Fluorides and Human Health*. Monograph Series No. 59. World Health Organization, Geneva, Switzerland, pp. 273–321.

Leslie, R. and Parberry, D.G. (1972) Growth of *Verticillium lecanii* on medium containing sodium fluoride. *Transactions of the British Mycological Society* 58, 351–352.

Less, L.N., McGregor, A., Jones, L.H.P., Cowling, D.W. and Leafe, E.L. (1975) Fluoride uptake by gas from aluminium smelter fume. *International Journal of Environmental Studies* 7, 153–160.

Lévy, L. and Strauss, R. (1973) Sur l'insolubilization de l'ion fluor à l'état de fluorure de calcium chez *Chara fragilis* Desvaux. *Comptes Rendus Hebdomadaires des Séances de l'Académie des Sciences, Serie D* 277, 181–184.

Lhoste, A.M. and Garrec, J.P. (1975) Étude biométrique des effets du fluor sur la structure des chloroplastes de maïs (*Zea mays* L.

var. Inra 260). *Journal de Microscopie et de Biologie Cellulaire* 24, 351–364.

Li, D., Duan, R., Wang, S., He, G., Li, P., Nie, Z. and Wen, T. (1999) Epidemiological and radiological study of skeletal fluorosis in Minzhi Town, Longli County, Guizhou Province, China. *Fluoride* 32, 55–59.

Li, J.X. and Cao, S.R. (1994) Recent studies on endemic fluorosis in China. *Fluoride* 27, 125–127.

Li, M.J. and Tang, S.Z. (1996) Fluoride, selenium, sulphur and carbon contents of coal in Hubei Province. *Fluoride* 29, 77–78.

Liang, J.Y., Shyu., T.H. and Lin, H.C. (1996) The aluminum complexes in the xylem sap of a tea plant. *Journal of the Chinese Agricultural Chemical Society* 34, 695–702.

Liébecq, C. and Osterrieth, P.M. (1963) Modification de la radiosensibilité de *Escherichia freundii* cultivée en présence de fluoroacétate. *Societê Internationale Physiologie et Biochimie* 70, 125–127.

Liébecq, C. and Osterrieth, P.M. (1965) Radiosensibilité des cultures de *Escherichia freundii* cultivée en présence de fluoroacétate. *International Journal of Radiation Biology* 9, 253–259.

Liébecq, C. and Peters, R.A. (1949) The toxicity of fluoroacetate and the tricarboxylic acid cycle. *Biochimica et Biophysica Acta* 3, 215–230.

Linero, A.A. and Baker, R.A. (1978) *Evaluation of Emissions and Control Techniques for Reducing Fluoride Emissions from Gypsum Ponds in the Phosphoric Acid Industry.* EPA 600/2–78–124, US Environmental Protection Agency, Research Triangle Park, North Carolina.

Livingstone, P.G. (1994) The use of 1080 in new Zealand. In: Seawright, A.A. and Eason, C.T. (eds) *Proceedings of the Science Workshop on 1080.* Miscellaneous Series 28, Royal Society of New Zealand, SIR Publishing, Wellington, pp. 1–9.

Loganathan, P., Hedley, G.C., Wallace, G.C. and Roberts, A.H.C. (2001) Fluoride accumulation in pasture forages and soils following long-term application of phosphorus fertilisers. *Environmental Pollution* 115, 275–282.

Lopes, C.F.F. (1999) *Fluoreto na atmosferica de Cubatão.* Technical Report, CETESB, São Paulo, Brazil.

Lorenc-Plucinska, G. and Oleksyn, J. (1982) Effect of hydrogen fluoride on photosynthesis, photorespiration and dark respiration in scotch pine (*Pinus sylvestris*). *Fluoride* 15, 149–156.

Louw, N., de Villiers, O.T. and Grobbelaar, N. (1970) The inhibition of aconitase activity from *Dichapetalum cymosum* by fluorocitrate. *South African Journal of Science* 66.

Lovelace, C.J. and Miller, G.W. (1967) *In vitro* effects of fluoride on tricarboxylic acid cycle dehydrogenases and oxidative phosphorylation. Part I. *Journal of Histochemistry and Cytochemistry* 15, 195–201.

Lovelace, J., Miller, G.W. and Welkie, G.W. (1968) The accumulation of fluoroacetate and fluorocitrate in forage crops collected near a phosphate plant. *Atmospheric Environment* 2, 187–190.

Lovelock, J.E. (1971) Atmospheric fluorine compounds as indicators of air movements. *Nature* 230, 379.

Luo, K., Xu, L., Li, R. and Xiang, L. (2002) Fluorine emissions from combustion of steam coal of North China Plate and Northwest China. *Chinese Science Bulletin* 47, 1346–1350.

Luštinec, J. and Pokorná, V. (1962) Alternation of respiratory pathways during the development of wheat leaf. *Biologia Plantarum (Praha)* 4, 101–109.

Luštinec, J., Pokorná, V. and Růžička, J. (1962) Activation of glycolysis and inhibition of glucose transport into leaves by fluoride. *Biologia Plantarum (Praha)* 4, 126–130.

Lynch, A.J., McQuaker, N.R. and Gurney, M. (1978) Calibration factors and estimation of atmospheric SO_2 and fluoride by use of solid absorbents. *American Chemical Society* 12, 169–173.

Lynch, D.W. (1951) Diameter growth of Ponderosa pine in relation to the Spokane-blight problem. *Northwest Science* 25, 157–163.

McClenahen, J.R. and Weidensaul, T.C. (1977) Geographic distribution of airborne fluorides near a point source in southeast Ohio. *Ohio Agricultural Research and Development Center Research Bulletin* 1093, 2–29.

McClure, F.J. (1949) Fluorine in foods. *Public Health Reports* 64, 1061–1074.

McCombie, H. and Saunders, B.C. (1946) Fluoroacetates and allied compounds. *Nature* 158, 382–388.

McCornack, A.A., Cochran, L.C., English, H. and Hemstreet, C.L. (1952) A black tip condition of peach fruits in California. *Plant Disease Reporter* 36, 99–100.

McCulloch, A. (1999) CFC and halon replacements in the environment. *Journal of Fluorine Chemistry* 100, 163–173.

McCulloch, A. (2000) Halocarbon greenhouse gas emissions during the next century. In: van Ham, J., Baede, A.P.M., Meyer, L.A. and

Ybema, R. (eds) *Non-CO₂ Greenhouse Gases: Scientific Understanding Control and Implementation.* Kluwer Academic Publishers, Noordwijkerhout, The Netherlands, pp. 223–230.

McCune, D.C. (1969) *On the Establishment of Air Quality Criteria, with Reference to the Effects of Atmospheric Fluoride on Vegetation.* Monograph No. 69–3, Air Quality Monographs, American Petroleum Institute, New York.

McCune, D.C. (1983) Interactions of fluorides with other air pollutants. In: Shupe, J.L., Peterson, H.B. and Leone, N.C. (eds) *Fluorides. Effects on Vegetation, Animals and Humans.* Paragon Press, Salt Lake City, Utah, pp. 121–126.

McCune, D.C. and Hitchcock, A.E. (1971) Fluoride in forage: factors determining its accumulation from the atmosphere and concentration in the plant. In: *Proceedings of the 2nd International Clean Air Congress.* Academic Press, New York, pp. 289–292.

McCune, D.C. and Weinstein, L.H. (1971) Metabolic effects of atmospheric fluorides on plants. *Environmental Pollution* 1, 169–174.

McCune, D.C., Weinstein, L.H., Jacobson, J.S. and Hitchcock, A.E. (1964) Some effects of atmospheric fluoride on plant metabolism. *Journal of the Air Pollution Control Association* 14, 465–468.

McCune, D.C., Hitchcock, A.E., Jacobson, J.S. and Weinstein, L.H. (1965) Fluoride accumulation and growth of plants exposed to particulate cryolite in the atmosphere. *Contributions from the Boyce Thompson Institute* 23, 1–12.

McCune, D.C., Hitchcock, A.E. and Weinstein, L.H. (1966) Effect of mineral nutrition on the growth and sensitivity of gladiolus to hydrogen fluoride. *Contributions from the Boyce Thompson Institute* 23, 295–300.

McCune, D.C., De Hertogh, A.A. and Weinstein, L.H. (1967) Effect of HF fumigation on ¹⁴C-glucose metabolism. In: *Abstracts of Papers for the 153rd Meeting, American Chemical Society.* American Chemical Society, Miami Beach, Florida. Spaulding-Moss, Boston, Maryland, abstract 15.

McCune, D.C., Weinstein, L.H., Mancini, J.F. and van Leuken, P. (1973) Effects of hydrogen fluoride on plant–pathogen interactions. In: *Proceedings of the Third International Clean Air Congress.* VDI-Verlag GmbH, Düsseldorf, pp. A146–A149.

McCune, D.C., MacLean, D.C. and Schneider, R.E. (1976) Experimental approaches to the effects of airborne fluoride on plants. In: Mansfield, T.A. (ed.) *Effects of Air Pollutants on Plants.* Cambridge University Press, Cambridge, pp. 31–43.

McCune, D.C., Silberman, D.H. and Weinstein, L.H. (1977) Effects of relative humidity and free water on the phytotoxicity of hydrogen fluoride and cryolite. In: *Proceedings, 4th International Clean Air Congress.* Japanese Union of Pollution Prevention Associations, Tokyo, pp. 116–119.

McCune, D.C., Ormrod, D.P. and Reinert, R.A. (1984) Effects of pollutant mixtures on vegetation. In: Lefohn, A.S. and Ormrod, D.P. (eds) *A Review and Assessment of the Effects of Pollutant Mixtures on Vegetation – Research Recommendations.* Publication EPA-600/3–84–037. US Environmental Protection Agency, Corvallis, Oregon.

McDonagh, M.S., Whiting, P.F., Wilson, P.M., Sutton, A.J., Chestnutt, I., Cooper, J., Misso, K., Bradley, M., Treasure, E. and Kleijnen, J. (2000) Systematic review of water fluoridation. *British Medical Journal* 321, 855–859.

McEwan, T. (1964a) Isolation and identification of the toxic principle of *Gastrolobium grandiflorum. Nature* 201, 827.

McEwan, T. (1964b) Isolation and identification of the toxic principle of *Gastrolobium grandiflorum. Queensland Journal of Agricultural Science* 21, 1–14.

McGannon, H.E. (1964) *The Making, Shaping and Treating of Steel,* 8th edn. United States Steel Corporation, Pittsburgh.

McIlroy, J.C. (1994) Susceptibility of target and non-target animals to 1080. In: Seawright, A.A. and Eason, C.T. (eds) *Proceedings of the Science Workshop on 1080.* Miscellaneous Series 28, Royal Society of New Zealand, SIR Publishing, Wellington, pp. 90–96.

MacIntire, W.H. (1957) Fate of air-borne fluorides and attendant effects upon soil reaction and fertility. *Journal of the Association of Official Agricultural Chemists* 40, 958–976.

MacIntire, W.H., Winterberg, S.H., Thompson, J.G. and Hatcher, B.W. (1942) Fluorine content of plants fertilized with phosphates and slags carrying fluorides. *Industrial and Engineering Chemistry* 34, 1469–1479.

MacIntire, W.H., Shaw, W.M., Robinson, B. and Sterges, A.J. (1948) Disparity in the leaching ability of fluorine from incorporations in phosphated and slagged soils. *Soil Science* 65, 321–329.

MacIntire, W.H., Winterberg, S.H., Clements, L.B., Jones, L.S. and Robinson, B. (1951a) Effect of fluorine carriers on crops and drainage waters. *Industrial and Engineering Chemistry* 43, 1797–1799.

MacIntire, W.H., Winterberg, S.H., Clements, L.B., Hardin, L.J. and Jones, L.S. (1951b) Crop and soil reactions to applications of hydrofluoric acid. *Industrial and Engineering Chemistry* 43, 1800–1803.

MacIntire, W.H., Shaw, W.M. and Robinson, B. (1955a) Behavior of incorporations of potassium and calcium fluorides in a 6-year lysimeter study. *Journal of Agriculture and Food Chemistry* 3, 772–777.

MacIntire, W.H., Sterges, A.J. and Shaw, W.M. (1955b) Fate and effects of hydrofluoric acid added to four Tennessee soils in a 4-year lysimeter study. *Journal of Agriculture and Food Chemistry* 3, 777–782.

MacIntire, W.H., Hardin, L.J. and Buehler, M.H. (1958) Fluorine in Maury County, Tennessee. *Tennessee Agricultural Experiment Station Bulletin*, Knoxville, Tennessee 279, 33 pp.

McKee, J.E. and Wolf, H.W. (1963) *Water Quality Criteria*, 2nd edn. Publication No. 3-A, State Water Quality Control Board, Resource Agency of California, Sacramento, California.

McLaughlin, M.J., Tiller, K.G., Naidu, R. and Stevens, D.P. (1996) Review: the behaviour and environmental impact of contaminants in fertilisers. *Australian Journal of Soil Research* 34, 1–54.

McLaughlin, M.J., Stevens, D.P., Keerthisinghe, G., Cayley, J.W.D. and Ridley, A.M. (2001) Contamination of soil with fluoride by long-term application of superphosphates to pastures and risk to grazing animals. *Australian Journal of Soil Research* 39, 627–640.

McLaughlin, S.B. and Barnes, R.L. (1975) Effects of fluoride on photosynthesis and respiration of some south-east American forest trees. *Environmental Pollution* 8, 93–96.

MacLean, D.C. (1982) Air quality standards for fluoride to protect vegetation: regional, seasonal, or other considerations. *Journal of the Air Pollution Control Association* 32, 82–84.

MacLean, D.C. (1983) Factors that modify the response of plants to fluoride. In: Shupe, J.L., Peterson, H.B. and Leone, N.C. (eds) *Fluorides: Effects on Vegetation, Animals and Humans*. Paragon Press, Salt Lake City, Utah, pp. 139–144.

MacLean, D.C. and Schneider, R.E. (1971) Fluoride phytotoxicity: its alteration by temperature. In: Englund, H.M. and Beery, W.T. (eds) *Proceedings. 2nd International Clean Air Congress*. Academic Press, New York, pp. 292–295.

MacLean, D.C. and Schneider, R.E. (1981) Effects of gaseous hydrogen fluoride on the yield of field-grown wheat. *Environmental Pollution* 24, 39–44.

MacLean, D.C., Roark, O.F., Folkerts, G. and Schneider, R.E. (1969) Influence of mineral nutrition on the sensitivity of tomato plants to hydrogen fluoride. *Environmental Science and Technology* 3, 1201–1204.

MacLean, D.C., Schneider, R.E. and McCune, D.C. (1973) Fluoride phototoxicity as affected by relative humidity. In: *Proceedings of the Third International Clean Air Congress*. VDI-Verlag GmbH, Dusseldorf, pp. A143–A145.

MacLean, D.C., Schneider, R.E. and McCune, D.C. (1976) Fluoride susceptibility of tomato plants as affected by magnesium nutrition. *Journal of the American Society for Horticultural Science* 101, 347–352.

MacLean, D.C., Schneider, R.E. and McCune, D.C. (1977) Effects of chronic exposure to gaseous fluoride on the field of field-grown bean and tomato plants. *Journal of the American Society for Horticultural Science* 102, 297–299.

MacLean, D.C., Schneider, R.E. and Weinstein, L.H. (1982) Fluoride-induced foliar injury in *Solanum pseudo-capsicum*: its induction in the dark and activation in the light. *Environmental Pollution (Series A)* 29, 27–33.

MacLean, D.C., Weinstein, L.H., McCune, D.C. and Schneider, R.E. (1984) Fluoride-induced suture red spot in 'Elberta' peach. *Environmental and Experimental Botany* 24, 353–367.

McMullin, J. (1935) Results and recommendations in the practical use of natural cryolite. *Northwest Fruit Grower* 7, 4–11.

McNulty, I.B. and Newman, D.W. (1957) Effects of atmospheric fluoride on the respiration rate of bush beans and gladiolus leaves. *Plant Physiology* 32, 121–124.

Madden, K.E. and Fox, B.J. (1997) Arthropods as indicators of the effects of fluoride pollution on the succession following sand mining. *Journal of Applied Ecology* 34, 1239–1256.

Madkour, S. and Weinstein, L.H. (1987) Effects of hydrogen fluoride on incorporation and transport of photoassimilates in soybean. *Environmental Toxicology and Chemistry* 6, 627–634.

Mahadevan, T.N., Meenakshy, V. and Mishra, U.C. (1986) Fluoride cycling in nature through precipitation. *Atmospheric Environment* 20, 1745–1749.

Mandl, R.H., Weinstein, L.H. and Keveny, M. (1973) A cylindrical open top chamber for the exposure of plants to air pollutants in the field. *Journal of Environmental Quality* 2, 371–376.

Mandl, R.H., Weinstein, L.H. and Keveny, M. (1975) Effects of hydrogen fluoride and sulphur dioxide alone and in combination on several species of plants. *Environmental Pollution* 9, 133–143.

Mandl, R.H., Weinstein, L.H., Dean, M. and Wheeler, M. (1980) The response of sweet corn to HF and SO_2 under field conditions. *Environmental and Experimental Botany* 20, 359–365.

Mann, H.C. (1992a) Normal superphosphate. In: Buonicore, A.J. and Davis, W.T. (eds) *Air Pollution Engineering Manual*. Air and Waste Management Association, Van Nostrand Reinhold, New York, pp. 576–578.

Mann, H.C. (1992b) Triple superphosphate. In: Buonicore, A.J. and Davis, W.T. (eds) *Air Pollution Engineering Manual*. Air and Waste Management Association, Van Nostrand Reinhold, New York, pp. 5479–5581.

Manning, W. and Feder, W. (1980) *Biomonitoring Air Pollutants with Plants*. Pollutant Monitoring Series, Applied Science, London.

Mansfield, T.A. and McCune, D.C. (1988) Problems of crop loss assessment when there is exposure to two or more gaseous pollutants. In: Heck, W.W., Taylor, O.C. and Tingey, D.T. (eds) *Assessment of Crop Loss from Air Pollutants*. Elsevier Applied Science, London, pp. 317–344.

Manskova, B. (1975) Influence of fluorine emission from an aluminum factory plant and its content in different developmental stages of European pine shoot moth *Rhyaconia buoliana* (Den. and Schiff.) (Lepidoptera). *Biologia Bratislava* 30, 355–360.

Marais, J.S.C. (1943) The isolation of the toxic principle 'potassium cymonate' from 'Gifblaar', *Dichapetalum cymosum* (Hook) Engl. *Onderstepoort Journal of Veterinary Science* 18, 203–206.

Marais, J.S.C. (1944) Monofluoroacetic acid, the toxic principle of 'Gifblaar' *Dichapetalum cymosum* (Hook) Engl. *Onderstepoort Journal of Veterinary Science* 20, 67–73.

Marchant, N.G. (1992) History of plant poisoning in Western Australia. In: Colegate, S.M. and Dorling, P.R. (eds) *Plant-associated Toxins: Agricultural, Phytochemical and Ecological Aspects*. CAB International, Wallingford, pp. 7–12.

Marier, J.R. (1977) Some current aspects of environmental fluoride. *Science of the Total Environment* 8, 253–265.

Marquis, R.E., Clock, S.A. and Mota-Meira, M. (2002) Fluoride and weak acids as modulators of microbial physiology. *FEMS Microbiology Reviews* 26, 493–510.

Marschner, H. (1995) *Mineral Nutrition of Plants*, 2nd edn. Academic Press, London, 889 pp.

Marshall, J., Groves, K. and Fallscheer, H. (1939) Cryolite in codling moth control and a new procedure for its application. *Washington State Horticultural Association Proceedings* 34, 123–131.

Martin v. Reynolds Metals Company (1958) 135 F Supp 379 (1955), aff'd, 258 F2d 321 (9th Cir), cert denied, 358 US 840.

Martin v. Reynolds Metals Company (1959) 221 Or 86, 342 P2d 790.

Martin v. Reynolds Metals Company (1964) 224 F Supp 9798 (1963), aff'd, 337 F2d 780 (9th Cir 1964).

Mascarenhas, J.P. and Machlis, L. (1964) Chemotrophic response of the pollen of *Antirrhinum majus* to calcium. *Plant Physiology* 39, 70–77.

Massey, L.M. (1952) Similarities between disease symptoms and chemically induced injury to plants. In: McCabe, L.C. (ed.) *Air Pollution. Proceedings US Technical Conference on Air Pollution*. McGraw-Hill, New York, pp. 48–52.

Matsushima, J. and Brewer, R.J. (1972) Influence of sulfur dioxide and hydrogen fluoride as a mix or reciprocal exposure on citrus growth and development. *Journal of the Air Pollution Control Association* 22, 710–713.

Maurizio, A. (1957) Factors affecting the toxicity of fluorine compounds to bees. *Bee World* 38, 314.

Maurizio, A. (1960) Bestimmung der letalen Dosis einger Fluorverbindungen für Bienen. Zugleich ein Beitrag zur Methodik der Giftwertbestimmung in Bienenversuchen. In: *Proceedings, Fourth International Congress of Crop Protection*, Vol. 2. Hamburg, pp. 1709–1713.

Maurizio, A. and Staub, M. (1956) Bienenwertgiftungen mitt Fluorhaltigen in der Schweiz. *Schweiz Bienen-Zeitung* 79, 476–486.

Mayer, D.F., Lunden, J.D. and Weinstein, L.H. (1988) Evaluation of fluoride levels and effects on honey bees (*Apis mellifera* L.) (Hymenoptera: Apidiae). *Fluoride* 21(3), 113–120.

Mead, R.J. and Segal, W. (1972) Fluoroacetic acid biosynthesis: a proposed mechanism. *Australian Journal of Biological Science* 25, 327–333.

Mead, R.J., Oliver, A.J. and King, D.R. (1979) Metabolism and defluorination of fluoroacetate by the brush-tailed possum (*Trichosurus vulpecula*). *Australian Journal of Biological Science* 32, 15–26.

Mead, R.J., Oliver, A.J., King, D.R. and Hubach, P.H. (1985a) The coevolutionary role of fluoroacetate in plant–animal interactions in Australia. *Oikos* 44, 55–60.

Mead, R.J., Twigg, L.E., King, D.R. and Oliver, A.J. (1985b) The tolerance to fluoroacetate of geographically separated population of the quokka (*Setonix brachyurus*). *Australian Zoologist* 21, 503–511.

Meiers, P. (2003) History of many aspects of fluoride. www.pmeiers.bei.t-online.de/

Mellanby, K. (2000) *Ecological Indicators for the Nation*. National Academy Press, Washington, DC.

Meyer, J.J.M., Grobbelaar, N. and Steyn, P.L. (1990) Fluoroacetate-metabolizing pseudomonad isolated from *Dichapetalum cymosum*. *Applied Environmental Microbiology* 56, 2152–2155.

Meyer, J.J.M., Grobbelaar, N., Vleggaar, R. and Louw, A.I. (1992) Fluoroacetyl-coenzyme A hydrolase-like activity in *Dichapetalum cymosum*. *Journal of Plant Physiology* 139, 369–372.

Mezzetti, A. and Sansavini, S. (1977) Esame dei danni prodotti su pesco da emissione di fluro e fattori concomitanti. *Genio Rurale* 5, 19–31.

Michel, J., Suttie, J.W. and Sunde, M.L. (1984) Fluorine deposition in bone as related to physiological state. *Poultry Science* 63, 1407–1411.

Middleton, J. (1845) On fluorine in bones, its sources, and its application to the determination of the geological age of fossil bones. *Quarterly Journal of the Geological Society*, 214–216.

Midgley, P.A. (1995) Alternatives to CFCs and their behaviour in the atmosphere. In: Hester, R.E. and Harrison, R.M. (eds) *Volatile Organic Compounds in the Atmosphere*. Issues in Environmental Science and Technology, No. 4. Royal Society of Chemistry, Cambridge, pp. 91–108.

Midgley, T. and Henne, A. (1930) Organic fluorides as refrigerants. *Industrial and Engineering Chemistry* 22, 542–547.

Midgley, P.A. and McCulloch, A. (1999) International regulations on halocarbons. In: Fabian, P. and Singh, O.M. (eds) *The Handbook of Environmental Chemistry*, Vol. 4, Part E. Springer-Verlag, Berlin, pp. 204–221.

Milhaud, G., Enriquez, B. and Clauw, M. (1984) Bioavailability of soil fluoride for sheep: preliminary study. *Recueil de Médecine Vétérinaire* 160, 377–380.

Miller, G.W. (1958) Properties of enolase in extracts from pea seed. *Plant Physiology* 33, 199–206.

Miller, J.E. and Miller, G.W. (1974) Effects of fluoride in fluoride on mitochondrial activity in higher plants. *Physiologia Plantarum* 32, 115–121.

Miller, V.L., Allmendinger, D.F., Johnson, F. and Polley, D. (1953) Lime papers and indicator plants in fluorine air pollution investigations. *Journal of Agriculture and Food Chemistry* 1, 526–529.

Mitterböck, F. and Führer, E. (1988) Effects of fluoride-polluted spruce leaves on nun moth caterpillars, *Lymantria monacha* (Lepidoptera, Lymantridae). *Journal of Applied Entomology* 105, 19–27.

Mizota, C. and Wada, K. (1980) Implications of clay mineralogy to the weathering and chemistry of AP horizons of Ando soils in Japan. *Geoderma* 23, 49–63.

Mohamed, A.H. and Kemner, P.A. (1970) Genetic effects of hydrogen fluoride gas on *Drosophila melanogaster*. *Fluoride Quarterly Reports* 3, 192–200.

Mohr, H. and Kragstrup, J. (1991) A histomorphometric analysis of the effects of fluoride on experimental ectopic bone formation in the rats. *Journal of Dental Research* 70, 957–960.

Molina, M.J. and Rowland, F.S. (1974) Stratospheric sink for chlorofluoromethanes: chlorine atom catalyzed destruction of ozone. *Nature* 249, 810–814.

Monteleone-Neto, R. and Castilla, E.E. (1994) Apparently normal frequency of congenital anomalies in the highly polluted town of Cubatão, Brazil. *American Journal of Medical Genetics* 52, 319–323.

Monteleone-Neto, R., Nrunoni, D., Laurenti, R., Jorge, M.H. de M., Davidson, S.L. and Lebrão, M.L. (1985) Birth defects and environmental pollution: the Cubatão example. In: Marois, M. (ed.) *Prevention and Mental Congenital Defects*. Part B: *Epidemiology, Early Detection and Therapy, and Environmental Factors*. Alan R. Liss, New York, pp. 65–68.

Moore, H.E. (1987) Gaseous fluoride emissions from gypsum settling and cooling ponds. *Florida Scientist* 50, 65–78.

Morton, G.O., Lancaster, J.E., Van Lear, G.E., Fulmor, W. and Meyer, W.E. (1969) The structure of nucelocidin. III (a new structure).

Journal of the American Chemical Society 91, 1535–1537.

Moshida, M. and Yoshida, M. (1974) Symptoms of fluorine intoxication on silkworms, especially the abnormal arthroidal membrane. In: *Japanese Society of Sericulture, 41st Annual Meeting.* p. 24.

Moss, J., Murphy, C.D., Hamilton, J.T.G., McRoberts, W.C., O'Hagan, D., Schaffrath, C. and Harper, D.B. (2000) Fluoroacetaldehyde: a precursor of both fluoroacetate and 4-fluorothreonine in *Streptomyces cattleya*. *Chemical Communications* 22, 2281–2282.

MRC (2002) *Water Fluoridation and Health*. Medical Research Council Working Group Report, Medical Research Council, London, UK.

Mueller, B. and Worseck, M. (1970) Damage to bees caused by arsenic- and fluorine-containing industrial flue gases. *Monatchefte für Veterinaermedicin* 25, 554–556.

Mukai, K. and Ishida, H. (1970) The alkaline filter paper method for surveying fluorides in the atmosphere. Paper presented to the American Institute of Mining, Metallurgy and Petroleum Engineers, Denver, Colorado, pp. 35–40.

Mun, A.I., Brazilevi, Z.A. and Budayeva, K.P. (1966) Geochemical behavior of fluorine in bottom sediments of continental basins. *Geochemistry International USSR* 3, 698–703.

Murray, F. (1981) Fluoride cycles in an estuarine ecosystem. *Science of the Total Environment* 17, 223–241.

Murray, F. (1982) Ecosystems as sinks for atmospheric fluorides. In: Murray, F. (ed.) *Fluoride Emissions: Their Monitoring and Effects on Ecosystems*. Academic Press, Sydney, pp. 191–205.

Murray, F. (1983) Response of grapevines to fluoride under field conditions. *Journal of the American Society for Horticultural Science* 108, 526–529.

Murray, F. and Wilson, S. (1988a) The joint action of sulphur dioxide and hydrogen fluoride on the yield and quality of wheat and barley. *Environmental Pollution* 55, 239–249.

Murray, F. and Wilson, S. (1988b) Effects of sulphur dioxide, hydrogen fluoride and their combination on three *Eucalyptus* species. *Environmental Pollution* 52, 265–279.

Murray, F. and Wilson, S. (1988c) Joint action of sulfur dioxide and hydrogen fluoride on growth of *Eucalyptus tereticornis*. *Environmental and Experimental Botany* 28, 343–349.

Murray, F. and Wilson, S. (1990) Yield responses of soybean, maize, peanut and navy bean exposed to SO$_2$, HF and their combination. *Environmental and Experimental Botany* 30, 215–223.

Murray, L.R. and Woolley, D.R. (1968) The uptake of fluoride by *Acacia georginae* F.M. Bailey. *Australian Journal of Soil Research* 6, 203–210.

Murray, L.R., McConnell, J.D. and Whitten, J.H. (1961) Suspected presence of fluoroacetate in *Acacia georginae* F.M. Bailey. *Australian Journal of Science* 24, 41–42.

Murray, M.M. and Wilson, D.C. (1946) Fluorine hazards with special reference to some social consequences of industrial processes. *Lancet* ii, 821–824.

Musselman, R.C. and Minnick, T.J. (2000) Nocturnal stomatal conductance and ambient air quality standards for ozone. *Atmospheric Environment* 34, 719–733.

Nagata, T., Hayatsu, M. and Kosuge, N. (2002) Identification of aluminium forms in tea leaves by [27]AI NMR. *Phytochemistry* 31, 1215–1218.

NAS (1971) *Biologic Effects of Air Pollutants: Fluorides*. National Academy of Sciences, Washington, DC.

NAS (1974) *Effects of Fluorides in Animals*. National Academy of Sciences, Washington, DC.

Nash, T.H. (1971) Lichen sensitivity to hydrogen fluoride. *Bulletin of the Torrey Botanical Club* 98, 103–106.

Nash, T.H. (1973) The effect of air pollution on other plants, particularly vascular plants. In: Ferry, B.W., Baddeley, M.S. and Hawksworth, D.L. (eds) *Air Pollution and Lichens*. Athlone Press, University of London, London.

Nash, T.H. (1976) Lichens as indicators of air pollution. *Naturwissenschaften* 63, 364–367.

Neeley, K.L. and Harbaugh, F.G. (1954) Effects of fluoride ingestion on a herd of dairy cattle in the Lubbock, Texas area. *Journal of the American Veterinary Medical Association* 124, 344–350.

Neuhold, J.M. and Sigler, W.F. (1960) Effects of sodium fluoride on carp and rainbow trout. *Transactions of the American Fish Society* 89, 358–370.

Newman, D.R. and Yu, M.-H. (1976) Fluorosis in black-tailed deer. *Journal of Wildlife Diseases* 12, 39–41.

Newman, J.R. (1984) Fluoride standards and predicting wildlife effects. *Fluoride* 17, 41–47.

Nicol, S. and Stolp, M. (1989) Sinking rates of cast exoskeletons of Antarctic krill *Euphasia superba* Dana and their role in the vertical flux of particulate matter and fluoride in the southern ocean. *Deep Sea Research. Part A. Oceanographic Research Papers* 36, 1753–1762.

NIDCR (n.d.) National Institute of Dental and Craniofacial Research. www.nidr.nih.gov/health/waterFluoridation.asp

Nilsson, J. and Grennfelt, P. (1988) *Critical Loads for Sulfur and Nitrogen.* Nordic Council of Ministers, Miljørapport 15, Stockholm.

NIOSH (1976) *Occupational Exposure to Hydrogen Fluoride.* US Department of Health, Education and Welfare, National Institute for Occupational Safety and Health Publication No. 76-143. Available at: http://www.cdc.gov/niosh/76-143.html

Nommik, H. (1953) Fluorine in Swedish agricultural products, soils and drinking water. *Acta Polytechnica* 127, 1–121.

NRC Committee (2000) *Ecological Indicators for the Nation.* Committee to Evaluate Indicators for Monitoring Aquatic and Terrestrial Environments, National Research Council, National Academy Press, Washington, DC.

NZ Review (2000) *Effects of Air Contaminants on Ecosystems and Recommended Critical Levels and Critical Loads.* Air Quality Technical Report No. 15, Ministry for the Environment, Wellington, New Zealand.

NZ Review (2002) *New Zealand Ambient Air Quality Guidelines Update.* Air Quality Report No. 32, Ministry for the Environment, Wellington, New Zealand (also at www.mfe.govt.nz).

Oelrichs, P.B. and McEwan, T. (1961) Isolation of the toxic principle in *Acacia georginae. Nature* 190, 808–809.

Oelschläger, W. (1971) Fluoride uptake in soil and its depletion. *Fluoride* 4, 80–84.

O'Hagan, D. and Harper, D.B. (1999) Fluorine-containing natural products. *Journal of Fluorine Chemistry* 100, 127–133.

Ohta, T., Wergedal, J.E., Matsuyama, T., Baylink, D.J. and William, L.K.H. (1995) Phenytoin and fluoride act in concert to stimulate bone formation and increase bone volume in adult male rats. *Calcified Tissue International* 56, 390–397.

Oliver, A.J., King, D.R. and Mead, R.J. (1977) The evolution of resistance to fluoroacetate intoxication in mammals. *Search* 8, 130–132.

Olsen, G.W., Burns, J.M., Mandel, J.H. and Zobel, L.R. (1999) Serum perfluorooctane sulfonate and lipid clinical chemistry tests in fluorochemical production employees. *Journal of Occupational and Environmental Medicine* 41, 799–806.

Omueti, J.A.I. and Jones, R.L. (1977a) Fluorine content of soil from Morrow plots over a period of 67 years. *Journal of the Soil Science Society of America* 41, 1023–1024.

Omueti, J.A.I. and Jones, R.L. (1977b) Fluoride adsorption by Illinois USA soils. *Journal of Soil Science* 28, 564–572.

Omueti, J.A.I. and Jones, R.L. (1980) Fluorine distribution with depth in relation to profile development in Illinois. *Journal of the Soil Science Society of America* 44, 247–249.

Ordin, L. and Altmann, A. (1965) Inhibition of phosphoglucomutase activity in oat coleoptiles by air pollutants. *Physiologia Plantarum* 18, 790–797.

Ormrod, D.P. (1982) Air pollutant interactions in mixtures. In: Unsworth, M.H. and Ormrod, D.P. (eds) *Effects of Gaseous Air Pollution in Agriculture and Horticulture.* Butterworths, London, pp. 307–332.

OSHA (undated) Occupational Safety and Health Guideline for Hydrogen Fluoride. http://www.osha.gov/SLTC/healthguidelines/hydrogenfluoride/recognition.html

Oshima, R.J. (1974) Viable system of biological indicators for monitoring air pollutants. *Journal of the Air Pollution Control Association* 24, 576–578.

Oskarsson, N. (1980) The interaction of volcanic gases and tephra – fluorine adhering to tephra of the 1970 Hekla eruption. *Journal of Volcanology and Geothermal Research* 8, 251–266.

Pack, M.R. (1966) Response of tomato fruiting to hydrogen fluoride as influenced by calcium nutrition. *Journal of the Air Pollution Control Association* 16, 541–544.

Pack, M.R. (1971) Effects of hydrogen fluoride on bean reproduction. *Journal of the Air Pollution Control Association* 21, 133–137.

Pack, M.R. (1972) Response of strawberry fruiting to hydrogen fluoride fumigation. *Journal of the Air Pollution Control Association* 22, 714–717.

Pack, M.R. and Sulzbach, C.W. (1976) Response of plant fruiting to hydrogen fluoride fumigation. *Atmospheric Environment* 22, 714–717.

Palomäki, V., Tynnyrinen, S. and Holopainen, T. (1992) Lichen transplantation in monitoring fluoride and sulphur deposition in the surroundings of a fertilizer plant and a strip mine at Siilinjarvi. *Annales Botanici Fennici* 29, 25–34.

Pankhurst, N.W., Boyden, C.R. and Wilson, J.B. (1980) The effect of fluoride effluent on marine organisms. *Environmental Pollution* 23, 299–312.

Paranjpe, M.G., Chandra, A.M.S., Qualls, C.W., McMurry, S.T., Rohrer, M.D., Whaley, M.M., Lochmiller, R.L. and McBee, K. (1994) Fluorosis in a wild cotton rat (*Sigmodon hispidus*) population inhabiting a petrochemical waste site. *Toxicologic Pathology* 22, 569–578.

Park, B.K., Kitteringham, N.R. and O'Neill, P.M. (2001) Metabolism of fluorine-containing drugs. *Annual Reviews of Pharmacology and Toxicology* 41, 443–470.

Paterson, D. and Kenworthy, J.B. (1982) An investigation of the effects of fluoride on selected moss species. In: Unsworth, M.H. and Ormrod, D.P. (eds) *Effects of Gaseous Air Pollution in Agriculture and Horticulture.* Butterworths, London, pp. 486–488.

Patra, R.C., Dwidedi, S.K., Bhardwaj, B. and Swarup, D. (2000) Industrial fluorosis in cattle and buffalo around Udaipur, India. *Science of the Total Environment* 253, 145–150.

Pattee, O.H., Wiemeyer, S.N. and Swineford, D.M. (1988) Effects of dietary fluoride on reproduction in eastern screech-owls. *Archives of Environmental Contamination and Toxicology* 17, 213–218.

Peakall, D.B. and Shugart, L.E. (1991) *Biomarkers: Research and Application in the Assessment of Environmental Health.* Springer-Verlag, Barcelona.

Perel'man, A.I. (1977) *Geochemistry of Elements in the Supergene Zone.* English translation by Teteruk-Schneider. Keter Publishing House, Jerusalem.

Perkins, D.F. (1992) Relationship between fluoride contents and loss of lichens near an aluminium works. *Water Air and Soil Pollution* 64, 503–510.

Perkins, D.F., Millar, R.O. and Neep, P.E. (1980) Accumulation of airborne fluoride by lichens in the vicinity of an aluminium reduction plant. *Environmental Pollution, Series A* 21, 155–168.

Perry, M.W. and Greenway, H. (1973) Permeation of uncharged organic molecules and water through tomato roots. *Annals of Botany* 37, 225–232.

Peters, R.A. (1952) Lethal synthesis. *Proceedings, Royal Society of London, Series B* 139, 143–170.

Peters, R.A. (1954) Biochemical light upon an ancient poison: a lethal synthesis. *Endeavour* 13, 147–154.

Peters, R.A. (1957) Mechanism of the toxicity of the active constituent of *Dichapetalum cymosum* and related compounds. *Advances in Enzymology* 18, 113–159.

Peters, R.A. and Hall, R.J. (1959) Further observations upon the toxic principle of *Dichapetalum toxicarium.* *Biochemical Pharmacology* 2, 25–36.

Peters, R.A. and Hall, R.J. (1960) Fluorine compounds in nature: the distribution of carbon–fluorine compounds in some species of *Dichapetalum.* *Nature* 187, 573–575.

Peters, R.A. and Shorthouse, M. (1964) Fluoride metabolism in plants of *Acacia georginae.* *Biochemistry Journal* 93, 20–21.

Peters, R.A. and Shorthouse, M. (1967) Observations on the metabolism of fluoride in *Acacia georginae* and some other plants. *Nature* 216, 80–81.

Peters, R.A. and Shorthouse, M. (1971) Identification of a volatile constituent formed by homogenates of *Acacia georginae* exposed to fluoride. *Nature* 231, 123–124.

Peters, R.A. and Shorthouse, M. (1972a) Fluorocitrate in plants and foodstuffs. *Phytochemistry* 11, 1337–1338.

Peters, R.A. and Shorthouse, M. (1972b) Formation of monofluorocarbon compounds by single cell cultures of *Glycine max* growing on inorganic fluoride. *Phytochemistry* 11, 1339.

Peters, R.A., Wakelin, R.W., Birks, F.T., Martin, A.J.P. and Webb, J. (1954) The toxic principle in seeds of *Dichapetalum toxicariuim* (ratsbane). *Biochemistry Journal* 58, xl.

Peters, R.A., Wakelin, R.W., Martin, A.J.P., Webb, J. and Birks, F.T. (1959) Observations upon the toxic principle in seeds of *Dichapetalum toxicarium.* *Biochemistry Journal* 71, 245–248.

Peters, R.A., Hall, R.J., Ward, P.F.V. and Sheppard, N. (1960) The chemical nature of the toxic compounds containing fluorine in the seeds of *Dichapetalum toxicarium.* *Biochemistry Journal* 77, 17–23.

Peters, R.A., Shorthouse, M. and Murray, L.R. (1964) Enolase and fluorophosphate. *Nature* 202, 1331–1332.

Peters, R.A., Murray, L.R. and Shorthouse, M. (1965a) Fluoride metabolism in *Acacia georginae* Gidyea. *Biochemistry Journal* 95, 724–730.

Peters, R.A., Shorthouse, M. and Ward, P.F.V. (1965b) The synthesis of the carbon–fluorine bond by *Acacia georginae* in vitro. *Life Sciences* 4, 749–752.

Pfeffer, A. (1962) Insektenschadlinge an Tannen im Bereich der Gasehatationen. *Zeitschrift für Angewante Entomologie* 51, 203–207.

Phipps, K.R., Orwoll, E.S., Mason, J.D. and Cauley, J.A. (2000) Community water fluoridation, bone mineral density, and fractures: prospective study of effects in older women. *British Medical Journal* 321, 860–864.

Pierce, A.W. (1952) Studies on fluorosis of sheep. 1. The toxicity of water-borne fluoride for sheep maintained in pens. *Australian Journal of Agricultural Research* 3, 326–340.

Pitman, M.G. (1982) Transport across plant roots. *Quarterly Reviews of Physics* 15, 481–554.

Poland, J.D. and Dennis, D.T. (1999) Treatment of plague. In: *WHO Plague Manual*. World Health Organization, Geneva, pp. 55–62.

Pollanschütz, J. (1969) Beobachtungen über die Empfindlichkeit verschiedener Baumarten gegenüber Immissionen von SO$_2$, HF und Magnesitstaub. In *Air Pollution, Proceedings of the 1st European Congress, Wageningen*. Centre for Agricultural Publishing and Documentation, Wageningen, pp. 371–377:

Polomski, J., Flühler, H. and Blaser, P. (1980) Behaviour of air-borne fluorides in soils. In: *Proceedings of a Symposium on Effects of Air Pollutants on Mediterranean and Temperate Forest Ecosystems*. General Technical Report PSW-43, Pacific Southwest Forest and Range Experiment Station, Forest Service, US Department of Agriculture, Berkeley, California, p. 246.

Poole, R.T. and Conover, C.A. (1973) Fluoride induced necrosis of *Cordyline terminalis* Kunth 'Baby Doll' as influenced by medium and pH. *Journal of the American Society for Horticultural Science* 98, 447–448.

Poole, R.T. and Conover, C.A. (1975) Fluoride induced necrosis of *Dracaena deremensis* Engler cv. Janet Craig. *HortScience* 10, 376–377.

Port, G.R., Davies, M.T. and Davison, A.W. (1998) Fluoride loading of a lepidopteran larva (*Pieris brassica*) fed on treated diets. *Environmental Pollution* 99, 233–239.

Posthumus, A.C. (1982) Biological indicators of air pollution. In: Unsworth, M.H. and Ormrod, D.P. (eds) *Effects of Gaseous Air Pollution in Agriculture and Horticulture*. Butterworths, London, pp. 27–42

Power, F.B. and Tutin, F. (1906) Chemical and physiological examination of the fruit of *Chailletia toxicaria*. *Journal of the American Chemical Society* 28, 1170–1183.

Powlesland, R.G., Knegtmans, J.W. and Styche, A. (2000) Mortality of North Island tomtits (*Petroica macrocephala*) caused by aerial 1080 possum control operations, 1997–1998, Pureora Forest Park. *New Zealand Journal of Ecology* 24, 161–168.

Preuss, P.W. and Weinstein, L.H. (1969) Studies on fluoro-organic compounds in plants II. Defluorination of fluoroacetate. *Contributions from the Boyce Thompson Institute* 24, 151–156.

Preuss, P.W., Lemmens, A.G. and Weinstein, L.H. (1968) Studies on fluoro-organic compounds in plants. I. Metabolism of 2-^{14}C-fluoroacetate. *Contributions from the Boyce Thompson Institute* 24, 25–32.

Preuss, P.W., Colavito, L. and Weinstein, L.H. (1970) The synthesis of monofluoroacetic acid by a tissue culture of *Acacia georginae*. *Experimentia* 26, 1059–1060.

Qian, D., Li, Z., Wang, J. and Gao, X. (1984) A study on fluoride and mulberry–silkworm ecosystem. *Journal of Environmental Sciences of China* 5, 7–11.

Rakowski, K.J. (1997) Hydrogen fluoride effects on plasma membrane composition and ATPase activity in needles of white pine (*Pinus strobus*) seedlings penetrated with 12 h photoperiod. *Trees (Berlin)* 11, 248–253.

Rakowski, K.J. and Zwiazek, J.J. (1991) Early effects of hydrogen fluoride on water relations, photosynthesis and membrane integrity on eastern white pine (*Pinus strobus*) seedlings. *Environmental and Experimental Botany* 32, 377–382.

Rakowski, K.J. and Zwiazek, J.J. (1992) Early effects of hydrogen fluoride on water relations, photosynthesis and membrane integrity in eastern white pine (*Pinus strobus*) seedlings. *Environmental and Experimental Botany* 33, 377–382.

Rakowski, K.J., Zwiazek, J.J. and Sumner, M.J. (1995) Hydrogen fluoride effects on plasma membrane composition. ATPase activity and cell structure in needles of eastern white pine (*Pinus strobus*) seedlings. *Trees (Berlin)* 9, 190–194.

Ramagopal, S., Welkie, G.W. and Miller, G.W. (1969) Fluoride injury of wheat roots and calcium nutrition. *Plant and Cell Physiology* 10, 675–685.

Rand, W.E. and Schmidt, H.J. (1952) The effect upon cattle of Arizona water of high fluorine content. *American Journal of Veterinary Research* 13, 50–61.

Reckendorfer, P. (1952) Ein Beitrag zur Microchemie des Rauchschadens durch Fluor. Die Wanderung des Fluor im pflanzlichen Gewebe. 1 Teil. Die unsichtbaren Schäden. *Pflanzenschutz-Berichte* 9, 33–35.

Reckendorfer, P. (1953) Ein Beitrag zur Microchemie des Rauchschadens durch Fluor. Die Wanderung des Fluor im pflanzlichen

Gewebe. II Teil. Die sichtbaren Schäden. *Pflanzenschutz-Berichte* 10, 112–124.

Reid Collins and Associates Ltd (1976) *Fluoride Emissions and Forest Growth*. Aluminum Company of Canada, Kitimat, British Columbia.

Reid Collins and Associates Ltd (1981) *Kitimat Valley Lichen Survey*. Aluminium Company of Canada, Kitimat, British Columbia.

Remde, A. and Debus, R. (1996) Biodegradability of fluorinated surfactants under aerobic and anaerobic conditions. *Chemosphere* 32, 1563–1574.

Renner, W. (1904a) A case of poisoning by the fruit of *Chailletia toxicaria* (ratsbane). *British Medical Journal*, 1314.

Renner, W. (1904b) Native poison, West Africa. *Journal of the African Society*, 109–111.

Renwick, J.A.A. and Potter, J. (1981) Effects of sulfur dioxide on volatile terpene emissions from balsam fir. *Journal of the Air Pollution Control Association* 31, 65–66.

Rich (1964) Our crops are like canaries (in reaction to toxic gases). *Frontiers of Plant Science* 16, 3.

Richards, B.L. and Cochran, L.C. (1957) *Virus and Viruslike Diseases of Stone Fruits in Utah*. Bulletin 384, Utah State Agricultural Experiment Station, Logan, Utah.

Riet Correa, F., del Carmen Mendez, M., Schild, A.L., Oliveira, J.A. and Zenebon, O. (1986) Dental lesions in cattle and sheep due to industrial pollution caused by coal combustion. *Pesquisa Veterinaria Brasileira* 6, 23–31.

Rimington, C. (1935) Chemical investigations of the 'gifblaar' *Dichapetalum cymosum* (Hood). Engl. Onderstepoort. *Journal of Veterinary Science and Animal Industry* 5, 81–95.

Roberts, B.A. and Thompson, L.K. (1980) Lichens as indicators of fluoride emission from a phosphorus plant, Long Harbour, Newfoundland, Canada. *Canadian Journal of Botany* 58, 2218–2228.

Roberts, B.A., Thompson, L.K. and Sidhu, S.S. (1979) Terrestrial bryophytes as indicators of fluoride emission from a phosphorus plant, Long Harbour, Newfoundland, Canada. *Canadian Journal of Botany* 57, 1583–1590.

Roberts, T.R. (1999) *Metabolic Pathways of Agrochemicals*. Part 1: *Herbicides and Plant Growth Regulators*. Royal Society of Chemistry, Cambridge.

Roberts, T.R. and Hutson, D.H. (1999) *Metabolic Pathways of Agrochemicals*. Part 2: *Insecticides and Fungicides*. Royal Society of Chemistry, Cambridge.

Robinson, E. (1957) Determining fluoride air concentrations by exposing limed filter papers. *American Industrial Hygiene Association Journal* 18, 145–148.

Robinson, W.O. and Edgington, G. (1946) Fluorine in soils. *Soil Science* 61, 341–353.

Roholm, K. (1937) *Fluorine Intoxication. A Clinical–Hygienic Study, With a Review of the Literature and Some Experimental Investigations*. H.K. Lewis, London.

Roques, A., Kerjean, M. and Auclair, D. (1980) Effets de la pollution atmosphérique par le fluor et le dioxyde de soufre sur l'appareil reproducteur femelle de *Pinus sylvestris* en forêt de Roumare (Seine-Maritime, France). *Environmental Pollution, Series A* 21, 191–201.

Ross, C.W., Wiebe, H.H. and Miller, G.W. (1962) Effect of fluoride on glucose catabolism in plant leaves. *Plant Physiology* 37, 305–309.

Royal Commission, Ontario (1968) *Report of the Committee Appointed to Inquire and Report upon the Pollution of Air, Soil and Water*. Frank Fogg, Queen's Printer, Ontario.

Samal, U.N. and Naik, B.N. (1989) Accumulation of fluoride in earthworms in the vicinity of an aluminium factory of Orissa. *Journal of Zoological Research* 2, 83–85.

Samal, U.N., Sahoo, L., Mohapatra, B.B. and Naik, B.N. (1990) Effect of fluoride pollutants on certain aquatic insects. *Geobios* 17, 31–33.

Sanada, M., Miyano, T., Iwadare, S., Williamson, J.M., Arison, B.H., Smith, J.L., Douglas, A.W., Liesch, J.M. and Inamine, E. (1986) Biosynthesis of fluorothreonine and fluoroacetic acid by the thienamycin producer, *Streptomyces cattleya*. *Journal of Antibiotics* 39, 259–265.

Sands, M., Nicol, S. and McMinn, A. (1998) Fluoride in Antarctic krill. *Marine Biology* 132, 591–598.

Saralakumari, D. and Ramakrishna, R.P. (1993) Endemic fluorosis in the village Ralla Anantapuram in Andhra Pradesh: an epidemiological study. *Fluoride* 26, 177–180.

Sato, M.I.Z., Valent, G.U., Coimbrão, C.A., Coelho, M.C.L.S., Sanchez, P.S., Alonso, C.D. and Martins, M.T. (1995) Mutagenicity of airborne particulate material from urban and industrial areas of São Paulo, Brazil. *Mutation Research* 335, 317–330.

Saunders, B.C. (1947) Toxic properties of ω-fluorocarboxylic acids and derivatives. *Nature* 160, 179–181.

Saunders, B.C. (1972) Chemical characteristics of the carbon–fluorine bond. In: *Carbon–Fluorine Compounds: Chemistry,*

Biochemistry, Biological Activity. Ciba Foundation Symposium 1971, Associated Scientific Publishers, London, pp. 9–32.

Schamshula, R., Un, P., Sugar, E. and Duppenthaler, J. (1988) The fluoride content of selected foods in relation to the fluoride concentration of water. *Acta Physiologica Hungarica* 72, 217–227.

Schiff, H.D., Fitzgerald, J., McCabe, M., Montanaro, D. and Shortell, V. (1981) *Correlation of Remote and Wet Chemical Sampling Techniques for Hydrogen Fluoride from Gypsum Ponds.* Report No. GCA-TR-80-76-G, US EPA, Washington, DC.

Schofield, H. (1999) Fluorine chemistry statistics: numbers of organofluorine compounds and publications associated with fluorine chemistry. *Journal of Fluorine Chemistry* 100, 7–11.

Scholl, G. (1971) Die immissionsrate von Fluor in Pflanzen als Maßstab für eine Immissionsbegrenzumg. *VDI-Berichte* 164, 39–45.

Schönbeck, H. (1969) Eine Methode zur Erfassung der biologischen Wirkung von Luftverunreinigungen durch transplanierte Flechten. *Staub-Reinhalt der Luft* 29, 14–18.

Schorr, J.R. (1977) *Source Assessment: Pressed and Blown Glass Manufacturing Plants.* US Environmental Protection Agency, Research Triangle Park, North Carolina.

Schroder, J.L., Basta, N.T., Rafferty, D.P., Lochmiller, R.L., Kim, S., Qualls, W. and McBee, K. (1999) Soil and vegetation fluoride exposure pathways to cotton rats on a petrochemical-contaminated landfill. *Environmental Toxicology and Chemistry* 18, 2028–2033.

Schultz, M., Kierdorf, U., Sedlacek, F. and Kierdorf, H. (1998) Pathological bone changes in the mandibles of wild red deer (*Cervus elaphus* L.) exposed to high environmental levels of fluoride. *Journal of Anatomy* 193, 431–442.

Schwela, D.H. (1979) An estimate of deposition velocities of several air pollutants on grass. *Ecotoxicology and Environmental Safety* 3, 174–189.

Seawright, A.A. and Eason, C.T. (1994) *Proceedings of the Science Workshop on 1080.* Miscellaneous Series 28, Royal Society of New Zealand, SIR Publishing, Wellington.

Semrau, K.T (1957) Emission of fluorides from industrial processes – a review. *Journal of the Air Pollution Control Association* 7, 92–108.

Sere, A., Kamgue, R.T., Assi, L.A. and Ba, A.C. (1982) *Spondianthus preussii* Engl. var. *preussii*, plante toxique pour le bétail

africain. *Revue d'Élévage et de Médecine Vétérinaire des Pays Tropicaux* 35, 73–82.

Sharpless, R.O. and Johnson, D.S. (1977) The influence of calcium on senescence changes in apple. *Annals of Applied Biology* 85, 450–453.

Shaw, W.M. (1954) Colorimetric determination of fluorine in water and soil extracts. *Analytical Chemistry* 26, 1212–1214.

Sherlock, J.C. (1984) Fluorides in foodstuffs and the diet. *Journal of the Royal Society of Health* 104, 34–36.

Shortt, J., Pandit, C. and Raghavachari, T. (1937) Endemic fluorosis in the Nellore district of South India. *Indian Medical Gazette*, 396.

Shupe, J.L. and Olson, A.E. (1983) Clinical and pathological aspects of fluoride toxicosis in animals. In: Shupe, J.L., Peterson, H.B. and Leone, N.C. (eds) *Fluorides. Effects on Vegetation, Animals and Humans.* Paragon Press, Salt Lake City, Utah, pp. 319–338.

Shupe, J.L., Miner, M.L., Harris, L.E. and Greenwood, D.A. (1962) Relative effects of feeding hay atmospherically contaminated by fluoride residue, normal hay plus calcium fluoride, and normal hay plus sodium fluoride to dairy heifers. *American Journal of Veterinary Research* 23, 777–787.

Shupe, J.L., Olson, A.E., Peterson, H.B. and Low, J.B. (1984) Fluoride toxicosis in wild ungulates. *Journal of the American Veterinary Medical Association* 185, 1295–1300.

Sidhu, S.S. (1979) Fluoride levels in air, vegetation and soil in the vicinity of a phosphorus plant. *Journal of the Air Pollution Control Association* 29, 1069–1072.

Sidhu, S.S. (1981) Patterns of fluoride accumulation in boreal forest species under perennial exposure to emissions from a phosphorus plant. *Atmospheric Pollution* 8, 425–432.

Sidhu, S.S. (1982) Fluoride deposition through precipitation and leaf litter in a boreal forest in the vicinity of a phosphorus plant. *Science of the Total Environment* 23, 205–214.

Silver, G.T. (1961) Notes on the chemical control of *Ectropis crepuscularia* Schiff, at Kitimat, BC. *Proceedings of the Entomological Society of British Columbia* 58, 13–16.

Singer, L. and Ophaug, R.H. (1983) Fluoride intakes of humans. In: Shupe, J.L., Peterson, H.B. and Leone, N.C. (eds) *Fluorides: Effects on Vegetation, Animals and Humans.* Paragon Press, Salt Lake City, Utah, pp. 62–82.

Singh, A., Chhabra, R. and Abrol, I.P. (1979) Effect of fluorine and phosphorus applied to a sodic soil on their availability and on yield and chemical composition of wheat. *Soil Science* 128, 90–97.

Skelly, J.M., Krupa, S.V. and Chevone, B.I. (1979) Field surveys. In: Heck, W.W., Krupa, S.V. and Linzon, S.N. (eds) *Handbook of Methodology for the Assessment of Air Pollution Effects on Vegetation.* Air Pollution Control Association, Pittsburgh, Pennsylvania, pp. 12-1–12-30.

Skye, E. (1979) Lichens as biological indicators of air pollution. *Annual Review of Phytopathology* 17, 325–342.

Slack, A.V. (1969) *Phosphoric acid.* Marcel Dekker, New York, 769 pp.

SMA (1990) *The Rain Forest of the Serra do Mar: Degradation and Reconstitution.* Document Series, Secretaria de Estado do Meio Ambiente, São Paulo, Brazil.

Smid, J.R. and Kruger, B.J. (1985) The fluoride content of some teas available in Australia. *Australian Dental Journal* 30, 25–28.

Smith, F.A. and Hodge, H.C. (1979) Airborne fluorides and man. Part I. *Critical Reviews in Environmental Control* 8, 293–371.

Smith, F.A., Gardner, D.E. and Yuile, C.L. (1977) Defluorination of fluoroacetate in the rat. *Life Sciences* 20, 1311–1318.

Soikkeli, S. (1981) Comparison of cytological injuries in conifer needles from several polluted industrial environments. *Annales Botanici Fennici* 18, 47–61.

Soikkeli, S. and Touvinen, T. (1979) Damage in mesophyll ultrastructure of needles of Norway spruce in two industrial environments in central Finland. *Annales Botanici Fennici* 16, 50–64.

Solberg, R.A. and Adams, D.F. (1956) Histological responses of some plant leaves to hydrogen fluoride and sulfur dioxide. *American Journal of Botany* 43, 755–760.

Solberg, R.A., Adams, D.F. and Ferchau, H.A. (1955) Some effects of hydrogen fluoride on the internal structure of *Pinus ponderosa* needles. In: *Proceedings of the 3rd National Air Pollution Symposium,* Pasadena, California, pp. 164–176.

Somani, L.L. (1977) Interactive effect of fluorine and salt contents of irrigation water on germination, growth and nodulation of Berseem (*Trifolium alexandrinum*). *Tropical Agriculture* 54, 219–222.

Sower, M.R., Clark, M.K., Jannausch, M.L. and Wallace, R.B. (1991) A prospective study of bone mineral content and fracture in communities with differential fluoride exposure. *American Journal of Epidemiology* 133, 649–660.

Specht, R.C. and MacIntire, W.H. (1961) Fixation, leaching, and plant uptake of fluorine from additions of certain fluorides in representative sandy soils: the fluorine problem in certain Florida areas. *Soil Science* 92, 172–176.

Spektor, D.M., Hofmeister, V.A., Artaxo, P., Brague, J.A.P., Echelar, F., Nogueira, D.P., Hayes, C., Thurston, G.D. and Lippmann, M. (1991) Effect of heavy industrial pollution on respiratory function in the children of Cubatão, Brazil – a preliminary report. *Environmental Health Perspectives* 94, 51–54.

Spierings, F.H.G. (1968) Effect of air pollution on crop yield. In: *1st International Congress of Plant Pathology.* London.

Spurr, E.B. (1994a) Impacts on non-target invertebrate populations of aerial application of sodium monofluoroacetate (1080) for brushtail possum control. In: Seawright, A.A. and Eason, C.T. (eds) *Proceedings of the Science Workshop on 1080.* Miscellaneous Series 28, Royal Society of New Zealand, SIR Publishing, Wellington, pp. 116–123.

Spurr, E.B. (1994b) Review of the impacts on non-target species of sodium monofluoroacetate (1080) in baits used for brushtail possum control in New Zealand. In: Seawright, A.A. and Eason, C.T. (eds) *Proceedings of the Science Workshop on 1080.* Miscellaneous Series 28, Royal Society of New Zealand, SIR Publishing, Wellington, pp. 124–133.

Städler, E. (1974) Host plant stimuli affecting host oviposition behavior of the eastern spruce budworm. *Entomologica Experimentalis et Applicata* 17, 176.

Steudle, E., Murrmann, M. and Peterson, C.A. (1993) Transport of water and solutes across maize roots modified by puncturing the endodermis. *Plant Physiology* 103, 335–349.

Stevens, D.P., McLaughlin, M.J. and Alston, A.M. (1997) Phytotoxicity of aluminium-fluoride complexes and their uptake from solution culture by *Avena sativa* and *Lycopersicon esculentum. Plant and Soil* 192, 81–93.

Stevens, D.P., McLaughlin, M.J. and Alston, A.M. (1998a) Phytotoxicity of the fluoride ion and its uptake from solution culture by *Avena sativa* and *Lycopersicon esculentum. Plant and Soil* 200, 119–129.

Stevens, D.P., McLaughlin, M.J. and Alston, A.M. (1998b) Phytotoxicity of hydrogen fluoride and fluoroborate and their uptake from solution culture by *Lycopersicon esculentum* and *Avena sativa. Plant and Soil* 200, 175–184.

Stevens, D.P., McLaughlin, M.J., Randall, P.J. and Keerthisinghe, G. (2000) Effect of fluoride supply on fluoride concentrations in five pasture species: levels required to reach phytotoxic or potentially zootoxic concentrations in plant tissue. *Plant and Soil* 227, 223–233.

Stewart, D., Treshow, M. and Harner, F.M. (1973) Pathological anatomy of conifer needle necrosis. *Canadian Journal of Botany* 51, 983–988.

Stewart, G.G. and Brunt, S.V. (1968) The effect of monofluoroacetic acid upon the carbohydrate metabolism of *Saccharomyces cerevisiae*. *Biochemical Pharmacology* 17, 2349–2355.

Stewart H., Jenkin, B., McCarthy, R. and Flinn, D. (1994) Public versus user perception of risks and benefits of plantation forestry in Victoria, Australia. In: Seawright, A.A. and Eason, C.T. (eds) *Proceedings of the Science Workshop on 1080*. Miscellaneous Series 28, Royal Society of New Zealand, SIR Publishing, Wellington, pp. 10–18.

Steyn, D.G. (1928) *A Summary of our Present Knowledge in Respect of Poisoning by* Dichapetalum cymosum. 13th and 14th Reports of the Director of Veterinary Education and Research, Onderstepoort.

Stokinger, H.E. (1949) Toxicity following inhalation of fluorine and hydrogen fluoride. In: Voegtlin, C. and Hodge, H.C. (eds) *Pharmacology and Toxicology of Uranium Compounds*. McGraw Hill, New York, pp. 1021–1057.

Streeton, J.A. (1990) *Air Pollution Health Effects and Air Quality Objectives in Victoria*. Environment Protection Authority, Victoria, Australia.

Stumm, W. and Morgan, J.J. (1981) *Aquatic Chemistry*, 2nd edn. John Wiley & Sons, New York.

Sultan, B.-A., Samson, F. and Robinson, J. (2000) Combustion atmosphere toxicity of polymeric materials intended for internal cables. In: *Proceedings of the 49th International Wire and Cable Symposium, 13–16 November 2000, Atlantic City, USA*. International Wire and Cable Symposium, Eatontown, New Jersey, pp. 231–242.

Sulzbach, C.W. and Pack, M.R. (1972) Effects of fluoride on pollen germination, pollen tube growth and fruit development of tomato and cucumber. *Phytopathology* 62, 1247–1253.

Suttie, J.S. (1969) Air quality standards for the protection of farm animals from fluorosis. *Journal of the Air Pollution Control Association* 19, 239–242.

Suttie, J.S., Dickie, R., Clay, A.B., Nielsen, P., Mahan, W.E., Baumann, D.P. and Hamilton, R.J. (1987) Effects of fluoride emissions from a modern primary aluminum smelter on a local population of white-tailed deer (*Odocoileus virginianus*). *Journal of Wildlife Diseases* 23, 135–143.

Suttie, J.W. (1977) Effects of fluoride on livestock. *Journal of Occupational Medicine* 19, 40–48.

Suttie, J.W. (1983) The influence of nutrition and other factors on fluoride tolerance. In: Shupe, J.L., Peterson, H.B. and Leone, N.C. (eds) *Fluorides. Effects on Vegetation, Animals and Humans*. Paragon Press, Salt Lake City, Utah, pp. 291–303.

Swaine, D.J. (1990) *Trace Elements in Coal*. Butterworths, London.

Swarup, D., Dey, S., Patra, R.C., Dwivedi, S.K. and Ali, S.L. (2001) Clinico-epidemiological observations of industrial bovine fluorosis in India. *Indian Journal of Animal Sciences* 71, 1111–1115.

Sykes, T.R., Quastel, J.H., Adam, M.J., Ruth, T.J. and Noujaim, A.A. (1987) The disposition and metabolism of fluorine-18 fluoroacetate in mice. *Biochemical Archives* 3, 317–324.

Symonds, R., Rose, W. and Reed, M. (1988) Contribution of Cl- and F-bearing gases to the atmosphere by volcanoes. *Nature* 334, 415–418.

Tahori, A.S. (1965) Investigation of the resistance pattern of a fluoroacetate-resistant house fly strain. In: Freeman, P. (ed.) *XIIth International Congress of Entomology*. London, p. 242.

Tahori, A.S. (1966) Changes in the resistance pattern of a fluoroacetate-resistant fly strain. *Journal of Economic Entomology* 59, 462–464.

Takmaz-Niscancioglu, S. (1983) Dynamics of fluoride uptake, movement and loss in higher plants. PhD thesis, University of Newcastle upon Tyne, UK.

Takmaz-Niscancioglu, S. and Davison, A.W. (1982) Loss of fluorides from grass swards and other surfaces. In: Unsworth, M.H. and Ormrod, D.P. (eds) *Effects of Gaseous Air Pollution in Agriculture and Horticulture*. Butterworths, London, 465 pp.

Takmaz-Nisancioglu, S. and Davison, A.W. (1988) Effects of aluminium on fluoride uptake by plants. *New Phytologist* 109, 149–155.

Tatera, B.S. (1970) Parameters which influence fluoride emissions from gypsum ponds. *Dissertation Abstracts* 3992-B.

Tatou-Kamgoue, R., Sylla, O., Pousset, J.L., Brunet, J.C. and Sere, A. (1979) Isolement et caractérisation des principes toxiques du *Spondianthus preussii* var. Glaber Engler. *Plantes Médicinales et Phytothérapie* 13, 252–259.

Taves, D.R. (1968) Evidence that there are two forms of fluoride in human serum. *Nature* 217, 1050–1051.

Taylor, F., Hetrick, D., Conrad, M., Parr, P. and Bledsoe, J. (1981) Uranium conservation and enrichment technologies: sources of atmospheric fluoride. *Journal of Environmental Quality* 10, 80–87.

Taylor, R.J. and Basabe F.A. (1984) Patterns of fluoride accumulation and growth reduction exhibited by Douglas fir in the vicinity of an aluminium reduction plant. *Environmental Pollution, Series A* 33, 221–235.

Tedder, J.M., Nechvatal, A. and Jubb, A.H. (1975) *Basic Organic Chemistry*. John Wiley & Sons, London.

Teller, A.J. and Hsieh, J.Y. (1992a) Fiberglass operations. In: Buonicore, A.J. and Davis, W.T. (eds) *Air Pollution Engineering Manual*. Air and Waste Management Association, Van Nostrand Reinhold, New York, pp. 766–770.

Teller, A.J. and Hsieh, J.Y. (1992b) Glass manufacturing. In: Buonicore, A.J. and Davis, W.T. (eds) *Air Pollution Engineering Manual*. Air and Waste Management Association, Van Nostrand Reinhold, New York, pp. 770–777.

Tessier, A.M. and Paris, R.R. (1974) On a poisonous Eurphorbiaceae: presence of substances belonging to curbitacins. *Annales Pharmaceutiques Françaises* 32, 177–182.

Thalenhorst, W. (1974) Investigations on the influence of fluor containing air pollutants upon the susceptibility of spruce plants to the attack of the gall aphid *Sacchiphantes abietis* (L.). *Zeitschrift für Pflanzenkrankheiten und Pflanzenschutz* 81, 717–727.

Thomas, J. and Gluskoter, H.J. (1974) Determination of fluoride in coal with the fluoride-ion selective electrode. *Analytical Chemistry* 46, 1321–1323.

Thomas, M.D. (1958) Air pollution with relation to agronomic crops. I. General status of research on the effects of air pollution and plants. *Agronomy Journal* 50, 545–550.

Thomas, M.D. (1961) Effects of air pollution on plants. In: *Air Pollution*. World Health Organization, Geneva, pp. 233–278.

Thomas, M.D. and Alther, E.W. (1966) The effects of fluoride on plants. In: Smith, F.A. (ed.) *Handbook of Experimental Pharmacology*. Vol. XX/1. *Pharmacology of Fluorides*. Springer-Verlag, New York, pp. 231–306.

Thomas, M.D. and Hendricks, W. (1956) Effects of air pollution on plants. In: Magill, P.L., Holdren, F.R. and Ackley, C. (eds) *The Pollution Handbook*. McGraw-Hill, New York, pp. 9.1–9.44.

Thompson, C.R., Taylor, O.C., Thomas, M.D. and Ivie, J.O. (1967) Effects of air pollutants on apparent photosynthesis and water use by citrus trees. *Environmental Science and Technology* 1, 644–650.

Thompson, L.K., Sidhu, S.S. and Roberts, B.A. (1979) Fluoride accumulation in soil and vegetation in the vicinity of a phosphorus plant. *Environmental Pollution* 18, 221–234.

Thorarinsson, S. (1967) The eruptions of Hekla in modern times. In: Einarsson, T., Kjartansson, G. and Thorarinsson, S. (eds) *The Eruption of Hekla 1947–1948*, Part 3. Societas Scientarum, H.F. Leiftur, Reykjavik, pp. 1–170.

Thorarinsson, S. (1979) On the damage caused by volcanic eruptions with special reference to tephra and gases. In: Sheets, P.D. and Grayson, D.K. (eds) *Volcanic Activity and Human Ecology*. Academic Press, New York, pp. 125–159.

Thorarinsson, S. and Sigvaldason, G.E. (1973) The Hekla eruption of 1970. *Bulletin Volcanologique* 36, 269–288.

Timmerman, A.D. (1967) Effects of hydrogen fluoride accumulation on plant cells microstructure and ultrastructure. PhD thesis, Texas A & M University, Texas.

Timperley, C.M. (2000) Highly toxic fluorine compounds. In: Banks, E.E. (ed.) *Fluorine Chemistry at the Millennium*. Elsevier, Amsterdam, pp. 499–536.

Tong, S.S.C., Morse, R., Bache, C.A. and Lisk, D.J. (1975) Elemental analysis of honey as an indicator of pollution. *Archives of Environmental Health* 30, 329–332.

Tonomura, K., Futai, F., Tanabe, O. and Yamaoka, T. (1965) Defluorination of monofluoroacetate by bacteria. I. Isolation of bacteria and their activity of defluorination. *Agricultural and Biological Chemistry (Japan)* 29, 124–128.

Trautwein, K., Bucher, R. and Kopp, C. (1968) Laboratory and field investigations of fluorine effects of bees. *Anim. Hyg. Freiberg.* 10 pp.

Treble, D.H., Lamport, D.T.A. and Peters, R.A. (1962) The inhibition of plant aconitate hydratase (aconitase) by fluorocitrate. *Biochemistry Journal* 85, 113–115.

Treshow, M. (1957) The effects of fluoride on the anatomy of Chinese apricot leaves. *Phytopathology* 46, 649.

Treshow, M. (1965) Response of some pathogenic fungi to sodium fluoride. *Mycologia* 57, 216–221.

Treshow, M. (1971) Fluorides as air pollutants affecting plants. *Annual Review of Phytopathology* 9, 21–44.

Treshow, M. (1975) Interactions of air pollutants and plant diseases. In: Mudd, J.B. and Kozlowski, T.T. (eds) *Responses of Plants to Air Pollution*. Academic Press, New York, pp. 307–334.

Treshow, M. and Pack, M. (1970) Fluoride. In: Jacobson, J.S. and Hill, A.C. (eds) *Recognition of Air Pollution Injury to Vegetation: A Pictorial Atlas*. Informative Report 1, TR-7 Agricultural Committee, Air Pollution Control Association, Pittsburgh, Pennsylvania, pp. D1–D77.

Treshow, M. and Transtrum, L.G. (1964) Similarity between leaf markings caused by air pollutants and other agents. 1. Moisture stress and fluoride expression. *Proceedings, Utah Academy of Science* 41, 49–52.

Treshow, M., Dean, G. and Harner, F.M. (1966) Stimulation of virus-induced local lesions by fluoride. *Phytochemistry* 56, 904.

Treshow, M., Dean, G. and Harner, F.M. (1967) Stimulation of tobacco mosaic virus-induced lesions on bean by fluoride. *Phytopathology* 57, 756–758.

Tromp, T.K., Ko, M.K.W., Rodriguez, J.M. and Sze, N.D. (1995) Potential accumulation of a CFC-replacement degradation product in seasonal wetlands. *Nature* 376, 327–330.

Tsuyuki, H. and Wold, F. (1964) Enolase: multiple molecular forms in fish muscle. *Science* 146, 535–537.

Twigg, L.E. (1994) Occurrence of fluoroacetate in Australian plants and tolerance to 1080 in indigenous Australian animals. In: Seawright, A.A. and Eason, C.T. (eds) *Proceedings of the Science Workshop on 1080*. Miscellaneous Series 28, Royal Society of New Zealand, SIR Publishing, Wellington, pp. 97–115.

Twigg, L.E. and King, D.R. (1991) The impact of fluoroacetate-bearing vegetation on native Australian fauna: a review. *Oikos* 61, 412–430.

Twigg, L.E. and Mead, R.J. (1990) Comparative metabolism of and sensitivity to fluoroacetate in geographically separated populations of *Tiliqua rugosa* Gray Scincidae. *Australian Journal of Zoology* 37, 617–626.

Twigg, L.E. and Socha, L.V. (1996) Physical versus chemical defense mechanisms in toxic *Gastrolobium*. *Ecology* 108, 21–28.

Twigg, L.E. and Socha, L.V. (2001) Defluorination of sodium monofluoroacetate by soil microorganisms from central Australia. *Soil Biology and Biochemistry* 33, 227–234.

Twigg, L.E., Mead, R.J. and King, D.R. (1986) Metabolism of fluoroacetate in the skink (*Tiliqua rugosa*) and the rat (*Rattus norvegicus*). *Australian Journal of Biological Science* 39, 1–16.

Twigg, L.E., King, D.R., Bowen, L.H., Wright, G.R. and Eason, C.T. (1996a) Fluoroacetate content of the toxic Australian plant genus, *Gastrolobium*, and its environmental persistence. *Natural Toxins* 4, 122–127.

Twigg, L.E., King, D.R., Bowen, L.H., Wright, G.R. and Eason, C.T. (1996b) Fluoroacetate found in *Nemcia spathulata*. *Australian Journal of Botany* 44, 411–412.

Twigg, L.E., Wright, G.R. and Potts, M.D. (1999) Fluoroacetate content of *Gastrolobium brevipes* in central Australia. *Australian Journal of Botany* 47, 877–880.

Twigg, L.E., Martin, G.R. and Lowe, T.J. (2002) Evidence of pesticide resistance in medium-sized mammalian pests: a case study with 1080 poison and Australian rabbits. *Journal of Applied Ecology* 39: 549–560.

UNEP (1987) *Montreal Protocol on Substances that Deplete the Ozone Layer*. United Nations Environmental Programme (UNEP), New York.

Unsworth, M.H. (1982) Exposure to gaseous pollutants and uptake by plants. In: Unsworth, M.H. and Ormrod, D.P. (eds) *Effects of Gaseous Air Pollution in Agriculture and Horticulture*. Butterworths, London, pp. 43–63.

USDA (2001) US Department of Agriculture APHIS Wildlife Services internet site, Tables of Animals Taken and Methods FY 2000: http://www.aphis.usda.gov/ws/tables/00table10t.rtf

US EPA (1980) *Reviews of the Environmental Effects of Pollutants: IX Fluoride*. Health Effects Research Laboratory, US Environmental Protection Agency, Cincinnati, Ohio.

US EPA (1992) *Hydrogen Fluoride Study: Report to Congress*. Section 301(n)(6) Clean Air Amendments of 1990, Office of Solid Waste and Emergency Response, Washington, DC.

US EPA (1995a) *Profile of the Electronics and Computer Industry*. EPA/310-R-95-002, EPA Office of Compliance Sector Notebook Project, Office of Enforcement and Compliance Assurance, Washington, DC.

US EPA (1995b) *Profile of the Iron and Steel Industry*. EPA/310-R-95-005, EPA Office of Compliance Sector Notebook Project, Office of Enforcement and Compliance Assurance, Washington, DC.

US EPA (1995c) *Profile of the Petroleum Refining Industry*. EPA/310-R-95-013, EPA Office of Compliance Sector Notebook Project, Office of Enforcement and Compliance Assurance, Washington, DC.

US EPA (1995d) *Reregistration Eligibility Decision (RED). Sodium Fluoroacetate*. EPA-738-F-95-025. Office of Prevention, Pesticides and Toxic Substances, US Environmental Protection Agency, Washington, DC.

US EPA (1997a) *Profile of the Fossil Fuel Electric Power Generating Industry*. EPA/310-R-97-007, Office of Enforcement and Compliance Assurance, Washington, DC.

US EPA (1997b) National Emission Standards for Hazardous Air Pollutants for Primary Aluminum Reduction Plants. 40 CFR Parts 9, 60, and 63. Final Rule. *Federal Register* 62(194), 52384–52428.

USGS (2001) *Rock Phosphate*. United States Geological Survey: www.minerals.usgs.gov/minerals/pubs/commodity/phosphate_rock/540302.pdf

USGS (2002) Fluorspar statistics. United States Geological Survey: www.minerals.usgs.gov. minerals/pubs/commodity/fluorspar/280302.pdf and Open-File Report 01–006 http://minerals.usgs.gov/minerals/pubs/og01–006.fluorspar.html

USHEW (1969) *Control Techniques for Particulate Air Pollutants*. NAPCA Publications AP-51, Department of Health, Education, and Welfare, Public Health Service, Environmental Health Service, National Air Pollution Control Administration, Washington, DC.

Valkovič, V. (1983) *Trace Elements in Coal*. CRC Press, Boca Raton, Florida.

van der Eerden, L.J. (1981) *Fluoride-accumulatie in gras: voorgangsverlag, no 1*. Report no. R256, Research Institute for Plant Protection, Wageningen, The Netherlands.

van der Eerden, L.J. (1991) Fluoride content in grass as related to atmospheric fluoride concentrations: a simplified model. *Agriculture, Ecosystems and Environment* 37, 257–273.

van Hook, C. (1974) Fluoride distribution in the Silver Bow, Montana area. *Fluoride* 9, 181–199.

Vartiainen, T. and Gynther, J. (1984) Fluoroacetic acid in guar gum. *Food Chemistry and Toxicology* 22, 307–308.

VDI (1987) VDI 2310, Maximale Immissions-verte für Fluoride zum Schutz der landwirtschaftlichen Nutztiere. In: *VDI-Handbuch Reinhaltung der Luft*, Vol. 1, Part 26. Verein Deutscher Ingenieure, Düsseldorf, Germany.

VDI (1989) VDI 2310, Maximale Immissions-Konzentrationen für Fluorwasserstoff. In: *VDI-Handbuch Reinhaltung der Luft*, Vol. 1, Part 3. Verein Deutscher Ingenieure, Düsseldorf, Germany.

Vejrup, K.V. and Lindblom, B. (2002) *Analysis of Perfluorooctanesulfonate Compounds in Impregnating Agents, Wax and Floor Polish Products*. Survey no. 17, Survey of Chemical Substances in Consumer Products, Danish Environmental Protection Agency, Danish Ministry of the Environment, Copenhagen, Denmark.

Venkateswarlu, P., Armstrong, W.D. and Singer, L. (1965) Absorption of fluoride and chloride by barley roots. *Plant Physiology* 40, 255–261.

Vickery, B. and Vickery, M.L. (1972) Fluoride metabolism in *Dichapetalum toxicarium*. *Phytochemistry* 11, 1905–1909.

Vickery, B. and Vickery, M.L. (1975) The synthesis and defluorination of monofluoroacetate in some *Dichapetalum species*. *Phytochemistry* 14, 423–427.

Vickery, B., Vickery, M.L. and Kaberia, F. (1979) A possible biomimetic synthesis of fluoroacetic acid. *Experientia* 35, 299–300.

Vike, F. and Håbjørg, A. (1995) Variation in fluoride content and leaf injury on plants associated with aluminum smelters in Norway. *Science of the Total Environment* 163, 25–34.

Vikøren, T. and Stuve, G. (1995) Bone fluoride concentrations in Canada geese (*Branta canadensis*) from areas with different levels of fluoride pollution. *Science of the Total Environment* 163, 123–128.

Vikøren, T. and Stuve, G. (1996a) Fluoride exposure in cervids inhabiting areas adjacent to aluminium smelters in Norway. II. Fluorosis. *Journal of Wildlife Disease* 32, 181–189.

Vikøren, T. and Stuve, G. (1996b) Fluoride exposure and selected characteristics of eggs and bones of the herring gull (*Larus argentatus*) and the common gull (*Larus canus*). *Journal of Wildlife Diseases* 32, 190–198.

Visscher, P.T., Culbertson, C.W. and Oremland, R.S. (1994) Degradation of trifluoroacetate in oxic and anoxic sediments. *Nature* 369, 729–731.

Vogel, J. and Ottow, J.C.G. (1991) Fluoride accumulation in different earthworm species near an industrial emission source in southern

Germany. *Bulletin of Environmental Contamination and Toxicology* 47, 515–520.

VPC (2000) *Controlled Pesticides.* Prepared by the Agricultural Compounds and Veterinary Medicines Group. Wellington, New Zealand.

Wahlstrom, V.L., Fugelsang, K.C. and Muller, C.J. (1996) *Effect of Fluoride on Fermentation Rate and Population Density of Fourteen Strains of* Saccharomyces *sp.* Publication No. 960102, California Agricultural Technology Institute, Sacramento, California.

Wainwright, M. and Supharungsun, S. (1984) Release by fungi of F⁻ from insoluble fluorides. *Transactions of the British Mycological Society* 82, 289–292.

Waldbott, G.L. and Cecilioni, V.A. (1969) Neighborhood fluorosis. *Clinical Toxicology* 2, 387–396.

Walker, J.R.L. and Lien, B.C. (1981) Metabolism of fluoroacetate by a soil *Pseudomonas* sp. and *Fusarium solani. Soil Biology and Chemistry* 13, 231–235.

Walsh, L.H. (1909) *South African Poisonous Plants.* T. Maskey Miller, Capetown.

Walton, K. (1984) Fluoride in fox bone near an aluminium reduction plant in Anglesey, Wales and elsewhere in the United Kingdom. *Environmental Pollution, Series B* 7, 273–280.

Walton, K.C. (1985) Fluoride in bones of small rodents near an aluminium reduction plant. *Water, Air and Soil Pollution* 26, 65–70.

Walton, K.C. (1986) Fluoride in moles, shrews and earthworms near an aluminium reduction plant. *Environmental Pollution, Series A* 42, 361–371.

Walton, K.C. (1987a) Fluoride in bones of small rodents living in areas with different pollution levels. *Water, Air, and Soil Pollution* 32, 113–122.

Walton, K.C. (1987b) Factors determining amounts of fluoride in woodlice *Oniscus asellus* and *Porcellio scaber,* litter and soil near an aluminium reduction plant. *Environmental Pollution* 46, 1–9.

Walton, K.C. (1988) Environmental fluoride and fluorosis in mammals. *Mammal Review* 18, 77–90.

Wang, G.X. and Cheng, G.D. (2001) Fluoride distribution in water and the governing factors of environment in arid north-west China. *Journal of Arid Environments* 49, 601–614.

Wang, J.X. and Bian, Y.M. (1988) Fluoride effects on the mulberry–silkworm system. *Environmental Pollution* 51, 11–18.

Wang, J.X., Qian, D.F., Li, Z.F., Gao, X.P., Ma, D.H. and Li, Q. (1980) Effects of fluoride in mulberry leaves on growth and development of the silkworm. *Environmental Quality China* 2, 33–38.

Wang L.F. and Huang, J.Z. (1995) Outline of control practice of endemic fluorosis in China. *Social Science and Medicine* 41, 1191–1195.

Warburg, O. and Christian, W. (1942) Isolierung und Kristallisation des Gärungsferment Enolase. *Biochemische Zentralblatt* 310, 385–421.

Ward, P.F.V., Hall, R.J. and Peters, R.A. (1964) Fluoro-fatty acids in the seeds of *Dichapetalum toxicarium. Nature* 201, 611–612.

Warrington, P.D. (1996) Ambient water quality criteria for fluoride. http://www.gov.bc.ca/wat/wq/BCguidelines/fluoride.html

Weeks, M. and Leicester, H. (1968) Discovery of the Elements, 7th Edn. Easton, Pennsylvania, 896 pp.

Wei, L.L. and Miller, G.W. (1972) Effects of HF on the fine structure of mesophyll cells from *Glycine max* Merr. *Fluoride* 5, 67–73.

Wei, M.W. (1992) Primary aluminum industry. In: Buonicore, A.J. and Davis, W.T. (eds) *Air Pollution Engineering Manual.* Van Nostrand Reinhold, New York, pp. 590–606.

Weidmann, S.M. and Weatherall, J.A. (1970) Distribution of fluorides: distribution in hard tissues. In: *Fluorides and Human Health.* Monograph Series No. 59, WHO, Geneva, pp. 104–128.

Weinstein, L.H. (1961) Effects of atmospheric fluorides on metabolic constituents of tomato and bean leaves. *Contributions from the Boyce Thompson Institute* 21, 215–231.

Weinstein, L.H. (1969) Discussion of A.C. Hill's paper on air quality standards for fluoride vegetation effects. *Journal of the Air Pollution Control Association* 19, 336.

Weinstein, L.H. (1977) Fluoride and plant life. *Journal of Occupational Medicine* 19, 49–78.

Weinstein, L.H. and Alscher-Herman, R. (1982) Physiological responses of plants to fluorine. In: Unsworth, M.H. and Ormrod, D.P. (eds) *Effects of Gaseous Air Pollution in Agriculture and Horticulture.* Butterworths, London, pp. 139–167.

Weinstein, L.H. and Bunce, H.W.F. (1981) Impact of emissions from an alumina reduction smelter on the forests of Kitimat, B.C.: a synoptic view. In: *74th Annual Meeting of the Air Pollution Control Association, Philadelphia, Pennsylvania, 21–26 June 1981,* Paper 81-44.4

Weinstein, L.H. and Hansen, K.S. (1988) Relative susceptibility of Brazilian vegetation to airborne fluoride. *Pesquisa Agropecuria Brasileira* 23, 1125–1137.

Weinstein, L.H. and Laurence, J.A. (1989) Indigenous and cultivated plants as bioindicators. In: *Biologic Markers of Air-Pollution Stress and Damage in Forests*. National Academy Press, Washington, DC, pp. 194–204.

Weinstein, L.H. and McCune, D.C. (1970) Field surveys, vegetation sampling, and air and vegetation monitoring. In: Jacobson, J.S. and Hill, A.C. (eds) *Recognition of Air Pollution Injury to Vegetation: A Pictorial Atlas*. Informative Report 1, TR-7 Agricultural Committee, Air Pollution Control Association, Pittsburgh, Pennsylvania, pp. G1–G4.

Weinstein, L.H. and McCune, D.C. (1971) Effects of fluoride on agriculture. *Journal of the Air Pollution Control Association* 21, 410–413.

Weinstein, L.H. and Mandl, R.H. (1971) The separation and collection of gaseous and particulate fluorides. *VDI-Berichte* 164, 53–63.

Weinstein, L.H., McCune, D.C., Mancini, J.F., Colavito, L.J., Silberman, D.H. and van Leuken, P. (1972) Studies on fluoro-organic compounds in plants. III. Comparison of the biosynthesis of fluoroorganic acids in *Acacia georginae* with other species. *Environmental Research* 5, 393–408.

Weinstein, L.H., McCune, D.C., Mancini, J.F. and van Leuken, P. (1973) Effects of hydrogen fluoride on the growth, development, and reproduction of the Mexican bean beetle. In: *Proceedings of the Third International Clean Air Congress*. VDI Verlag GmbH, Düsseldorf, pp. A150–A153.

Weinstein, L.H., Laurence, J.A., Mandl, R.H. and Wälti, K. (1990) Use of native and cultivated plants as bioindicators of pollution damage. In: Wang, W., Gorsuch, J.W. and Lower, S.R. (eds) *Plants for Toxicity Assessment*. ASTM, Philadelphia, Pennsylvania, pp. 117–126.

Weinstein, L.H., Davison, A.W. and Arndt, U. (1998) Fluoride. In: Flagler, R.B. (ed.) *Recognition of Air Pollution Injury to Vegetation. A Pictorial Atlas*, 2nd edn. Air and Waste Management Association, Pittsburgh, Pennsylvania, pp. 4-1–4-27.

Weismann, L. and Svartarakova, L. (1974) Toxicity of sodium fluoride on some species of harmful insects. *Biologia (Bratislava)* 29, 847–852.

Weismann, L. and Svartarakova, L. (1975) Paralysing effect of calcium chloride on the toxic action of sodium fluoride in larvae of the Colorado potato beetle. *Biologia (Bratislava)* 30, 841–845.

Weismann, L. and Svartarakova, L. (1976) Toxicity of sodium fluoride for the spider mite *Tetranychus utticae* Koch. *Biologia (Bratislava)* 31, 125–132.

Wenzel, W. and Blum, E.H. (1992) Fluorine speciation and mobility in F-contaminated soils. *Soil Science* 153, 357–364.

Whitford, G.M. (1992) Acute and chronic fluoride toxicity. *Journal of Dental Research* 71, 1249–1254.

Whittem, J.H. and Murray, L.R. (1963) The chemistry of pathology of Georgina River poisoning. *Australian Veterinary Journal* 39, 168–173.

WHO (1970) *Fluorides and Human Health*. Monograph Series No. 59, WHO, Geneva, 364 pp.

WHO (1984) *Fluorine and Fluorides*. Environmental Health Criteria 36, World Health Organization, Geneva.

WHO (1994) *Fluorides and Oral Health. Report of a WHO Expert Committee on Oral Health Status and Fluoride Use*. WHO Technical Report Series 846, World Health Organization, Geneva.

WHO (2002a) Environmental Health Criteria 227, World Health Organization, Geneva.

WHO (2002b) *Guidelines for Drinking Water Quality*. http://www.who.int/water_sanitation_health/GDWQ/draftchemicals/fluoride2003.pdf

WHO (n.d.) *WHO and Water Supply in India*. http://www.whoindia/org/UNIC.htm

Wiebe, H.H. and Poovaiah, B.W. (1973) Influence of moisture, heat, and light stress on hydrogen fluoride fumigation injury to soybeans. *Plant Physiology* 52, 542–545.

Wilson, W.L., Campbell, M.W., Eddy, L.D. and Poppe, W.H. (1967) Calibration of limed filter paper for measuring short-term hydrogen fluoride doses. *American Industrial Hygiene Association Journal* 27, 254–259.

WMO (1989) *Scientific Assessment of Stratospheric Ozone*. Research and Monitoring Project Report No. 20, Vol. II, World Meteorological Organization, Geneva, Switzerland.

Wolting, H.G. (1975) Synergism of hydrogen fluoride and leaf necrosis on freesias. *Netherland Journal of Plant Pathology* 81, 71–77.

Wong, D.H., Kirkpatrick, W.E., King, D.R. and Kinnear, J.E. (1992) Defluorination of sodium monofluoroacetate (1080) by microorganisms isolated from Western Australian soils. *Soil Biology and Biochemistry* 24, 833–838.

Wright, D.A. and Thompson, A. (1978) Retention of fluoride from diets containing materials produced during aluminium smelting. *British Journal of Nutrition* 40, 139–147.

Wright, D.A., Davison, A.W. and Johnson, M.S. (1978) Fluoride accumulation by long-tailed mice (*Apoldemus sylvaticus*) and field voles (*Microtus agrestis* L.). *Environmental Pollution* 17, 303–310.

Yang, S.F. and Miller, G.W. (1963a) Biochemical studies on the effect of fluoride on higher plants. 1. Metabolism of carbohydrates, organic acids and amino acids. *Biochemistry Journal* 88, 505–509.

Yang, S.F. and Miller, G.W. (1963b) Biochemical studies on the effect of fluoride on higher plants. 2. The effect of fluoride on sucrose-synthesizing enzymes from higher plants. *Biochemistry Journal* 88, 510–516.

Yoshida, Y. (1975) Experimental study on mulberry leaf poisoning by air pollution due to factories' exhaust. *Shiga Prefecture Experiment Station Proceedings* 32, 93–97.

Yu, M.-H. and Miller, G.W. (1967) Effect of fluoride on the respiration of leaves from higher plants. *Plant Cell Physiology* 8, 483–493.

Yu, M.-H. and Miller, G.W. (1970) Gas chromatographic identification of fluoroorganic acids. *Environmental Science and Technology* 4, 492–495.

Yu, M.-H., Miller, G.W. and Lovelace, J. (1971) Gas chromatographic analysis of fluoroorganic acids in plants and animal tissue. In: Englund, H.M. and Beery, W.T. (eds) *Proceedings, 2nd International Clean Air Congress*. Academic Press, New York, pp. 156–158.

Zahavi, M., Tahori, A.S. and Kindler, S.H. (1964) Studies on the biochemistry of fluoroacetate resistance in house flies. *Israel Journal of Chemistry* 2, 320–321.

Zahavi, M., Tahori, A.S. and Mager, J. (1968) Studies on the biochemical basis of susceptibility and resistance of the housefly to

fluoroacetate. *Biochimica et Biophysica Acta* 153, 787–798.

Zhang, C.L. and Huang, H.B. (1998) The stunted fruit disorder – a physiological anomaly in mango caused by a fluorine pollution. *Journal of Horticultural Science and Biotechnology* 73, 513–516.

Zhang, X.P., Deng, W. and Yang, X.M. (2002) The background concentrations of 13 soil trace elements and their relationships to parent materials and vegetation in Xizang (Tibet), China. *Journal of Asian Earth Sciences* 21, 167–174.

Zhang, Y. and Cao, S.R. (1996) Coal burning induced endemic fluorosis in China. *Fluoride* 29, 207–211.

Zheng, B., Ding, Z., Huang, R., Zhu, J., Yu, X., Wang, A., Zhou, D., Mao, D. and Su, H. (1999) Issues of health and disease relating to coal use in southwestern China. *International Journal of Coal Geology* 40, 119–132.

Zimmerman, P.W. and Hitchcock, A.E. (1946) Fluorine compounds given off by plants. *American Journal of Botany* 33, 233.

Zimmerman, P.W., Hitchcock, A.E. and Gwirtsman, J. (1957) Fluorine in food with special reference to tea. *Contributions from the Boyce Thompson Institute* 19, 49–53.

Zipkin, I., Lucas, S.M., Lavender, D.R., Fullmer, H.M., Schiffman, E. and Corcoran, B.A. (1970) Fluoride and calcification of rat aorta. *Calcified Tissue Research* 6, 173–182.

Zubovic, P., Oman, C., Coleman, S., Bragg, L., Kerr, P., Kozey, K., Simon, F., Rowe, J., Medlin, J. and Walker, F. (1979) *Chemical Analysis of 617 Coal Samples from the Eastern United States*. Open-File Report No. 79-665, US Geological Survey.

Index